Other Titles in This Series

171 **L. A. Bunimovich, B. M. Gurevich, and Ya. B. Pesin, Editors,** S
Dynamical Systems

170 **S. P. Novikov, Editor,** Topics in Topology and Mathematical Ph

169 **S. G. Gindikin and E. B. Vinberg, Editors,** Lie Groups and Lie Algebras: E. B. Dynkin's Seminar

168 **V. V. Kozlov, Editor,** Dynamical Systems in Classical Mechanics

167 **V. V. Lychagin, Editor,** The Interplay between Differential Geometry and Differential Equations

166 **O. A. Ladyzhenskaya, Editor,** Proceedings of the St. Petersburg Mathematical Society, Volume III

165 **Yu. Ilyashenko and S. Yakovenko, Editors,** Concerning the Hilbert 16th Problem

164 **N. N. Uraltseva, Editor,** Nonlinear Evolution Equations

163 **L. A. Bokut′, M. Hazewinkel, and Yu. G. Reshetnyak, Editors,** Third Siberian School "Algebra and Analysis"

162 **S. G. Gindikin, Editor,** Applied Problems of Radon Transform

161 **Katsumi Nomizu, Editor,** Selected Papers on Analysis, Probability, and Statistics

160 **K. Nomizu, Editor,** Selected Papers on Number Theory, Algebraic Geometry, and Differential Geometry

159 **O. A. Ladyzhenskaya, Editor,** Proceedings of the St. Petersburg Mathematical Society, Volume II

158 **A. K. Kelmans, Editor,** Selected Topics in Discrete Mathematics: Proceedings of the Moscow Discrete Mathematics Seminar, 1972–1990

157 **M. Sh. Birman, Editor,** Wave Propagation. Scattering Theory

156 **V. N. Gerasimov, N. G. Nesterenko, and A. I. Valitskas,** Three Papers on Algebras and Their Representations

155 **O. A. Ladyzhenskaya and A. M. Vershik, Editors,** Proceedings of the St. Petersburg Mathematical Society, Volume I

154 **V. A. Artamonov et al.,** Selected Papers in K-Theory

153 **S. G. Gindikin, Editor,** Singularity Theory and Some Problems of Functional Analysis

152 **H. Draškovičová et al.,** Ordered Sets and Lattices II

151 **I. A. Aleksandrov, L. A. Bokut′, and Yu. G. Reshetnyak, Editors,** Second Siberian Winter School "Algebra and Analysis"

150 **S. G. Gindikin, Editor,** Spectral Theory of Operators

149 **V. S. Afraĭmovich et al.,** Thirteen Papers in Algebra, Functional Analysis, Topology, and Probability, Translated from the Russian

148 **A. D. Aleksandrov, O. V. Belegradek, L. A. Bokut′, and Yu. L. Ershov, Editors,** First Siberian Winter School "Algebra and Analysis"

147 **I. G. Bashmakova et al.,** Nine Papers from the International Congress of Mathematicians, 1986

146 **L. A. Aĭzenberg et al.,** Fifteen Papers in Complex Analysis

145 **S. G. Dalalyan et al.,** Eight Papers Translated from the Russian

144 **S. D. Berman et al.,** Thirteen Papers Translated from the Russian

143 **V. A. Belonogov et al.,** Eight Papers Translated from the Russian

142 **M. B. Abalovich et al.,** Ten Papers Translated from the Russian

141 **H. Draškovičová et al.,** Ordered Sets and Lattices

140 **V. I. Bernik et al.,** Eleven Papers Translated from the Russian

139 **A. Ya. Aĭzenshtat et al.,** Nineteen Papers on Algebraic Semigroups

138 **I. V. Kovalishina and V. P. Potapov,** Seven Papers Translated from the Russian

137 **V. I. Arnol′d et al.,** Fourteen Papers Translated from the Russian

136 **L. A. Aksent′ev et al.,** Fourteen Papers Translated from the Russian

135 **S. N. Artemov et al.,** Six Papers in Logic

134 **A. Ya. Aĭzenshtat et al.,** Fourteen Papers Translated from the Russian

133 **R. R. Suncheleev et al.,** Thirteen Papers in Analysis

(See the AMS catalog for earlier titles)

Sinai's Moscow Seminar on Dynamical Systems

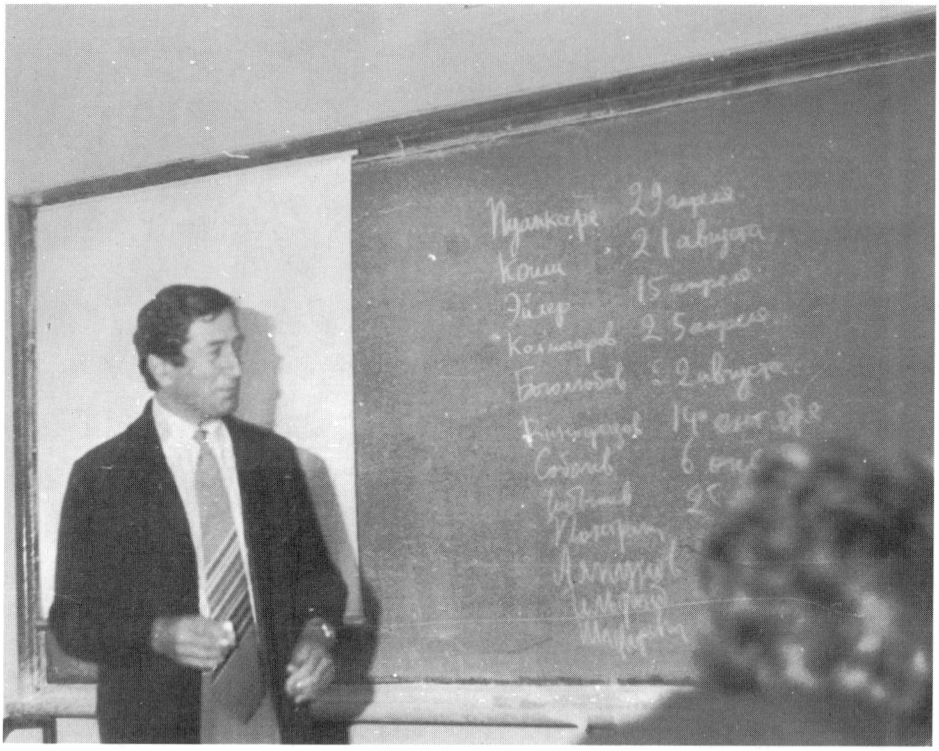

Ya. G. Sinai (1985)

American Mathematical Society

TRANSLATIONS

Series 2 • Volume 171

Advances in the Mathematical Sciences – 28

(*Formerly Advances in Soviet Mathematics*)

Sinai's Moscow Seminar on Dynamical Systems

L. A. Bunimovich
B. M. Gurevich
Ya. B. Pesin
Editors

American Mathematical Society
Providence, Rhode Island

Translation edited by N. PESIN

1991 *Mathematics Subject Classification.* Primary 28Dxx, 58F11; Secondary 58Fxx.

ABSTRACT. The book consists of papers by the participants of the Seminar on Dynamical Systems and Applications held by Ya. G. Sinai at the Moscow State University for more than thirty years. The book is useful to researchers and advanced graduate students working in dynamical systems, ergodic theory, global analysis, and statistical mechanics.

Library of Congress Card Number 91-640741
ISBN 0-8218-0426-1
ISSN 0065-9290

∞ The paper used in this book is acid-free and falls within the guidelines
established to ensure permanence and durability.
Printed on recycled paper.

10 9 8 7 6 5 4 3 2 1   01 00 99 98 97 96

Contents

Amer. Math. Soc. Transl.
(2) Vol. **171**, 1996

Foreword

For everyone who has ever dealt with the theory of stochasticity of dynamical systems, Professor Sinai's name has been surrounded by a mystical oreol.
"Cambridge University Press in the advertisement of Sinai's first book."

Anyone who has met Yakov Grigorievich Sinai will be extremely surprised to know that September 21, 1995, is his 60th birthday. He does not fit the stereotypical image of a 60 year old. For instance, a standard "Sinai situation" occurred last summer on the eve of the Congress of Mathematical Physics in Paris. Most senior and prominent participants were assigned to the quietest and most comfortable hotel. When someone asked the organizers about Sinai, the answer was, "He is too young, only people in their mid-fifties and older stay there." Even those of us, who have known him for more than 30 years can hardly believe that he is 60. However, we can testify that Sinai holds all the necessary documents to claim this surprising age.

The present volume was put together by his friends, students, and close co-workers, who carry the spirit of Sinai's seminars through space-time to commemorate the anniversary. Each paper in this volume has an author or a coauthor who was Sinai's student and/or an active participant of his celebrated seminar over the last 33 years.

It has puzzled more than one generation of mathematicians, especially in the West, "How could Sinai possibly have so many old students while being so young himself?" Indeed, his first students were only a few years younger than he.

Sinai is a descendent of the family that has played a prominent role in Russia's scientific and cultural history even before the 1917 October Revolution. Perhaps only a few mathematicians know that the famous geometer Veniamin Fedorovich Kagan was Sinai's grandfather.

There have been many outstanding seminars in Moscow's mathematical history. In addition to always exceeding the highest mathematical standards, Sinai's seminar on dynamical systems was also something special. All mathematicians who visited the seminar experienced the excitement, pleasant atmosphere and a bit of good humor (see, e.g., Halmos' memoirs). Another very special feature of the seminar was the enormous optimism and enthusiasm that was always radiated by Sinai. Many, many times during talks that seemingly had nothing to do with the theme of the seminar, we heard him say, "It is our business." Many times it did indeed become our business. Sinai managed to build so many bridges between the theory of dynamical systems and numerous other areas that scores of scientists who use those bridges often do not even know who actually built them. It is because of his miraculous intuition and unusually wide interests that a purely mathematical seminar on dynamical systems became the place where prominent physicists, chemists, biologists, etc. were invited to speak. Often physicists who did not agree on certain issues would come to the seminar to explain their views and seek advice, as opposed to the usual practice where physicists leave to mathematicians only obvious results that need technical proof.

It was not easy to "survive" at Sinai's seminar. Upon arriving, students were usually shocked by their lack of ability to understand anything. But there was something very attractive in this nurturing atmosphere where mathematics was treated as an art and the participants witnessed the creation of a masterpiece. It was no surprise that some people used to attend just to be a part of this wonderful atmosphere. In the late seventies, some seminars had lost their leaders to a wave of immigration and many gifted mathematicians remaining in the USSR began attending Sinai's seminar. The seminar provided much relief for the participants who, in the great majority, were never allowed to work as mathematicians in the former Soviet Union. Instead they worked in various institutions with funny names in Moscow or even outside Moscow and did mathematics in their spare time. However, Thursday evenings were a holy time. People began to hang around room 14-04 much earlier than when the seminar would begin and they often stayed late afterwards.

This spirited influence has affected not only the "dynamical people" in Moscow, former Soviet Union, and Europe, but also in the USA. Someone from Russia, who was allowed a short visit to the US was told by a famous American dynamicist: "We were shocked when the first Soviet dynamicists immigrated in the late seventies. How is it possible to leave such a place, where life is so exciting?" Unfortunately, life outside the seminar was full of hardship.

The seminar devoted to Sinai's 50th birthday was very memorable. There were several short lectures on various cute and beautiful problems, then some songs and poems were performed which were written especially for that day. It was even attended by many of his students who were no longer active in mathematics, but always considered the seminar as one of the best times in their lives.

Many participants who were not students and did not work at the Moscow University had a problem getting into the building. It was especially difficult for foreign mathematicians to attend Sinai's seminar while visiting Moscow. Usually one or two participants helped them get around the guards. There were several standard tricks developed to do that. However, Sinai himself often improvised. For instance, when a Mexican mathematician was visited in June 1986, Sinai went to the guard and told him: "This man is a very important person in Mexico and he promises to send a soccer ball from the World Cup as a gift to the Moscow University." This incident is very typical of Sinai and his attitude towards Soviet rules and regulations.

Once a year, the meeting seminar was moved to the countryside to face a bloody competition with Kirillov's representation theory seminar. Everyone had to play soccer and volleyball even if they did not know how. Now Sinai's team is playing all over the world; however, Thursdays in 14-04 will always be one of our fondest memories.

Sinai and his seminar exerted enormous influence on the whole area of dynamical systems and essentially shaped modern ergodic theory. We are very happy that Yakov

Grigorievich continues to surprise us with his enthusiasm and a fountain of bright new ideas. We send him our very best wishes and hope to see Sinai's "business" prospering many more years.

The editors

Amer. Math. Soc. Transl.
(2) Vol. **171**, 1996

Hierarchical Coding and Normal Sequences

O. N. Ageev and A. M. Stepin

For Yakov Sinai on his 60th birthday

In this paper we consider coding of symbolic dynamical systems and its relation to the corresponding measures of maximal entropy with statistical behavior of trajectories. More precisely, for the codes without overlapping (so-called S-codes) and the corresponding hierarchical coding procedure we prove the uniqueness of the measure of maximal entropy, give a description of this measure as an image of certain Bernoulli measure, and provide an explicit construction of normal sequences (with respect to the measure of maximal entropy).

Let A be a finite alphabet, $\Omega = A^{\mathbb{Z}}$ the set of sequences $\omega = \dots\omega_{-1}\omega_0\omega_1\dots$, $\omega_i \in A$, endowed with the standard product topology. The triple $(\Omega, \mathcal{B}, \mu)$, where $\mathcal{B}$ is the σ-algebra of Borel subsets in Ω and μ a shift-invariant probability measure on $\mathcal{B}$, defines a stationary random process. Let $(\Omega', \mathcal{B}', \mu')$ be another process (with possibly another alphabet A'). A shift-equivariant measurable mapping $\varphi : \Omega \to \Omega'$ with the property $\varphi^*\mu = \mu'$ is called a coding. If such a mapping φ is one-to-one for μ'-a.e. $\omega' \in \Omega'$ (i.e., there exists measurable $\psi : \Omega' \to \Omega$ such that $\psi\varphi = \mathrm{id}$ for μ-a.e. $\omega \in \Omega$), the coding φ is called invertible (isomorphism).

In **[1]**, Ornstein found the solution of the classification problem (up to an isomorphism) for Bernoulli automorphisms. In **[2]** Keane and Smorodinsky solved the classification problem for the class of finitary codings for Bernoulli automorphisms. In **[3]**, Del Junco found a generalization of this result. Namely, he constructed a coding with finite anticipation for Bernoulli processes of the same entropy. Since in general a description of classification up to an isomorphism may look quite complicated, some particular coding procedures are of interest. One of these procedures is that of hierarchical coding (see, for example, **[5]**).

A mapping $\varphi : \Omega \to \Omega'$ is called hierarchical coding if for some subset $\Omega_1 \subset \Omega$, $\mu(\Omega_1) = 1$, and any $\omega \in \Omega_1$ there exists a sequence of cylindrical sets $E_i^{\omega} = \{\eta \in \Omega_1 : \eta_k = \omega_k \text{ if } m_i(\omega) \le k \le n_i(\omega)\}$ satisfying the conditions: a) $E_i^{\omega} \searrow \{\omega\}$, b) $[\varphi(\eta)]_k$ (the kth coordinate of $\varphi(\eta)$) does not depend on $\eta \in E_i^{\omega}$ for $m_i(\omega) \le k \le n_i(\omega)$.

1991 *Mathematics Subject Classification*. Primary 28D20; Secondary 58F03.

The work of the second author was supported by the Russian Foundation for Fundamental Research (project 93-01-00239) and ISF (grant M1E000).

A specific example of such a coding appeared in Meshalkin's paper **[4]**, where the first nontrivial example of an isomorphism for Bernoulli processes with the same entropy was constructed. We mention the papers **[5]**, **[6]**, and **[7]** among those devoted to the analysis of the properties of hierarchical coding, and notice that there exists a criterion that allows us to verify whether the image of Bernoulli (or Markovian) measure under hierarchical coding is a Bernoulli (resp. Markovian) measure **[8]**.

Hierarchical codings admit effective description in purely combinatorial terms. To do this we recall and introduce relevant notions and notations. Fix an alphabet $A = \{a_i : i = 1, \dots, r\}$. The number of letters in a word $W = a_{i_1} \dots a_{i_\ell}$ over A is called the length of W and is denoted by $|W|$; empty word has zero length. The word V_1V_2 obtained by attaching the word V_2 to the right of the word V_1 is called the concatenation of V_1 and V_2. The insertion of a word U into a word V is said to be the word of the form V_1UV_2, where $V = V_1V_2$ and V_1, V_2 are nonempty words. A word U enters a word W if $W = V_1UV_2$ for some (possibly empty) words V_1 and V_2. We write $\mathrm{Fr}(W, U)$ for the frequency of occurence of the word U in the word W. The cylindrical set

$$\{x \in A^{\mathbb{Z}} : x_0 \dots x_{|B|-1} = B\}$$

with the base B is denoted by $\overline{B}$.

For any set of words $K = \{B_i\}$ over the alphabet A we denote by $\mathcal{E}(K)$ (resp. $\widetilde{\mathcal{E}}(K)$) the closure of K with respect to operations of concatenation and inserting of words (resp. concatenation only).

Denote

$$\Omega_K = \bigcap_{n=1}^{\infty} \bigcup_{C_j \in \mathcal{E}(K)} \bigcup_{n \le s \le |C_j|-1} T^s \overline{C_j},$$

where T is the shift transformation in $\Omega = A^{\mathbb{Z}}$. A Borel set Ω_K consists of sequences ω with the following property: for any pair of integers M and N, $M < N$, there exist $m, n \in \mathbb{Z}$ such that $m < M$, $n > N$ and

$$\omega_m \dots \omega_n \in \mathcal{E}(K).$$

It follows that Ω_K is a T-invariant subset. Let $M(K)$ be the collection of T-invariant Borel probability measures supported by Ω_K.

Suppose that a set K' of words over alphabet A' and a mapping $\varphi : K \to K'$ are such that there exists an extension $\varphi_1 : \mathcal{E}(K) \to \mathcal{E}(K')$ of φ satisfying the following conditions:

1. $|\varphi_1(W)| = |W|$ for $W \in \mathcal{E}(K)$;
2. $\varphi_1(UV) = \varphi_1(U)\varphi_1(V)$ for $U, V \in \mathcal{E}(K)$;
3. $\varphi_1(U_1WU_2) = U_1'\varphi_1(W)U_2'$ for $U = U_1U_2$, $W \in \mathcal{E}(K)$, where $U_1'U_2' = \varphi_1(U)$ and $|U_1'| = |U_1|$;
4. if $U = U_2W$, $V = WU_1$, where $U, V \in \mathcal{E}(K)$ then $\varphi_1(U) = U_1'W'$, $\varphi_1(V) = W'V_1'$ and $|W'| = |W|$.

The mapping φ_1 can be naturally extended to a stationary mapping $\varphi_2 : \Omega_K \to \Omega_{K'}$ (this means that φ_2 is equivariant with respect to shifts). If $\mu \in M(K)$, then φ_2 is an hierarchical coding of $(\Omega, \mathcal{B}, \mu)$ and $(\Omega', \mathcal{B}', \varphi_2^*\mu)$.

It turns out that any hierarchical coding can be obtained by the scheme described above. The set K of words in the above coding procedure is called the code and the mapping $\varphi : K \to K'$ defining hierarchical coding $\Omega_K \to \Omega_{K'}$ is also called the coding.

Consider a code K satisfying the conditions:

1. if $V \in K$ is of the form $V_1 U V_2$ and $|V| > |U|$, then $U \notin K$;
2. we have $\{U : \exists\ V \in K,\ \exists\ U_1\,(|U_1| > 0)\, UU_1 = V\} \cap \{U : \exists\ V \in K\ \exists\ U_2\,(|U_2| > 0)\, U_2 U = V\} = \varnothing$.

Such a code is called an S-code or a code without overlapping. The straightforward verification shows that the set $\{B_i\}$ is an S-code if and only if for every pair $U, V \in \mathcal{E}(\{B_i\})$ of the form $U = U_1 W$, $V = W V_1$ the overlap W belongs to $\mathcal{E}(\{B_i\})$.

Recall that a sequence $\omega \in A^{\mathbb{Z}}$ is called regular for a shift-invariant measure μ on $\mathcal{B}$ if for any word B over alphabet A we have

$$\lim_{\substack{m \to -\infty \\ n \to +\infty}} \operatorname{Fr}(\omega_m \ldots \omega_n, B) = \mu(\overline{B}).$$

It is known that μ-a.a. sequences are regular with respect to an ergodic measure μ.

Given a symbolic dynamical system, i.e., a shift-invariant Borel subset $M \subset A^{\mathbb{Z}}$, we call a sequence $\omega \in M$ normal if ω is regular for the measure of maximal entropy supported by M (if such a measure exists). The notion of normality of numbers with respect to decimal expansion (i.e., for $\{0, 1, \ldots 9\}^{\mathbb{N}}$) was introduced by E. Borel **[9]**. He also established the existence of normal numbers using a measure-theoretical approach. The first explicit example of normal numbers was given by Champernowne **[10]**. Earlier Lebesgue and Sierpinksi suggested effective constructions of normal numbers (see **[11]**). Other interesting constructions of normal sequences for one-sided topological Bernoulli shifts can be found in **[12]**.

We notice the recent publication **[13]** with contemporary view on the problem of normality. Given a collection of words $K = \{B_i\}$, we denote by φ_K the generating function for K, i.e.,

$$\varphi_K(z) = \sum_{n=0}^{\infty} \#\{j : |B_j| = n\} z^n,$$

and by R_K its radius of convergence. Let $\widetilde{\Omega}_K$ be a one-sided analog of Ω_K for the concatenation closure $\widetilde{\mathcal{E}}(K)$ of K. It is shown in **[13]** that for the prefix code K satisfying the condition $R_{\widetilde{\mathcal{E}}(K)} < R_K$, the sequences of the form $B_1 B_2 B_3 \ldots$, are normal for $\widetilde{\Omega}_K$, where B_i is the concatenation of all words from $\{U \in \widetilde{\mathcal{E}}(K) : |U| = i\}$.

In the present paper for S-code K we study the shift-invariant measure of maximal entropy supported by Ω_K and construct corresponding normal points in Ω_K.

Let $B_j^{\pm}$, $j = 1, 2, \ldots$, be the concatenations of some words from $\{W \in \mathcal{E}(K) : |W| = j\}$ such that every word $W \in \mathcal{E}(K)$ of length j enters $B_j^- B_j^+$ only once. Define C_j inductively by the formula $C_{j+1} = B_{j+1}^- C_j B_{j+1}^+$, where C_0 is empty word. Choose a point $u \in \Omega_K$ such that there exist sequences $m_i \to -\infty$ and $n_i \to +\infty$ for which the following conditions hold:

1. $u_{m_i} \ldots u_{n_i} = C_i$,
2. $\displaystyle\liminf_{i \to \infty} \frac{\min\{|n_i|, |m_i|\}}{\max\{|n_i|, |m_i|\}} > 0.$

Our objective is to find conditions on K that guarantee normality of the point $u \in \Omega_K$. For that, we need some properties of generating functions for S-codes, that are stated in the following proposition. We call S-code trivial if all its words are of length one.

PROPOSITION. *Let K be an S-code.*

1. *The following equality holds in a neighborhood of zero:*

$$\varphi_{\mathcal{E}(K)}(t) = \varphi_K(t \cdot \varphi_{\mathcal{E}(K)}(t)). \tag{1}$$

2. *Let* $r_0 := \lim_{t \to R_{\mathcal{E}(K)}} t \cdot \varphi_{\mathcal{E}(K)}(t)$. *Then* $r_0 = R_{\mathcal{E}(K)}\varphi_{\mathcal{E}(K)}(R_{\mathcal{E}(K)})$ *and* $r_0 \le R_K$.
3. *If K is nontrivial, then* $r_0 < \infty$.

Claim 1 follows from Theorem 2 in **[5]**. The first part of Claim 2 follows from the power series representation of $\varphi_{\mathcal{E}(K)}(t)$.

We prove the second part of Claim 2 by contradiction. Since the function $\psi(t) = t\varphi_{\mathcal{E}(K)}(t)$ is monotone on $[0, R_{\mathcal{E}(K)})$, there exists a number $t_0 < R_{\mathcal{E}(K)} \le R_K$ such that $\psi(t_0) = R_K$. Consider the holomorphic inverse function $\psi^{-1}(\tau)$ in a neighborhood of R_K. The continuation of $\varphi_K(\tau)$ to the neighborhood of R_K by the formula

$$\varphi_K(\tau) = \varphi_{\mathcal{E}(K)}(\psi^{-1}(\tau))$$

yields the required contradiction because R_K is a singular point for $\varphi_K(\tau)$.

Finally, suppose that $r_0 = \infty$ and K is nontrivial. Then the right-hand side of (1) grows faster than the left-hand side as $t \to R_{\mathcal{E}(K)}$. However, according to the Claim 2 equality (1) must be valid for any t, $|t| < R_{\mathcal{E}(K)}$. This proves Claim 3.

Set $s_0 = r_0/R_{\mathcal{E}(K)}$. If the code K is nontrivial, then, passing to the limit in (1) as $t \to R_{\mathcal{E}(K)}$, we see that s_0 is a root of the equation

$$\varphi_K(R_{\mathcal{E}(K)}t) = t. \tag{2}$$

For the majority of codes s_0 is a multiple root of this equation. In what follows we shall assume this property, since otherwise there is no measure of maximal entropy supported by Ω_K.

To a nontrivial S-code $K = \{W_i\}$ we associate the *canonical* code

$$K_0 = \{\underbrace{a_0 \dots a_0}_{n_i} a_i\}$$

with the same number of words and the Bernoulli measure ν_0 given by the formula

$$\nu_0(\bar{a}_0) = \frac{1}{s}, \quad \nu_0(\bar{a}_i) = s^{n_i} R^{n_i+1}_{\mathcal{E}(K_0)},$$

where $n_i = |W_i| - 1$, s is a multiple positive root of (2). Notice that the condition $\nu_0(\Omega_{K_0}) = 1$ follows from the choice of s (see **[7]**).

It turns out that the measure of maximal entropy for K admits the following description generalizing the result in **[5]**.

THEOREM 1. *The measure of maximal entropy for a nontrivial S-code K is unique and coincides with the image of Bernoulli measure ν_0 under the hierarchical coding corresponding to the mapping $K_0 \to K$, where K_0 is the canonical code for K. Furthermore,*

$$\sup_{\mu \in M(K)} h_\mu(T) = h_{\nu_0}(T) = -\log R_{\mathcal{E}(K)}.$$

The proof follows from the pattern suggested in **[5]**. The coding $\varphi : K \to K'$ naturally defines a bijection $M(K) \to M(K')$ preserving entropy with respect to the shift transformation. Hence it suffices to prove the claim of the theorem for the canonical code K_0. (Here the alphabet $A = \{a_i\}$ is not assumed to be finite.)

First we solve the problem for the subclass of $M(K_0)$ consisting of Bernoulli measures. We shall use the following facts (see [5], [7]):

$$\mu(\Omega_{K_0}) = 1 \iff \mu(\overline{a}_0) = \sum_{i>0} n_i \mu(\overline{a}_i), \tag{3}$$

$$\sup_{\mu \in M(K)} h_\mu(T) \le -\log R_{\mathcal{E}(K)}. \tag{4}$$

The equality $h_{\nu_0}(T) = -\log R_{\mathcal{E}(K)}$ can be verified directly. Using this equality and calculating local maxima of the function

$$-\sum \mu(\overline{a}_i) \log \mu(\overline{a}_i) \tag{5}$$

under the conditions that $\sum \mu(\overline{a}_i) = 1$ and $\mu(\overline{a}_0) = \sum_{i>0} \mu(\overline{a}_i) n_i$, we can prove the uniqueness of the global maximum of the function (5).

Returning to the general case, fix an ergodic measure $\mu \in M(K_0)$ and define the Bernoulli measure ν in $A^{\mathbb{Z}}$ with the one-dimensional distribution $\nu(\overline{a}_i) = \mu(\overline{a}_i)$. Clearly, $h_\nu(T) \ge h_\mu(T)$ and the equality is attained exactly when $\nu = \mu$.

However, in general $\nu \notin M(K_0)$. Since the code K_0 is canonical, for any $W \in \mathcal{E}(K_0)$ we have

$$\mathrm{Fr}(W, a_0) = \sum_{j>0} n_j \,\mathrm{Fr}(W, a_j).$$

Therefore for any point $\omega = \dots \omega_{-K} \dots \omega_0 \dots \omega_\ell \dots \in \Omega_{K_0}$ there exist sequences $m_i \to -\infty$ and $n_i \to +\infty$ such that

$$\mathrm{Fr}(\omega_{m_i} \dots \omega_{n_i}, a_0) = \sum_{j>0} n_j \,\mathrm{Fr}(\omega_{m_i} \dots \omega_{n_i}, a_j).$$

By the ergodic theorem, for μ-almost all $\omega \in \Omega_{K_0}$ we have

$$\lim_{i \to \infty} \mathrm{Fr}(\omega_{m_i} \dots \omega_{n_i}, a_j) = \mu(\overline{a}_j)$$

for any j and hence

$$\mu(\overline{a}_0) \ge \sum_{j>0} n_j \mu(\overline{a}_j).$$

Consider the Bernoulli measure ν' with the one-dimensional distribution

$$\begin{aligned} \nu'(\overline{a}_i) &= \nu(\overline{a}_i) \qquad \text{for } i > 1, \\ \nu'(\overline{a}_1) &= \nu(\overline{a}_1) + c, \\ \nu'(\overline{a}_0) &= \nu(\overline{a}_0) = c, \end{aligned}$$

where

$$c = \frac{1}{n_1 + 1} \left[\nu(\overline{a}_0) - \sum_{j>0} n_j \nu(\overline{a}_j) \right].$$

By (3) we have

$$\nu'(\Omega_{K_0}) = 1.$$

In view of convexity of the function $-x \log x$ we get $h_{\nu'}(T) \ge h_\nu(T)$.

If μ is nonergodic, then, using the ergodic decomposition, representation of $h_\mu(T)$ as an integral of entropies on ergodic components, and uniqueness of the maximal entropy measure for the subclass of ergodic measures in $M(K_0)$, we have

$$h_\mu(T) < h_{\nu_0}(T).$$

REMARK. There exist S-codes having no measures of maximal entropy.

Now we fix an S-code K over a finite alphabet $A = \{a_i\}$. Let $\varphi_K(z) = 1 + \sum b_K z^K$, $\varphi_{\mathcal{E}(K)}(z) = 1 + \sum\limits_{K>0} c_K z^K$.

LEMMA 1. *If ν is the measure of maximal entropy in the class $M(K)$, then it is ergodic and for any $B \in \mathcal{E}(K)$*

$$\nu(\overline{B}) = R_{\mathcal{E}(K)}^{|B|}.$$

The proof for a nontrivial code K immediately follows from Theorem 1. If the code K is trivial, then $\Omega_K = A_1^{\mathbb{Z}}$, where $A_1 \subset A$. It is clear that ν is a Bernoulli measure and $\nu(\overline{a}_i) = 1/|A_1|$ for $a_i \in A_1$. This proves Lemma 1.

LEMMA 2. *We have*

$$C_n = \frac{1}{(n+1)!} \lim_{z\to 0} \frac{d^n(\varphi_K^{n+1}(z))}{dz^n}.$$

PROOF. Setting $\psi(z) = \overline{z}\varphi_{\mathcal{E}(K)}(z)$ and using (1), we obtain the following functional relation:

$$z(\psi) = \psi\left[1 + \sum_{K>0} b_K \psi^K\right]^{-1}.$$

The inverse function of $z(\psi)$ is holomorphic in a neighborhood of zero. Coefficients of power expansion for the inverse function can be obtained by formulas

$$d_0 = 0, \quad d_n = \frac{1}{n!} \lim_{\psi\to 0} \frac{d^{n-1}}{d\psi^{n-1}} \left(\frac{\psi}{z(\psi)}\right)^n.$$

THEOREM 2. *Let K be an S-code and $u \in \Omega_K$ a sequence of the form $u = \dots B_i^- \dots B_1^- B_1^+ \dots B_i^+ \dots$, where $B_i^\pm = C_1(i,\pm)\dots C_{p(i,\pm)}(i,\pm)$, $C_\ell(i,\pm) \in \{W \in \mathcal{E}(K) : |W| = i\}$. A point U is normal for K if and only if the limit*

$$\lim_{K\to\infty} \frac{C_{Kd}}{C_{(K+1)d}} \left(= R_{\mathcal{E}(K)}^d\right)$$

exists; here $d = GCD\{i : b_i > 0\}$.

PROOF. Set $B_i = B_i^- B_i^+$, $C_i^- = B_i^- \dots B_1^-$, $C_i^+ = B_1^+ \dots B_i^+$, $C_i = C_i^- C_i^+$. We remark from the very beginning that the frequency of the occurences of an arbitrary word B into C_i such that B is not contained completely in some $C_\ell(p,\pm)$, tends to zero as $i \to \infty$. Therefore, in what follows we may not take into account these exceptional occurences.

For $\mathcal{D} \in \mathcal{E}(K)$ we have equalities (here $p = |\mathcal{D}|$)

$$|C_i| \operatorname{Fr}(C_i, \mathcal{D}) = \sum_{j=p}^{i} \operatorname{Fr}(B_j, \mathcal{D})|B_j| = \sum_{j=p}^{i} C_{j-p}(j-p+1),$$

$$|C_i| = \sum_{j=1}^{i} |B_j| = \sum_{j=1}^{i} j\, C_j.$$

Clearly $C_K = 0$ if K is not a multiple of d and $C_{\ell d} \neq 0$ for any ℓ greater than some ℓ_0.

If u is a normal point, then by Lemma 1 we have

$$\begin{aligned} R^{\ell d}_{\mathcal{E}(K)} &= \lim_{i\to\infty} \left(\sum_{j=1}^{i} j C_j \right)^{-1} \sum_{j=\ell d}^{i} C_{j-\ell d}(j^{-\ell d+1}) \\ &= \lim_{k\to\infty} \left(\sum_{j=1}^{K} C_{jd}\, jd \right)^{-1} \sum_{j=1}^{K} [(j-\ell)d+1] C_{(j-\ell)d} \\ &= \lim_{k\to\infty} \left(\sum_{j=1}^{K} j C_{jd} \right)^{-1} \sum_{j=1}^{K-\ell} j C_{jd} \end{aligned}$$

for any $\ell > \ell_0$. Since $0 < R_{\mathcal{E}(K)} < 1$, it follows that for $\alpha_j = jC_{jd}$ we obtain consecutively

$$\lim_{K\to\infty} \left(\sum_{i=1}^{K} \alpha_i \right)^{-1} \alpha_{K-\ell} \neq 0, \qquad \lim_{K\to\infty} \alpha_K^{-1} \alpha_{K-\ell} \neq 0.$$

Hence, we see that the limit of the ratio α_{K-1}/α_K exists as $K \to \infty$.

Conversely, if $\lim\limits_{K\to\infty} C_{Kd}/C_{(K+1)d}$ exists, then as above for any $\mathcal{D} \in \mathcal{E}(K)$

$$\lim_{i\to\infty} \operatorname{Fr}(C_i, \mathcal{D}) = \lim_{K\to\infty} \frac{C_{Kd}}{C_{(K+p)d}} = R^{pd}_{\mathcal{E}(K)}, \tag{6}$$

where $pd = |\mathcal{D}|$. Consider the sequence of measures

$$\nu_n = \frac{1}{|C_n|} \sum_{i=-|C_n^-|}^{|C_n^+|+1} \delta_{T^i(u)},$$

where δ_x is the Dirac measure supported at x. According to (6) and Lemma 1, for any $\mathcal{D} \in \mathcal{E}(K)$ we have

$$\lim_{n\to\infty} \nu_n(\overline{\mathcal{D}}) = \nu(\overline{\mathcal{D}}),$$

where ν is the measure of maximal entropy in $M(K)$.

We claim that $\lim \nu_n = \nu$. Assuming the contrary, we choose a subsequence ν_{n_K} such that

$$\lim_{K\to\infty} \nu_{n_K} = \widetilde{\nu} \neq \nu;$$

(a priori, not necessarily $\widetilde{\nu}(\Omega_K) = 1$). Fix an arbitrary word $\mathcal{D}$ over the alphabet A. We call a word $G \in \mathcal{E}(K)$ bordering for $\mathcal{D}$ if $G = \mathcal{D}, \mathcal{D}\mathcal{D}_2$ (possibly $\mathcal{D}_1 = \varnothing$ or $\mathcal{D}_2 = \varnothing$). Let $\{\mathcal{D}_i\}$ be the set of those bordering words for $\mathcal{D}$ that are not represented in the form $\mathcal{D}^1\mathcal{D}^0\mathcal{D}^2$ with $\mathcal{D}^0$ being bordering word for $\mathcal{D}$ and $|\mathcal{D}^1| + |\mathcal{D}^2| > 0$. We have

$$\widetilde{\nu}(\overline{\mathcal{D}}) \geq \sum_i n_i \widetilde{\nu}(\overline{\mathcal{D}}_i) = \sum_i n_i \nu(\overline{\mathcal{D}}_i) = \nu(\overline{\mathcal{D}}),$$

where n_i is the number of occurences of $\mathcal{D}$ into $\mathcal{D}_i$. Hence we come to the contradiction.

For sequences $n_i \to +\infty$ and $m_i \to -\infty$ we consider the corresponding measures

$$\mu_i = \frac{1}{n_i - m_i} \sum_{j=m_i}^{n_i-1} \delta_{T^j(u)}.$$

We claim that $\mu_i \to \nu$ as $i \to \infty$. If this were not true there would exist a subsequence μ_{n_K} such that $\mu_{n_K} \to \mu_0 \neq \nu$. Since

$$|C_{i+d}| : |C_i| < 2\, R^{-d}_{\mathcal{E}(K)}$$

for any sufficiently large i we can find a number a such that $\mu_0(\overline{B}) < a\nu(\overline{B})$ for any cylindric set $\overline{B}$. Since μ_0 is absolutely continuous with respect to ν, we have $\mu_0 = \nu$.

THEOREM 3. *If $\varphi''_K(r_0) < \infty$ for an S-code K, then the sequence u in Theorem 2 is normal for K.*

The proof is based on asymptotic estimates of certain integrals. Note first that the function $\varphi''_K(z)$ is continuous on the circle $|z| = r_0$. Denote $\varphi_K(r_0) = d_1$, $\varphi''_K(r_0) = d_2$ $(d_i > 0)$. The fact that s_0 is a multiple root of (2) can be written as follows:

$$\varphi_K(r_0) = \varphi'_K(r_0) \cdot r_0.$$

For the function $\widetilde{\varphi}_K(z) = \varphi_K(r_0 z)/d_1$ we have $\widetilde{\varphi}_K(1) = \widetilde{\varphi}'_K(1) = 1$, $\widetilde{\varphi}''_K(1) = r_0^2/d_1 = d_3$.

It follows from Lemma 2 that

$$\begin{aligned} C_{nd} &= \frac{1}{2\pi i(nd+1)} \int_{|z|=r_0} \frac{\varphi_K^{nd+1}(z)dz}{z^{nd+1}} \\ &= \frac{d_1^{nd+1}}{r_0^{nd} 2\pi i(nd+1)} \int_{|z|=1} \frac{\widetilde{\varphi}_K^{nd+1}(z)}{z^{nd+1}} dz = \frac{1}{2\pi(nd+1)r_0^{nd}} \int_{-\pi}^{\pi} e^{\psi_n(x)} dx, \end{aligned}$$

where $\psi_n(x) = (nd+1)\ell n \widetilde{\varphi}_K(e^{ix}) - ixnd$.

Now we estimate the contribution of the integral

$$I(n) = \int_{-\lambda}^{\lambda} e^{\psi_n(x)} dx$$

to the asymptotics of c_{nd} (the choice of λ will be made later).

The Taylor formula for $\psi_n(x)$ gives

$$\psi_n(x) = ix - \frac{1}{2}(nd+1)d_3\, x^2 + O(x^2).$$

Consider the integrals

$$I_1(C) = \frac{1}{\sqrt{nd_3d}} \int_{|\mu|<C} e^{\psi_n\left(\frac{\mu}{\sqrt{nd_3d}}\right)} d\mu,$$

$$I_2(C) = \frac{1}{\sqrt{nd_3d}} \int_{C<|\mu|<\lambda\sqrt{nd_3d}} e^{\psi_n\left(\frac{\mu}{\sqrt{nd_3d}}\right)} d\mu.$$

For any $C > 0$ we have

$$\sqrt{nd_3d}\ I_1(C) \sim \int_{|\mu|<C} e^{-\frac{1}{2}\mu^2} d\mu$$

as $n \to \infty$. Now choose λ such that for $|x| < \lambda$ and any $n \in \mathbb{N}$ the inequality

$$\operatorname{Re} \psi_n''(x)/(nd+1) < -\frac{d_3}{2}$$

holds (recall that $\psi_n''(0)/(nd+1) = -d_3$).

Further, since $\psi_n(x) = ix + \frac{1}{2}\psi_n''(\theta(x))x^2$, where $|\theta(x)| \leq |x| < \lambda$, we have

$$\sqrt{nd_3d}\ |I_2(C)| = \int_{C<|\mu|<\lambda\sqrt{nd_3d}} e^{\operatorname{Re}\psi_n\left(\frac{\mu}{\sqrt{nd_3d}}\right)} d\mu \leq \int_{C<|\mu|<\lambda\sqrt{nd_3d}} e^{-\frac{1}{4}\mu^2} d\mu.$$

Therefore,

$$I(n) \sim \sqrt{\frac{2\pi}{nd_3d}} \quad \text{as} \quad n \to \infty.$$

The function $e^{\psi_n(x)}$ is invariant with respect to the translation $x \mapsto x + \frac{2\pi}{d}$. It is also continuous and attains its maximal absolute value exactly at the points $2\pi K/d$, $K \in \mathbb{Z}$. Hence,

$$C_{nd} \sim \frac{d_1^{nd+1}}{(nd+1)r_0^{nd}} \sqrt{\frac{d}{2\pi\, nd_3}}.$$

The proof is completed by using Theorem 2.

We mention that Theorem 3 remains true under the assumption that fractional α-derivative, $\alpha > 1$, of the function φ_K at the point r_0 is finite. Some analogs of the results obtained in the paper are also valid for hierarchical coding of one-sided shifts.

Set $K' = \{U \in \mathcal{E}(K) : U = U_1U_2,\ U_1 \in \mathcal{E}(K) \Longrightarrow U_2\overline{\in}\mathcal{E}(K)\}$. It can be verified that

$$\widetilde{\mathcal{E}}(K') = \mathcal{E}(K), \qquad R_{K'} = R_{\mathcal{E}(K)}.$$

Therefore Theorem 3 can be considered as a refinement for $A^{\mathbb{Z}}$ of a result from **[13]**.

References

1. D. S. Ornstein, *Two Bernoulli shifts with the same entropy are isomorphic*, Adv. Math. **4** (1970), no. 3, 337–352.
2. M. Keane and M. Smorodinsky, *Bernoulli schemes of the same entropy are finitarily isomorphic*, Ann. Math. **109** (1979), no. 2, 397–406.
3. A. del Junco, *Bernoulli shifts of the same entropy are finitarily and unilaterally isomorphic*, Ergodic Theory Dynamical Systems **10** (1990), 687–716.
4. L. D. Meshalkin, *A case of isomorphism of Bernoulli schemes*, Dokl. Akad. Nauk SSSR, **128** (1959), no. 1, 41–44. (Russian)
5. R. I. Grigorchuk and A. M. Stepin, *On coding of Markov sources*, Ergodic Theory and Related Topics, Proc. Conf. Vitte/Hiddensee (GDR) 1981, Berlin, 1982, 207–229.

6. A. H. Zaslavskiĭ, *On the isomorphism problem for stationary processes*, Teor. Verojatnost. i Primenen. **9** (1964), no. 2, 318–326; English transl. in Theory Probab. Appl. **9** (1964).
7. A. H. Livšic, *The isomorphism problem for Bernoulli schemes*, Teor. Verojatnost. i Primenen. **19** (1974), no. 2, 409–416; English transl. in Theory Probab. Appl. **19** (1974).
8. O. N. Ageev, *Combinatorial and approximation methods in ergodic theory*, Ph. D. Thesis, Moscow State Univ., 1988. (Russian)
9. E. Borel, *Lecons sur la theorie des functions*, 3rd ed., Paris, 1928.
10. D. C. Champernowne, *The construction of the decimals normal in the scale of ten.*, J. London Math. Soc. **8** (1933), 254–260.
11. *Normal number*, Mathematical Encyclopedia, vol. 3, "Sovetskaya Encyclopediya", Moscow, 1982, pp. 1070–1071; English transl., Kluwer, Dordrecht.
12. A. G. Postnikov, *Ergodic aspects of the theory of congruences and of the theory of Diophantine approximations*, Trudy Mat. Inst. Steklov. **82** (1966), 3–112; English transl. in Proc. Steklov Inst. Math. **82** (1967).
13. A. Bertrand-Mathis, *Points generiques de Champernowne sur certain systemes codes; applications aux shifts*, Ergodic Theory Dynamical Systems **8** (1988), 35–51.

DEPARTMENT OF MATHEMATICS, MOSCOW STATE UNIVERSITY, MOSCOW 119899, RUSSIA

Amer. Math. Soc. Transl.
(2) Vol. **171**, 1996

On the Number of Flattening Points on Space Curves

V. Arnold

To Ya. Sinai admiringly

§1. Introduction

The results of this paper belong to the symplectic and contact topology, but to explain these results I start from an ellipse in the Euclidean 3-space. This space curve is infinitely degenerate: its torsion vanishes identically. Consider a generic smooth deformation of the ellipse (small with the derivatives[1]). We obtain a generic space curve, close to the ellipse. The result, applied to this case, claims that *the number of the flattening points (where the torsion of the deformed curve vanishes) is at least four*, (provided that the deformation is sufficiently small, taking the derivatives into account).

The smallness of the derivatives is crucial here: it is easy to construct a helix spiralling around the initial ellipse without any flattening points at all.

Our main result extends the above lower bound of the number of the flattening points to curves in higher-dimensional spaces.

Below a *curve* in $\mathbb{R}^m$ always means a smooth immersion $f : S^1 \to \mathbb{R}^m$. An immersion is *good* if the derivatives of f of orders $(1, \dots, m-1)$ are linearly independent at any point. A generic immersion is good.

DEFINITION 1. A *flattening point* of a good immersion is a point where the derivatives of f of orders $(1, \dots, m)$ are linearly dependent.

DEFINITION 2. A *generalized ellipse* in $\mathbb{R}^{2n+1}$ is an immersion defined by a vector Fourier polynomial

$$f(t) = \sum_{k \le n} a_k \cos(kt) + b_k \sin(kt),$$

the vectors $(a_0, \dots, a_n, b_1, \dots, b_n)$ being linearly independent.

THEOREM 1. *Every immersion of a circle into $\mathbb{R}^{2n+1}$ has at least $2n+2$ geometrically different flattening points, provided that it is not too far from a generalized ellipse (taking the derivatives into account).*

1991 *Mathematics Subject Classification*. Primary 53C15; Secondary 53A99.

[1]The reader can easily calculate the order of the derivatives required to be small in the statements below.

REMARK 1. This fact is a special case of an unusual generalization of the Morse inequalities in the spirit of the theory of the Lagrangian and Legendrian collapses in symplectic and contact topology (see **[1–4]**). For simplicity, in this paper we consider only the particular case of the affine or projective space curves; the proofs of the general theorems in symplectic and contact geometry extending Theorem 1 to these cases, are similar to those given below.

The flattening points of a generic space curve are the worse (pointwise) generic Lagrangian singularities of its Gauss mapping (associating to every oriented normal of the curve in the direction of this normal). If the curve is a generalized ellipse, its Gauss mapping is infinitely degenerate. After a small deformation the Lagrangian singularities of the Gauss mapping become generic. Our theorem provides a lower bound for the number of the pointwise generic singularities into which the infinitely degenerate singularity is decomposed after a generic small perturbation.

REMARK 2. For the infinitesimally small deformations Theorem 1 is a rather simple corollary of a theorem of Sturm–Kellog–Tabachnikov **[5]**, bounding from below the number of zeros of the function

$$f(t) = \sum_{k \geq N} a_k \cos(kt) + b_k \sin(kt)$$

on the circle $\{t \bmod 2\pi\}$ by the number $2N$ of the zeros of its first harmonic.

This theorem is a direct generalization of the Morse inequality for functions on the circle (to which the theorem is equivalent for $N = 1$).

The case $N = 2$ of this Sturm-type theorem is related to the four vertices theorem of the plane curves geometry and to its generalizations in symplectic and contact topology (see **[1–5]**).

REMARK 3. It would be interesting to understand how large the deviation of the space curve from a generalized ellipse may be in Theorem 1 (similar questions remain open also in the theories of the Lagrange and Legendre collapses, see **[1–4]** for the corresponding conjectures).

The situation here is similar to that in the sixties in the theory of the fixed points of the symplectomorphisms and of the Lagrange intersections.

In these theories the infinitesimal perturbations case is directly reducible to the Morse theory (for instance, the fixed points of the exact symplectomorphisms are the critical points of the Hamilton functions). The 1965 extension from the infinitesimal perturbations to the finite (but not too large) perturbations in these theories has been provided by the theory of generating functions (see **[6]**).

Later, in a series of brilliant papers by Conley and Zehnder **[7]**, Floer **[8]**, Chaperon **[9]**, Chekanov **[10]**, Laudenbach and Sikorav **[11]** (and many others) the perturbation smallness conditions have been removed (in many cases). The resulting theory can be considered as a far reaching generalization of the Morse theory to the multivalued functions (called in symplectic and contact geometry the Lagrange and Legendre submanifolds).

In the Lagrange and Legendre collapse theory, to which the present paper belongs, the role of the Morse theory is played by the Sturm theory. In this case the infinitesimal version minorations are provided by the Sturm–Kellog–Tabachnikov theorem. The results of **[1–4]** and of the present paper also reduce to the Sturm theory the study of the finite (but not too large) perturbations.

The final extension of these results to the case of the very large perturbations (in the spirit of the Chekanov theorem on the quasicritical points of the quasifunctions see **[12]**) depends on the (yet missing) "multivalued version of the Sturm theory", waiting for the courageous researchers (even in the simple particular case of space curves considered in this paper).

§2. The infinitesimal study

Denote the perturbed immersion by

$$R(t) = r(t) + \varepsilon z(t),$$

where $r(t)$ is a generalized ellipse, ε being a small parameter.

The flattening condition is the equation

$$R_1 \wedge R_2 \wedge \cdots \wedge R_{2n+1}(t) = 0, \qquad R_i = d^i R/dt^i,$$

for the unknown point $t \in S^1$. The infinitesimal flattening condition takes into account only terms of order 1 (and of course of order 0) in ε. To write this equation explicitly, we use the basis $(a_0, \dots, b_n)$, formed by the vectors, defining the generalized ellipse. For $\varepsilon = 0$ the component of the direction of a_0 is missing in all the vectors R_j. Hence all the zero order terms vanish, and the first order terms in ε are linear combinations of the functions

$$u_i(t) = d^i u/dt^i, \quad i = 1, \dots, 2n+1,$$

where the real-valued function $u(t)$ is the component of the perturbing vector $z(t)$ in the direction of the basic vector a_0.

Hence, the infinitesimal flattening condition has the form $(Lu)(t) = 0$, where L is some linear homogeneous differential operator of order $2n+1$ and where u is the a_0-component of the perturbation z. The other components of the perturbation have no effect on the flattening points (in the first approximation in ε). The order of the operator is indeed $2n+1$, since the generalized ellipse is a *good* curve.

Operator L can be computed explicitly, as a polynomial with respect to the operator $\partial = d/dt$.

LEMMA 1. $L = \partial(\partial^2 + 1)(\partial^2 + 4) \dots (\partial^2 + n^2)$.

PROOF. Let $z(t) = u(t)a_0$, where u is a linear combination of the functions $(1, \cos t, \sin t, \dots, \cos(nt), \sin(nt))$. The perturbed curve is then still a generalized ellipse. All its points are the flattening points. Hence all the $2n+1$ functions mentioned above are the solutions of the linear differential equation $Lu = 0$ of order $2n+1$. Hence they form the fundamental system of its solutions. This fundamental system defines the equation $Lu = 0$ unambiguously. Lemma 1 follows.

THEOREM 2. *Function Lu has at least $2n+2$ zeros on the circle $\{t \mod 2\pi\}$, whatever the choice of the function u on the circle is.*

PROOF. The Fourier series of function Lu does not contain the harmonics of orders smaller than $n+1$ (according to Lemma 1). Hence Theorem 2 follows from the Sturm-Kellog-Tabachnikov theorem (a proof of which can be found below, in §5).

§3. Equation for flattening points on finitely perturbed curves

Write the perturbed immersion in the form of a vector $R(t)$ with components

$$z(t), \cos t + u_1(t), \sin t + v_1(t), \ldots, \cos(nt) + u_n(t), \sin(nt) + v_n(t)$$

in the basis $(a_0, a_1, b_1, \ldots, a_n, b_n)$.

At the flattening points, the determinant formed by the components of the derivatives $R_i(t)$ of orders $i = 1, \ldots, 2n+1$ vanishes.

This determinant is a linear combination of the elements of its first row, i.e., a linear combination of the derivatives of function z. We can write the flattening condition in the form of the equation $L(t)z(t) = 0$ (on t), where L is a differential operator of order $2n+1$ (depending on u, v).

LEMMA 2. *Let the perturbations (u, v) be sufficiently small (taking the derivatives into account). Then the differential operator L is close (taking the derivatives into account) to the operator $L_0 = \partial(\partial^2+1)\ldots(\partial^2+n^2)$ from §2 and the differential equation $Lu = 0$ has a fundamental system of $2n+1$ 2π-periodic solutions (close to $1, \cos t, \sin t, \ldots, \cos(nt), \sin(nt)$).*

PROOF. The fundamental system of solutions of the equation $Lu = 0$ is formed by the $2n+1$ functions

$$1, \cos t + u_1(t), \sin t + v_1(t), \ldots, \cos(nt) + u_n(t), \sin(nt) + v_n(t).$$

§4. Flattening condition and Chebyshev systems

It is useful for the sequel to consider the Lagrange conjugate operator L^* of the operator L of §3.

LEMMA 3. *Let the perturbations (u, v) be sufficiently small (taking the derivatives into account). Then the differential operator L^*, conjugate to L, is close (taking the derivatives into account) to the operator $-L_0$ and the differential equation $L^*w = 0$ has a fundamental system of $2n+1$ 2π-periodic solutions (close to $1, \cos t, \sin t, \ldots, \cos(nt), \sin(nt)$).*

PROOF. Integrating by parts, one observes that $L_0^* = -L_0$ and hence L^* is close to $-L_0$. The dimensions of the kernels of the conjugate operators L and L^* in the space of the 2π-periodic functions are equal (since this is true for L_0 and since L may be obtained from L_0 by a homotopy, which does not change the index of an operator). According to Lemma 2, these dimensions are equal to $2n+1$.

DEFINITION. A system of N functions on a circle is called a *Chebyshev system* if the sum of the multiplicities of the zeros of every nontrivial linear combination of these functions is smaller than N.

EXAMPLE. $\{1, \sin t, \cos t\}$ is a Chebyshev system of 3 functions.

The sequel is based on the following simple remark.

REMARK. *The set of the Chebyshev systems (of a given number of functions) on a circle is open (taking the derivatives into account while defining the topology).*

Indeed, this follows directly from the semicontinuity of the sum of the multiplicities, since the projective space of the nontrivial linear combinations (modulo a common multiplier) is compact.

LEMMA 4. *Let the perturbations (u, v) be sufficiently small (taking the derivatives into account). Then the fundamental system of the solutions of the differential equation $L^*w = 0$ is a Chebyshev system.*

PROOF. The unperturbed fundamental system $\{1, \cos t, \sin t, \dots, \cos(nt), \sin(nt)\}$ is a Chebyshev system. Indeed, the sum of the multiplicities of the real intersection points of the plane curves $x^2 + y^2 = 1$ and $P(x, y) = 0$ (where P is a real polynomial of degree n) does not exceed $2n$.

According to Lemma 3, the perturbed fundamental system (of solutions of the differential equation $L^*w = 0$) is close to the above unperturbed system. According to the remark above the perturbed system is a Chebyshev system, as required.

§5. Lower bound on the number of flattening points

Now we prove Theorem 1 formulated in §1.

LEMMA 5. *Let the perturbations (u, v) be sufficiently small (taking the derivatives into account). Then the function $L(t)z(t)$ on the circle $\{t \bmod 2\pi\}$ has at least $2n + 2$ zeros.*

PROOF. Suppose that the number of zeros is smaller. For simplicity, we start with the case, where all zeros are simple, their number being $2n$.

Choose a nontrivial linear combination of the functions of the fundamental system of solutions of the equation $L^*w = 0$ vanishing at these $2n$ points. Such a combination exists, since a system of $2n$ linear homogeneous equations with $2n+1$ unknowns always has a nontrivial solution.

This combination g has no other zeros, while these $2n$ zeros of g are simple (since the fundamental system is a Chebyshev system, according to Lemma 4).

The product $(L(t)z(t))g(t)$ does not change sign (and vanishes exactly at the above $2n$ points). Here its integral does not vanish.

But we also have

$$\int (Lz)g\,dt = \int z(L^*g)\,dt = 0,$$

since g is a solution of the equation $L^*g = 0$.

The contradiction proves Lemma 5 in the case where the equation $L(t)z(t) = 0$ has $2n$ simple roots t.

To obtain the contradiction in the general case, choose those zeros of the function Lz, where it changes the sign. The number $2s$ of these zeros is even and is smaller then $2n + 2$, whence $s \le n$.

Choose a solution g of the differential equation $L^*g = 0$ that has $2s$ simple zeros at the $2s$ zeros of Lz chosen above, and $n - s$ double zeros at some other points. Such a function g exists and has no other zeros (since the fundamental system of solutions of the differential equation $L^*w = 0$ is a Chebyshev system, according to Lemma 4). The product $(Lz)g$ does not change the sign, in contradiction with the orthogonality of Lz and g guaranteed by the differential equation $L^*g = 0$. Lemma 5 (and hence Theorem 1) is proved.

The proof used only the Chebyshev property of the fundamental system of solutions of the conjugate differential equation $L^*w = 0$. The Sturm–Kellog–Tabachnikov theorem is proved too.

REMARK 1. The above proof of Theorem 1 restricts only the projection of the curve from $\mathbb{R}^{2n+1}$ to some $2n$-dimensional space. No restrictions are imposed on the last component (along the projection direction).

In the case of the curves in the ordinary 3-space the restriction is the requirement that the projection to a plane should be convex (positively curved).

§6. Convex curves in projective spaces

DEFINITION. An immersion of a circle into the projective space $\mathbb{R}P^m$ is called a *convex curve*, if every hyperplane $\mathbb{R}P^{m-1}$ intersects it, at most, at m points (taking the multiplicities into account).

EXAMPLE 1. A convex curve in $\mathbb{R}P^2$ is an embedding of a circle into an ordinary plane $\mathbb{R}^2$ whose curvature does not vanish.

EXAMPLE 2. A generalized ellipse defines a convex curve in $\mathbb{R}P^{2n}$.

EXAMPLE 3. An example of a convex curve in $\mathbb{R}P^{2n-1}$ is provided by the embedding $x_1 = \cos t, \dots, x_n = \cos(2n-1)t$; $y_1 = \sin t, \dots, y_n = \sin(2n-1)t$ of the circle $\{t \bmod \pi\}$ (where $x_1, \dots, y_n$ are the homogenuous coordinates). The convex curves in odd-dimensional spaces are not contractable.

EXAMPLE 4. The only convex curve in $\mathbb{R}P^1$ is this projective line itself.

REMARK. A convex curve in the even-dimensional space $\mathbb{R}P^{2n}$ can be defined by a Chebyshev system of $2n+1$ functions on the circle. These functions are the homogeneous coordinates of some lift of the curve from $\mathbb{R}P^{2n}$ to $\mathbb{R}^{2n+1} \setminus 0$.

A convex curve in an even-dimensional space does not intersect some hyperplane. Hence one can choose a lift for which one of the functions of the corresponding Chebyshev system is identically 1.

A convex curve in an odd-dimensional space $\mathbb{R}P^{2n-1}$ corresponds to a Chebyshev system of $2n$ sections of the Möbius bundle (which is the nontrivial line bundle over the circle). To construct such a system, consider the double covering of the convex curve. The double covering of a noncontractable circle in $\mathbb{R}P^{2n-1}$ can be lifted to $\mathbb{R}^{2n} \setminus 0$. We can choose the lift for which any two points covering the same point of the initial curve are lifted to opposite points of $\mathbb{R}^{2n}$. The coordinates of the lifted point are functions on the covering circle. On the original curve they define the sections of the Möbius bundle.

If the original curve in the projective space was convex, these $2n$ sections of the Möbius bundle form a Chebyshev system; every nontrivial linear combination of these sections has at most $2n-1$ zeros.

Thus, the theory of the convex curves in projective spaces coincides with the theory of the Chebyshev systems of the sections of the line bundles over a circle.

I list here some properties of the convex curves (which are more or less known):

1. The set of the convex curves in the projective space of any dimension is open and pathwise connected (M. Z. Shapiro).

2. Every boundary curve of the set of the convex curves in the projective space of any dimension has biflattening points, where the last but one torsion vanishes (M. Z. and B. Z. Shapiro, S. Anisov).

3. A convex curve in an even-dimensional projective space is an affine curve (it does not intersect some hyperplane). A convex curve in an odd-dimensional projective space

is noncontractable: it intersects some hyperplane at exactly one point, the intersection being transversal (S. Anisov).

4. The curve projectively dual to a convex curve is convex (M. E. Kazarian).

REMARK. Plane convex curves can be defined as curves having no inflection points intersecting their tangent lines only at the tangency points. The space convex curves also have no flattening points and do have only one common point with every their oscilating hyperplane. I do not know whether these two properties imply convexity.

§7. Flattening points on the projective curves having convex projections

In a projective space of any dimension N consider a curve whose projection to a hyperplane (from some point) is convex.

EXAMPLE. A small space perturbation of a convex curve of a hyperplane has this property.

REMARK. This example is essentially the only example, since an appropriate projective transformation moves a curve having a convex projection to a hyperplane, making it arbitrarily close to its convex projection.

Our main result is the following

THEOREM. *The number of the flattening points of a space projective curve having a convex projection to a hyperplane is higher than the dimension of the projective space.*

EXAMPLE. Every curve in the projective plane that projects regularly to the projective line has at least 3 inflection points.

This Möbius theorem holds also for all the noncontractable embedded curves in $\mathbb{R}P^2$ (see, e.g., [**13**]).

CONJECTURE. *The number of the inflection points is at least* 3, *provided that the curve can be connected to the standard embedding of a projective line by a regular homotopy without positive selftangency moments* (*i.e., provided that the corresponding Legendrian knot in the space of the oriented line elements of* $\mathbb{R}P^2$ *does not change its type*).

The proof of the theorem follows the arguments of §5. The convexity of the curve implies the Chebyshev property both for the fundamental system of solutions of the linear operator L introduced in §4 and also for the fundamental system of solutions of the Lagrange dual operator L^*. Namely, this operator L^* corresponds to the curve, projectively dual to the original curve (Khesin, Ovsienko [**14**]).

The proof of this rather strange duality theorem, presented below, is due to Kazarian.

§8. The Lagrange duality and the projective duality

Consider a projective curve $f : \mathbb{R} \to \mathbb{R}P^n$. Let $\tilde{f} : \mathbb{R} \to \mathbb{R}^{n+1} \setminus 0$ be its arbitrary lift. The curve f has no flattening if and only if the coordinates of $\tilde{f}$ form a fundamental system of solutions of some linear operator $L = (d/dt)^{n+1} + \cdots$.

This operator depends, in general, a) on the parametrization of the curve; b) on the choice of the lift (which is defined up to the multiplication by a nowhere vanishing function).

Let $\tilde{f}^* : \mathbb{R} \to \mathbb{R}^{n+1*}$ be the curve, whose components form a fundamental system of solutions of the Lagrange dual operator L^*, and let $f^* : \mathbb{R} \to \mathbb{R}P^{n*}$ be the corresponding projective curve.

DUALITY THEOREM. *The curve f^* is projectively dual to f and, in particular, does not depend on the choices* a) *and* b) *above.*

PROOF. Write the integration by parts formula

$$\int_\alpha^\beta (Ly)z\,dt - \int_\alpha^\beta (L^*z)y\,dt = B(y,z;\beta) - B(y,z;\alpha).$$

If $Ly = 0$ and $L^*z = 0$, the bilinear form $B(y,z;t)$ does *not* depend on the parameter t. Hence B defines a bilinear pairing of the spaces S and S^* of the solutions of the equations $Ly = 0$ and $L^*z = 0$.

LEMMA 1. *This pairing is nondegenerate. For any t and for any $k + \ell = n + 1$ the linear space $F_t^k \subset S$ of these solutions of the equation $Ly = 0$, which have at t a zero of order at least k, is the B-orthogonal complement of the space $F_t^\ell \subset S^*$ of those solutions of the equation $L^*z = 0$, which have at t a zero of order at least ℓ.*

PROOF OF LEMMA 1. Calculate the matrix of the form $B : S \times S^* \to \mathbb{R}$ in the basis formed by the solutions y_k, z_ℓ of the equations $Ly = 0$ and $L^*z = 0$ for which at t all the derivatives of orders $0 \le i \le n$ vanish, except the derivative of order k (respectively ℓ), which equals one (here $0 \le k, \ell \le n$).

The explicit integration by parts shows that this matrix is triangular, namely $b_{k,\ell} = 0$ for $k + \ell > n$. Moreover, the diagonal elements $(k + \ell = n)$ do not depend on L. One may, for instance, take $L = \partial^{n+1}/\partial t^{n+1}$. Thus one finds that the diagonal elements are ± 1, which proves the Lemma 1.

The space F_t^k has codimension k in S. For instance, consider the lines F_t^n. They are some points of $\mathbb{R}P^n = (S \setminus 0)/(\mathbb{R} \setminus 0)$. All such points form a curve $s : \mathbb{R} \to \mathbb{R}P^n$. I shall call it the *solutions curve* of L. The solutions corresponding to the points of this curve, have a zero of order n at t.

LEMMA 2. *The oscilating hyperplane of the solution curve is the projection f_t^1 to the projective space $\mathbb{R}P^n$ of the hyperplane $F_t^1 \subset S$ consisting of the solutions vanishing at t.*

The proof is a straightforward (but not too short) calculation.

Lemmas 1 and 2 imply

LEMMA 3. *The solution curves of the Lagrange dual operators L and L^* are projectively dual* (*in the sense of B*).

It means that the oscilating hyperplane of the solution curve of L is the B-orthocomplement of the solution curve of L^*.

But by Lemma 2 this oscilating plane is f_t^1. By Lemma 1 it is B-dual to the point F_t^{*n} of the solution curve of L^* in the space $\mathbb{R}P^{n*} = (S^* \setminus 0)/(\mathbb{R} \setminus 0)$.

The choice of a fundamental system of solutions $(y_0, \dots, y_n)$ of $Ly = 0$ identifies the space of solutions S with the dual space $(\mathbb{R}^{n+1})^\vee$ to the space $\mathbb{R}^{n+1}$ in which the curve $\tilde{f}$ lives.

Namely, a solution $y \in S$ is bijectively represented in the form $c_0 y_0 + \cdots + a_n y_n$. The point $c \in (\mathbb{R}^{n+1})^\vee$ is the image of $y \in S$.

LEMMA 4. *The solution curve in $(S \setminus 0)/(\mathbb{R} \setminus 0)$ is projectively dual to the curve f in $(\mathbb{R}^{n+1} \setminus 0)/(\mathbb{R} \setminus 0)$.*

PROOF. Consider the oscilating plane to the solution curve at point t. According to Lemma 2, the corresponding solutions vanish at t. Identifying S with $\mathbb{R}^{n+1\text{v}}$, as explained above, we interprete this hyperplane as the set of those c, for which $c_0y_0(t)+\cdots+c_ny_n(t)=0$. In the dual space $\mathbb{R}^{n+1}$, this hyperplane defines the line with the direction $\tilde{f}(t)$, whence the curve projectively dual to s is f, as required.

Combining Lemmas 3 and 4, we obtain a triple of the projective dualities

$$f \overset{4}{\longleftrightarrow} s \overset{3}{\longleftrightarrow} s^* \overset{4}{\longleftrightarrow} f^*,$$

which proved the theorem.

COROLLARY. *A curve projectively dual to a convex curve is convex.*

PROOF. The generalized ellipse (in the projective space of even or odd dimension, see Example 3 in §6) is convex. The projectively dual curves are projectively equivalent to the original ones in this case, according to the duality theorem, since $L^*=\pm L$. Hence in these examples the projectively dual curves are convex.

The general result follows, since the convexity may only be lost at a biflattening. Since the space of the convex curves is connected, we may attain any convex curve from an ellipse by a regular path. The dual curves along this path have no flattenings and hence their convexity persists.

REMARK. As Kazarian observed, the projective duality $f^{**}=f$ implies rather strange identities for the Wronskians, such as

$$W(W(f,g),W(f,h))=fW(f,g,h).$$

§9. Example: duality of the plane curves

In the case of the convex curves on the Euclidean plane $\mathbb{R}^2$ the above theory leads to the following nice formulas.

First consider the convex curve $\{u(t),v(t)\}$ parametrized by the arc length t. Let $K(t)$ be its curvature.

PROPOSITION 1. *The equation $Ly=0$, whose fundamental system of solutions is $(1,u,v)$, has the form $K^3y^{\text{l}}-K^{\text{l}}y^{\text{ll}}+Ky^{\text{lll}}=0$ (where $\text{l}=d/dt$).*

Now choose as the parameter the azimuth φ of the normal to the curve.

PROPOSITION 2. *In the angular coordinate φ the above equation $Ly=0$ is*

$$(K\dot{y})^{\cdot\cdot}+K\dot{y}=0,$$

where the dot denotes $d/d\varphi$.

Integrating by parts, we obtain

PROPOSITION 3. *The dual equation $L^*z=0$ has the form*

$$(K\ddot{z}+Kz)^{\cdot}=0.$$

REMARK. According to the Minkovski theorem (or to the Pascal hydrodynamical results) the function $K(\varphi)$ satisfies the conditions

$$\int K^{-1} \sin \varphi d\varphi = 0, \quad \int K^{-1} \cos \varphi d\varphi = 0.$$

Therefore, the dual equation has a three-dimensional space of 2π-periodic solutions.

PROPOSITION 4. *A fundamental system of solutions of the dual equations $L^* z = 0$ is formed by the functions*

$$z_1 = \cos \varphi, \quad z_2 = \sin \varphi, \quad z_3 = w\varphi,$$

w being a solution of the equation

$$\ddot{w} + w = F(\varphi), \quad \text{where} \quad F = K^{-1}.$$

The function w can be computed by the variation of constants. The resulting function is 2π-periodic (thanks to the Minkowski conditions):

$$w = \sin \varphi \int_{\varphi}^{\alpha} F(\theta) \cos \theta \, d\theta - \cos \varphi \int_{\varphi}^{\alpha} F(\theta) \sin \theta \, d\theta.$$

PROPOSITION 5. *The above formula defines on the circle $\{\varphi \bmod 2\pi\}$ a function which is positive everywhere except at the point $\varphi = 0$.*

PROOF. We have

$$\int_{\varphi}^{\alpha} F(\theta) \cos \theta d\theta = M(\varphi)\bar{u}(\varphi), \quad \int_{\varphi}^{\alpha} F(\theta) \sin \theta d\theta = M(\varphi)\bar{v}(\varphi),$$

where $(\bar{u}(\varphi), \bar{v}(\varphi))$ is the mass center of the arc $(u = \cos \theta, v = \sin \theta, \ 0 \leq \theta \leq \varphi)$ of density $F(\theta) = K^{-1}(\theta)$, $M(\varphi)$ being the mass of this arc. Thus we represent w in the form

$$w(\varphi) = M(\varphi) \begin{vmatrix} \bar{u}(\varphi) & \cos \varphi \\ \bar{v}(\varphi) & \sin \varphi \end{vmatrix}.$$

If $0 < \varphi \leq \pi$, the mass center lies in the sector $0 < \varphi < \pi$.

If $\pi < \varphi < 2\pi$, the mass center lies of the arc $\{\theta : 0 \leq \theta \leq \varphi\}$ is opposed to the mass center of the complementary arc $\{\theta : \varphi \leq \theta \leq 2\pi\}$. In both cases the direction to the mass center of the arc $\{\theta : 0 \leq \theta \leq \varphi\}$ has the azimuth ψ in the sector $\varphi - \pi < \psi < \varphi$, whence $w > 0$.

PROPOSITION 6. *The dual equation has a positive solution.*

PROOF. Consider $W(\varphi) = w(\varphi) + \varepsilon \cos \varphi$, where $\varepsilon > 0$ is sufficiently small.

Now represent the resulting projective curve f^* as a curve in the dual Euclidean plane. Propositions 4–6 imply

PROPOSITION 7. *The dual curve f^* is defined in the polar coordinates on the Euclidean plane by the equations*

$$p = r \cos \varphi, \quad q = r \sin \varphi, \quad r = W^{-1}.$$

PROPOSITION 8. *The curve from Proposition 7 is convex.*

PROOF. The convexity condition for a star-like curve has the form $A > 0$, where $A = (r - \ddot{r})r + 2\dot{r}^2$. Indeed, we first define

$$A = \det \begin{vmatrix} \dot{p} & \ddot{p} \\ \dot{q} & \ddot{q} \end{vmatrix}$$

and then express it in terms of $r(\varphi)$.

Now for $r = W^{-1}$ and $\ddot{W} + W = K^{-1}$ a (rather strange) calculation shows that $A = K^{-1}W^{-3} > 0$.

The duality theorem of §8 implies that this curve is projectively dual to the original curve in the plane (u, v). Our explicit calculations show moreover the Euclidean duality: on the Euclidean plane (u, v) the equation $uq(\varphi) - vp(\varphi) = 1$ defines the tangent line to the original convex curve.

§10. Proof of the minoration of the number of flattenings of the curves having a convex projection

Consider a curve in $\mathbb{R}P^{2n+1}$ whose projection from 0 to $\mathbb{R}P^{2n}$ is convex. The projected curve does not intersect some hyperplane H in $\mathbb{R}P^{2n}$. The hyperplane in $\mathbb{R}P^{2n+1}$ containing 0 and H does not intersect our curve.

Choose an affine coordinate system $(x_1, \dots, x_{2n}; z)$ in the complement to this hyperplane such that $\mathbb{R}P^{2n}$ is the horizontal plane $z = 0$ with coordinates $(x_1, \dots,$ $x_{2n})$, the direction towards 0 being vertical $(x = 0)$. The convexity of the curve means that the $2n+1$ functions $(1, x_1 = f_1(t), \dots, x_{2n} = f_{2n+1}(t))$ form a Chebyshev system on the circle.

This system is a fundamental system of solutions for a linear differential equation $Ly = 0$. Consider the Lagrange dual equation $L^*z = 0$. Its fundamental system is a Chebyshev system. Indeed, it corresponds to a curve in $\mathbb{R}P^{2n*}$ dual to the original convex curve in $\mathbb{R}P^{2n}$ (according to the Duality Theorem of §8).

The dual curve of a convex curve is convex according to the Corollary of this theorem. Thus the curve f^* corresponding to the equation $L^*z = 0$ is convex. Hence its fundamental system of solutions is a Chebyshev system.

The rest of the proof is the same as in §5, where the curve was supposed to be close to a generalized ellipse. The closeness to the ellipse was only used for the proof of the Chebyshev property of the fundamental system of solutions of the dual equation. The theorem is thus proved for the curves in the odd-dimensional projective spaces.

In the even-dimensional case the arguments are similar if one works on the double covering of the initial curve.

DEFINITION. A *Möbius function* on a circle $\{t \bmod 2\pi\}$ is a 2π-periodic function satisfying the condition $f(t + \pi) = -f(t)$. The $2n$ homogenuous coordinates of the lift of the double covering of a projective noncontractable curve in $\mathbb{R}P^{2n-1}$ can be chosen to be Möbius functions. They form a fundamental system of solutions of the differential equation $Ly = 0$ acting on the space of the Möbius functions. The convexity of the curve means that this system is a Chebyshev system (the number of zeros on $\{t \bmod \pi\}$ being smaller than the dimension of the system).

The pairing B is well defined for the Möbius functions, since the product of two Möbius functions is an ordinary function. The positiveness of the product of two Möbius functions $(Lz)g$ is well defined.

The fundamental system of solutions of the Lagrange dual equation forms a Chebyshev system of Möbius functions and the proof ends as in §5.

REMARK. One might conjecture that in the flattening points counting problems (as well as in the wave front reversion problem and in many other similar problems of the symplectic and contact topology) the really crucial restriction on the perturbations can be formulated in terms of the metamorphosis of the corresponding Legendrian knots (like in the Chekanov's Legendrian Morse theory, see **[10, 12]**). These conjectures, however, are not settled even in the simpler situation of the symplectic and contact geometry versions of the 4 vertices theorem, studied in **[1–4]** and of the Möbius theorem on the 3 inflection point (see **[13]**).

In any case, like the 4 vertices theorem and the Möbius theorem, the flattening points theory and the Chebyshev system theory belong to the symplectic and contact topology rather than to the Euclidean or projective geometry, apparently entering in the definitions of these objects.

References

1. V. I. Arnold, *Topological invariants of plane curves and caustics*, University Lecture Series, vol. 5, Amer. Math. Soc., Providence, RI, 1984 pps..
2. ———, *Invariants and perestroikas of plane fronts*, Trudy Mat. Inst. Steklov. **209** (1995) (to appear).
3. ———, *Sur les propriétés topologiques des projection Lagrangiennes en géométrie symplectique des caustiques,*, Cahiers Math. de la Decision, no. 9320, 14/6/93, Univ. Paris-Dauphine, 1–9.
4. ———, *On topological properties of Legendre projections in contact geometry of wave fronts*, Algebra i Analiz **6** (1994), no. 3, 1–16; English transl. in (St.-Petersburg Math. J. **6** (1995).
5. S. L. Tabachnikov, *Around four vertices*, Russian Math. Surveys **45** (1990), no. 1, 229–230.
6. V. I. Arnold, *Sur une propriété topologique des applications globalement canoniques de la mécanique classique*, C.R. Acad. Sci. Paris, Sec. A **261** (1965), 3719–3722.
7. C. Conley and E. Zehnder, *The Birkhoff-Lewis fixed point theorem and a conjecture of V. I. Arnold*, Invent. Math. **73** (1983), 33–49.
8. D. Floer, *Proof of the Arnold conjecture and generalizations to certain Kähler manifolds*, Duke Math. J. **53** (1986), 1–32.
9. M. Chaperon, *Quelque questions de géométrie symplectique [d'aprés, entre autres, Poincaré Arnold, Conley et Zehnder]*, Astérisque **105–106** (1983), 231–249.
10. Yu. V. Chekanov, *Legendre Morse theory*, Uspehi Mat. Nauk **42** (1987), no. 4, 139–141. (Russian)
11. F. Laudenbach and J.-C. Sikorav, *Persistence d'intersection avec la section nulle au course d'une isotopie hamiltonienne dans le fibré cotangent*, Invent. Math., **82** (1985), no. 2, 349–358.
12. V. I. Arnold, *First steps of symplectic topology*, Russian Math. Surveys **41** (1986), no. 6, 1–21.
13. ———, *Ramified covering $\mathbb{C}P^2 \to S^4$, hyperbolicity and projective topology*, Sibirsk. Mat. Zh. **29** (1988), no. 4, 36–47; English transl. in Siberian Math. J. **29** (1988).
14. V. Ovsienko and B. Khesin, *Symplectic leaves of the Geffaud-Dickey Brackets and homotopy classes of nondegenerate curves*, Functional. Anal. Appl. **24** (1990), no. 1, 62–69.

STEKLOV MATHEMATICAL INSTITUTE, MOSCOW, RUSSIA

Amer. Math. Soc. Transl.
(2) Vol. **171**, 1996

Invariant Tori and Symplectic Topology

M. Bialy and L. Polterovich

§1. Introduction and an overview

1.1. Introduction. As it was understood in the sixties (see **[A]**), many qualitative questions of classical mechanics should be considered in the framework of *symplectic topology*, a far-reaching generalization of the theory of curves and areas on surfaces. During the past 10 years exciting progress in symplectic topology was made due to the development of powerful techniques based on nonlinear functional analysis and algebraic geometry (we mention here the pioneer works **[C-Z, G]**, see **[H-Z, M-Sa]** for further references).

In the present paper we discuss some applications of modern symplectic topology to the theory of invariant tori of Hamiltonian systems. We introduce a special class of Hamiltonians on the cotangent bundle of the n-dimensional torus and concentrate on the following subjects[1]:

- dynamics on invariant tori and topology of Lagrangian submanifolds;
- converse KAM-theory and bi-invariant geometry on the group of Hamiltonian diffeomorphisms.

This paper is based on works **[E-P2]** (with Yakov Eliashberg) and **[B-P2]**, and contains further developments of their results.

1.2. Acknowledgments. The Tori were given to us by Sinai (cf. Exodus) when we worked under his supervision about 10 years ago. We express our gratitude to Professor Ya. G. Sinai for introducing us to this beautiful field.

We also thank F. Lalonde and D. McDuff for the explaination of some important ideas that are used in §3.

1.3. A class of Hamiltonians. Consider the cotangent bundle $T^*\mathbb{T}^n$ of the n-dimensional torus endowed with the canonical coordinates $(p, q \bmod 1)$. Denote by B^n the Euclidean unit ball $\{p|\ |p| \leq 1\}$, and by $B^*\mathbb{T}^n$ the unit ball bundle $\{(p,q)|\ |p| \leq 1\}$.

A smooth function $H : \mathbb{R}^n \to [-1;\infty)$ (respectively, $H : B \to [-1;0]$) is called *unimodal* if it is proper (respectively, vanishes on the boundary ∂B) and (in both cases!) has a unique critical point y, which is nondegenerate and satisfies $H(y) = -1$.

1991 *Mathematics Subject Classification*. Primary 58F99; Secondary 53C15.

[1]Various applications related to multi-dimensional Birkhoff's theory will not be discussed here, see **[B-P1, Au-Laf]** for the exposition of these results.

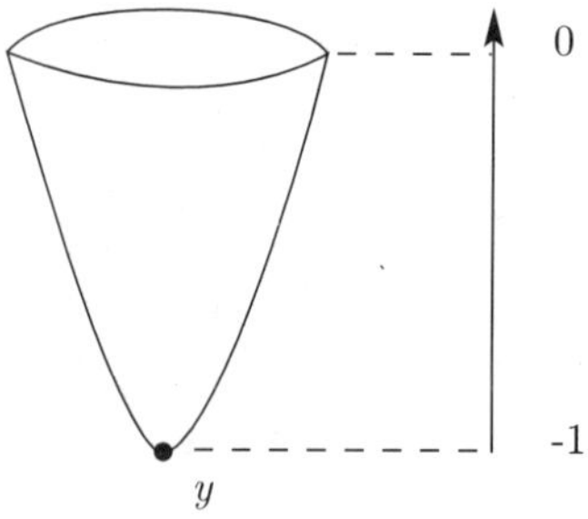

Clearly, y is a point of minimum of H. The graph of a unimodal function is drawn in Figure 1.

We say that a smooth function $H(p, q)$ on $T^*\mathbb{T}^n$ or on $B^*\mathbb{T}^n$ *belongs to the class* $\mathcal{F}$ if it is fiberwise unimodal, that is, for all $q \in \mathbb{T}^n$ the function $H(\cdot, q)$ is unimodal. The minima set of such a function is denoted by $\Sigma(H)$. Obviously, $\Sigma(H)$ is a smooth section of the cotangent bundle.

A function H is called *optical* if its restriction to every fiber of the cotangent bundle is strictly convex: $H_{pp} > 0$.

EXAMPLE 1.3.A. Let g be a Riemannian metric on $\mathbb{T}^n$. Then the Hamiltonian $H_g(p, q) = \frac{1}{2}|p|_g^2 - 1$ generates the corresponding geodesic flow on $T^*\mathbb{T}^n$. This function is optical, belongs to $\mathcal{F}$, and the minima set $\Sigma(H_g)$ coincides with the zero section.

1.4. Invariant tori: basic questions. Let H be a function of the class $\mathcal{F}$. Below we study those invariant tori L of the Hamiltonian flow generated by H that satisfy the following conditions:

- L is embedded and Lagrangian with respect to the standard symplectic structure $\omega = dp \wedge dq$ on $T^*\mathbb{T}^n$;
- L is homotopic to the zero section;
- L lies in a regular energy level $\{H = \text{const}\}$.

Such invariant tori appear naturally in the context of integrable Hamiltonian systems (Liouville-Arnold theorem) and their small perturbations (Kolmogorov-Arnold-Moser theorem).

Results discussed below arose from an attempt to understand the behavior of invariant tori in systems that are not necessarily close to integrable ones. The following two questions are of particular interest.

QUESTION 1.4.A (homotopical dynamics). Let ξ be the restriction of the Hamiltonian vector field to an invariant torus L. Is it true that ξ is homotopic to a "constant" vector field through nonsingular vector fields on L?

QUESTION 1.4.B (converse KAM). What are the obstructions to the existence of invariant tori in a given energy level?

Some answers to these questions are given in 1.5 and 1.6 below. Interestingly enough, they strongly depend on the property of the minima set $\Sigma(H)$ being Lagrangian.

1.5. Homotopic universality of quasi-periodic motion. Invariant tori that arise in integrable systems and KAM theory carry quasi-periodic motion. Hence the question 1.4.A above can be formulated as follows: is dynamics on a general invariant torus "homotopic" to the quasi-periodic dynamics?

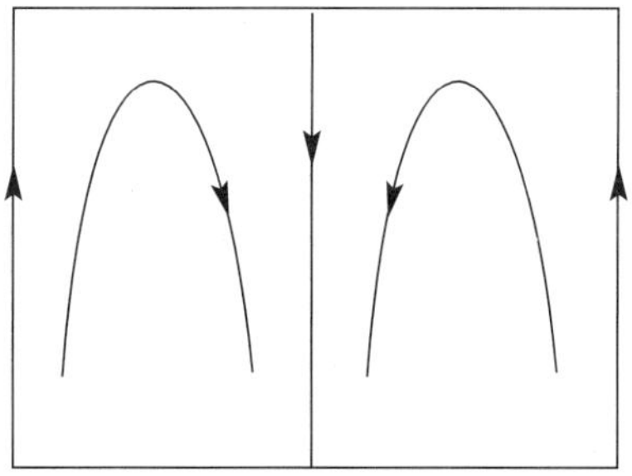

Recall that the homotopy classes of nonsingular vector fields on the n-dimensional torus $\mathbb{T}^n = \mathbb{R}^n/\mathbb{Z}^n$ are in one-to-one correspondence with the homotopy classes of maps $\mathbb{T}^n \to S^{n-1}$. In particular, the set of different classes is infinite. We say that a nonsingular vector field on $\mathbb{T}^n$ is *homotopically trivial* if it is homotopic to a translation invariant vector field. One can show that the class of homotopically trivial vector fields is the unique Diff $(\mathbb{T}^n)$-invariant homotopy class. An example of a homotopically *nontrivial* vector field on the 2-torus is given in Figure 2. Here the vector field $v(x)$ makes a full turn around the vertical direction while the point x moves along the horizontal cycle. Such fields have rather unpleasant dynamical properties (at least from the common point of view accepted in classical mechanics). For instance, they must have closed orbits (see **[R]** for the proof of this and many other interesting results), and cannot preserve a smooth measure (this follows immediately from the Laudenbach theorem cited in the proof of Theorem 1.5.A below).

For a function H from $\mathcal{F}$ denote by ξ_H the Hamiltonian vector field.

THEOREM 1.5.A (cf. **[P1]**). *Let $H \in \mathcal{F}$ be an optical Hamiltonian function, and L an embedded Lagrangian torus which is homotopic to the zero section and lies in a regular level of H. If the minima set $\Sigma(H)$ is Lagrangian, then the restriction of the Hamiltonian field ξ_H to L is homotopically trivial.*

REMARK 1.5.B. In the case when $\Sigma(H)$ is not Lagrangian, one can achieve an arbitrary homotopy type of dynamics. Indeed, let $v(q)$ be *an arbitrary* vector field on $\mathbb{T}^n$ of unit Euclidean length. Set $H(p,q) = \frac{1}{2}(p + v(q))^2 - 1$. This Hamiltonian clearly belongs to $\mathcal{F}$ and the level $\{H = -\frac{1}{2}\}$ contains the zero section Z. We can identify Z with the initial torus $\mathbb{T}^n$. Then the restriction of the Hamiltonian vector field to Z coincides with $v(q)$.

PROOF OF THEOREM 1.5.A. A remarkable theorem by F. Laudenbach (see **[La]**) states that a vector field v on $\mathbb{T}^n$ is homotopically trivial provided there exists a closed 1-form λ on $\mathbb{T}^n$ such that $\lambda(v) > 0$ at every point. Assume without loss of generality that $\Sigma(H)$ is the zero section. Then the 1-form $\lambda = pdq$ is closed on L, since $d\lambda = \omega$ and L is Lagrangian. Moreover, $\lambda(\xi_H) = pH_p > 0$ since H is strictly convex on every fiber and assumes a minimum at 0. Hence $\xi_H|_L$ is homotopically trivial in view of Laudenbach's theorem. □

In general it is unknown whether the assertion of Theorem 1.5.A remains true without additional restrictions on H that imply the existence of a 1-form λ on the energy level such that $d\lambda = \omega$ and $\lambda(\xi_H) > 0$. However, in the *low-dimension case* $n = 2$ the following fairly general result holds.

THEOREM 1.5.C (cf. **[E-P2]**). *Let $H \in \mathcal{F}$ be a Hamiltonian function such that $\Sigma(H)$ is Lagrangian. Let $L \subset T^*\mathbb{T}^2$ be an embedded Lagrangian torus which is homotopic to the zero section and lies in a regular level of H. Then the Hamiltonian field on L is homotopically trivial.*

This theorem is closely related to the symplectic rigidity phenomenon for embedded Lagrangian submanifolds. Its proof is based on the technique developed in the paper **[E-P2]**, where a similar result for invariant tori in the symplectic linear space was obtained. More details can be found in §2 below.

1.6. Hofer's geometry and invariant tori of optical flows. In the present section we work on the unit ball bundle $M = B^*\mathbb{T}^n$. Let $H \in \mathcal{F}$ be an optical function that has the following additional property:

there exists $E_0 > 0$ such that for all $E \in [0; E_0]$ the level
set $\{H = -E\}$ is given by $\{p_1^2 + p_2^2 = \text{const}\}$.

Let L be a Lagrangian torus lying in a level set of H. Then it is automatically invariant under the Hamiltonian flow g_H^t of H. Suppose in addition that L is homotopic to the zero section and the restriction of the flow g_H^t to L has an invariant measure which is positive on open sets. According to the multi-dimensional version of the Birkhoff's second theorem obtained in **[P2, B-P1]**, the torus L is *a section* of the cotangent bundle. In view of this discussion we arrive at the following natural question of the converse KAM theory. Define $E(H)$ as the supremum of those $E \in [0; 1]$ for which the energy level $\{H = -E\}$ contains a Lagrangian section. Clearly, $E(H) \geq E_0$. However, the KAM-theory implies that $E(H)$ is *strictly greater* than E_0. How to estimate the value of $E(H)$ from above? Notice that if the minima set $\Sigma(H)$ is Lagrangian, then $E(H) = 1$. Below we present a nontrivial bound for $E(H)$ in the case when $\Sigma(H)$ is not Lagrangian. Such an estimate appears from *Hofer's geometry* on the group of Hamiltonian diffeomorphisms.

Hofer's geometry is defined as follows. Let $\mathfrak{d}$ be the Lie algebra (with respect to the Poisson brackets) of smooth functions on M that vanish on the boundary and are constant on hypersurfaces $\{p_1^2 + p_2^2 = \text{const}\}$ near the boundary. Denote by D the corresponding Lie group of Hamiltonian diffeomorphisms. By definition, every element of D is the time-one map of the Hamiltonian flow generated by a function $H(t, x)$, $t \in [0; 1]$, $x \in M$, that satisfies the following condition:

$$H(t, \cdot) \in \mathfrak{d} \quad \text{for all } t \in [0; 1].$$

Define the norm $\|\cdot\|$ on $\mathfrak{d}$ by the formula

$$\|H\| = \max_M H - \min_M H.$$

This norm is invariant under the adjoint action of D on $\mathfrak{d}$, that is

$$\|H \circ \varphi\| = \|H\| \quad \text{for all } H \in \mathfrak{d}, \varphi \in D.$$

Hence it defines a bi-invariant Finsler metric on D, which is called *Hofer's metric*. The length of a path $\gamma(t), t \in [a; b]$, generated by a Hamiltonian function $H(t, x)$, is calculated as follows:

$$\text{length}(\gamma) = \int_a^b \|H(t, \cdot)\| dt.$$

The length structure on D defines a bi-invariant (pseudo) distance d on D as follows:

$$d(\varphi, \psi) = \inf_{\gamma} \operatorname{length}(\gamma),$$

where $\varphi, \psi \in D$ and the infimum is taken over all smooth paths γ connecting φ and ψ. It is a deep fact that d is a *genuine* distance function, that is, $d(\varphi, \psi) \neq 0$ provided $\varphi \neq \psi$ (see **[H2, 3, P3, L-M1]**).

For a function $H \in \mathfrak{d} - \{0\}$ consider the path $\{g_H^t\}_{t\in\mathbb{R}}$ on D. Define a quantity $\mu(H)$, which measures asymptotic nonminimality of this path, in the following way:

$$\mu(H) = \lim_{t\to+\infty} \frac{d(\mathbb{1}, g_H^t)}{t\|H\|}$$

(note that the limit exists since the function $t \mapsto d(\mathbb{1}, g_H^t)$ is subadditive in view of the triangle inequality). Clearly $\mu(H) \leq 1$. If $\{g_H^t\}$ is *a minimal geodesic*, that is $d(g_H^{t_1}, g_H^{t_2}) = \|H\| \cdot |t_1 - t_2|$ for all $t_1, t_2 \in \mathbb{R}$, then $\mu(H) = 1$.

Interestingly enough, the energy value $E(H)$ which is responsible for the destruction of invariant tori can be estimated in terms of the asymptotic nonminimality $\mu(H)$.

THEOREM 1.6.A. *Let $H \in \mathfrak{d}$ be an optical Hamiltonian of the class $\mathcal{F}$ and $\Sigma(H)$ the minima set of H.*

1. *The inequality $E(H) \leq \mu(H)$ holds;*
2. *If $\Sigma(H)$ is Lagrangian, then $E(H) = \mu(H) = 1$ and the path $\{g_H^t\}_{t\in\mathbb{R}}$ is a minimal geodesic on D;*
3. *If $\Sigma(H)$ is not Lagrangian, then $\mu(H) < 1$, and hence the path $\{g_H^t\}$ is not minimal.*

For the proof of parts (1) and (2) of Theorem 1.6.A we refer the reader to **[B-P2, 1.4.C]**. The last assertion is a new one, and we prove it in §3.

§2. Topology of Lagrangian tori and homotopical dynamics

In the present section we discuss several results that illustrate the so-called symplectic rigidity phenomenon for embedded Lagrangian tori in $T^*\mathbb{T}^2$. As an application, we prove Theorem 1.5.C (see 2.5 below). In order to make the exposition more clear, we neither give the shortest proof nor define the intermediate notions in maximal generality.

2.1. Unknottedness of Lagrangian tori in $T^*\mathbb{T}^2$. Consider the space of all smooth embedded tori in $T^*\mathbb{T}^2$ that are homotopic to the zero section. This space has an infinite number of connected components with respect to the C^∞-topology. In other words, there is an infinite number of different *knot types* $\mathbb{T}^2 \to T^*\mathbb{T}^2$ in the homotopy class of the zero section. However, only one of them can be represented by a Lagrangian torus!

THEOREM 2.1.A **[E-P1]**. *Every embedded Lagrangian torus in $T^*\mathbb{T}^2$ that is homotopic to the zero section is, in fact, isotopic to the zero section.*

REMARK 2.1.B. It is still unknown if such a torus is isotopic to the zero section through *Lagrangian tori*.

COROLLARY 2.1.C. *Let $L \subset \{p_1^2 + p_2^2 < 1\}$ be an embedded Lagrangian torus homotopic to the zero section. Then the complement $T^*\mathbb{T}^2 - L$ can be retracted to the 3-torus $\{p_1^2 + p_2^2 = 1\}$.*

This follows immediately from 2.1.A.

2.2. Giroux's theorem. In what follows we denote by Z the zero section of $T^*\mathbb{T}^2$. Notice that $T^*\mathbb{T}^2 - Z$ can be retracted to the 3-torus $\{p_1^2 + p_2^2 = 1\}$. Consider the homotopy classes $\mathbb{T}^2 \to T^*\mathbb{T}^2 - Z$ represented by those tori that are homotopic to a section. Clearly, they are classified by mappings from $\mathbb{T}^2$ to the unit circle in the (p_1, p_2)-plane. It turns out that only one of these classes contains an embedded Lagrangian torus!

THEOREM 2.2.A ([**Gi**], see also [**E-P2**]). *Let $L \subset T^*\mathbb{T}$ be an embedded Lagrangian torus that is homotopic to the zero section and does not intersect the zero section. Then L is homotopic to the flat section $\{p_1 = 1, p_2 = 0\}$ inside $T^*\mathbb{T}^2 - Z$.*

2.3. Two additional structures. Recall that (p_1, p_2, q_1, q_2) are canonical coordinates on $T^*\mathbb{T}^2$ and the symplectic form ω is written as $dp_1 \wedge dq_1 + dp_2 \wedge dq_2$. We identify all tangent spaces $T_{(.)}T^*\mathbb{T}^2$ with the linear space $\mathbb{R}^4$. In what follows, we use two additional structures:

- the complex structure $J : \mathbb{R}^4 \to \mathbb{R}^4$ given in (p, q)-coordinates by the matrix

$$\begin{pmatrix} 0 & -\mathbb{1} \\ \mathbb{1} & 0 \end{pmatrix};$$

- the Euclidean metric $dp_1^2 + dp_2^2 + dq_1^2 + dq_2^2$.

They are related to the symplectic structure ω as follows: $(\xi, \eta) = \omega(\xi, J\eta)$ for all $\xi, \eta \in \mathbb{R}^4$. In the examples below we illustrate the significance of these notions.

EXAMPLE 2.3.A. A surface $L \subset T^*\mathbb{T}^2$ is Lagrangian if and only if $J(T_xL)$ is orthogonal to T_xL for all $x \in L$.

EXAMPLE 2.3.B. The Hamiltonian field ξ_H of a function $H : T^*\mathbb{T}^2 \to \mathbb{R}$ equals $J \operatorname{grad} H$.

2.4. Framings. In order to handle homotopical dynamics on an invariant Lagrangian torus it is convenient to pass from nonvanishing tangent vector fields to nonvanishing normal vector fields, or *framings*.

Given a Lagrangian torus $L \subset T^*\mathbb{T}^2$, we denote by $\mathcal{T}(L)$ the set of homotopy classes of nonvanishing tangent vector fields on L and by $\mathcal{N}(L)$ the set of homotopy classes of nonvanishing normal vector fields along L. In view of 2.3.A, the complex structure J yields the following correspondence j between $\mathcal{N}(L)$ and $\mathcal{T}(L)$:

$$j([v]) = [Jv]$$

for every framing v. Clearly, j is a one-to-one correspondence.

Both sets $\mathcal{T}(L)$ and $\mathcal{N}(L)$ have distinguished elements. Of course, for $\mathcal{T}(L)$ it is just the class of a homotopically trivial vector field. In order to describe *distinguished framings* we have to carry out some minor additional work.

Assume henceforth that the torus L lies in $\{p_1^2 + p_2^2 < 1\}$. This can always be achieved by a suitable homothety along the fibers of the cotangent bundle.

For a framing v of L consider a small shift $L + \varepsilon v$ of L along v. Clearly, $L + \varepsilon v$ is contained in the complement to L. Applying Corollary 2.1.C, we see that $L + \varepsilon v$ is homotopic inside $T^*\mathbb{T}^2 - L$ to a section of the cotangent bundle that lies in the 3-torus $\{p_1^2 + p_2^2 = 1\}$.

DEFINITION 2.4.A. A framing v is called *distinguished* if $L+\varepsilon v$ is homotopic inside $T^*\mathbb{T}^2 - L$ to a flat Lagrangian section $\{p_1 = 1, p_2 = 0\}$.

The following result is a straightforward consequence of Theorem 1.1.A in **[E-P2]**.

THEOREM 2.4.B (cf. **[E-P2]**). *Let $L \subset T^*\mathbb{T}^n$ be an embedded Lagrangian torus homotopic to the zero section. The map $j : \mathcal{N}(L) \to \mathcal{T}(L)$ takes the class of distinguished framings to the class of homotopically trivial tangent vector fields.*

REMARK 2.4.C. Distinguished framings, as well as the map $j : \mathcal{N}(L) \to \mathcal{T}(L)$, can be defined in a somewhat similar manner for a wider class of embedded 2-tori, namely for *totally real* tori. Recall that a surface L is totally real with respect to the complex structure J if for all $x \in L$ the space $J(T_xL)$ is *transversal* (but in general, not orthogonal) to T_xL. One can show that the assertion of Theorem 2.4.B is *not* true for totally real tori (cf. **[P1]**), and hence describes a specifically symplectic phenomenon.

2.5. Proof of Theorem 1.5.C. Now we are ready to prove Theorem 1.5.C. Let $H \in \mathcal{F}$ be a Hamiltonian function such that the set $\Sigma(H)$ is a Lagrangian section. Without loss of genericity we may assume that $\Sigma(H)$ coincides with the zero section Z. Let $L \subset \{H = h\}$, $h > -1$ be an embedded Lagrangian torus homotopic to the zero section. Set $W = \{H \leq h\}$, and $N = \partial W = \{H = h\}$. We assume that $W \subset \{p_1^2 + p_2^2 < 1\}$. Denote by v the field grad H along L. The rest of the proof is divided into two steps.

1) The field v is orthogonal to N and hence is a framing of L. We claim that this framing is a distinguished one (see 2.4.A). Indeed, v is directed outside W and thus a small shift $L + \varepsilon v$ is contained in $T^*\mathbb{T}^2 - W$. Obviously, $T^*\mathbb{T}^2 - W$ is a deformation retract of $T^*\mathbb{T}^2 - Z$. Moreover, L, and therefore $L + \varepsilon v$, is homotopic to a flat section $K = \{p_1 = 1, p_2 = 0\}$ in $T^*\mathbb{T}^2 - Z$ in view of Giroux's theorem 2.2.A. Since K is contained in $T^*\mathbb{T}^2 - W$, we obtain that $L + \varepsilon v$ is homotopic to K in $T^*\mathbb{T}^2 - W$, and therefore in $T^*\mathbb{T}^2 - L$. By definition, this means that v is a distinguished framing, and our claim follows.

2) Since $v = \operatorname{grad} H|_L$ is a distinguished framing, we get from 2.4.B that $Jv = Js \operatorname{grad} H|_L$ is a homotopically trivial vector field on L. But $J \operatorname{grad} H|_L$ equals the Hamiltonian vector field $\xi_H|_L$ (see 2.3.B). This completes the proof. □

2.7. Techniques. Proofs of Theorems 2.1.A., 2.2.A, and 2.4.B above concerning the rigidity of embedded Lagrangian tori are based on the powerful technique of pseudo-holomorphic curves in symplectic manifolds, which was introduced by M. Gromov in **[G]**. We refer the reader to a survey **[E-P3]** for related discussion.

§3. Estimation of distances in Hofer's geometry

In this section we prove part (3) Theorem 1.6.A. We use notations introduced in 1.6.

3.1. A symplectic invariant of functions. For a function $H \in \eth - \{0\}$ define

$$\delta(H) = \inf_{\varphi \in D} \frac{\|H + H \circ \varphi\|}{2\|H\|}.$$

Notice that $\delta(H) \leq 1$. Moreover the equality $\delta(H) = 1$ implies that for every $\varphi \in D$

$$\|H + H \circ \varphi\| = 2\|H\|,$$

and hence the function H must satisfy the following very strong conditions:

$$\varphi(\text{minset } H) \cap \text{minset } H \neq \varnothing,$$
$$\varphi(\text{maxset } H) \cap \text{maxset } H \neq \varnothing$$

for all $\varphi \in D$.

Here minset H (resp. maxset H) denotes the set of all points of minimum (resp. maximum) of a function H.

3.2. Computations for functions of class $\mathcal{F}$.

THEOREM 3.2.A. *Let $H \in \mathfrak{d}$ be a function of class $\mathcal{F}$ and $\Sigma(H)$ the minima set of H. Then*

1. *If $\Sigma(H)$ is Lagrangian, then $\delta(H) = 1$;*
2. *If $\Sigma(H)$ is not Lagrangian, then $\delta(H) < 1$.*

PROOF. If $\Sigma(H)$ is a Lagrangian section, then $\varphi(\Sigma(H)) \cap \Sigma(H) \neq \varnothing$ for all $\varphi \in D$ in view of Arnold's conjecture (see **[G, H, La-S1]** for the proof).

If $\Sigma(H)$ is not Lagrangian, then it follows from **[P4]** (see also **[La-S2]**) that there exists $\varphi \in D$ such that $\varphi(\Sigma(H)) \cap \Sigma(H) = \varnothing$. The statement of Theorem 3.2.A follows immediately in view of the discussion in 3.1. □

3.3. A curve-shortening procedure.

THEOREM 3.3.A. *For every $H \in \mathfrak{d} - \{0\}$ the following inequality holds*

$$\mu(H) \leq \delta(H).$$

PROOF. The proof is based on the combination of the Sikorav trick **[S]** and the Lalonde-McDuff shortening procedure **[L-M2]**. Set $g_H^t = g^t$ and take arbitrary $\varphi \in D$ and $T > 0$.

Decompose g^{2T} in the following manner:

$$g^{2T} = B \circ A,$$

where

$$B = g^T \varphi g^T \varphi^{-1}, \quad A = \varphi g^{-T} \varphi^{-1} g^T.$$

Since the metric d is bi-invariant, the triangle inequality implies that

$$d(\mathbb{1}, A) \leq 2d(\mathbb{1}, \varphi),$$

and hence also

$$d(\mathbb{1}, g^{2T}) \leq d(\mathbb{1}, A) + d(\mathbb{1}, B) \leq d(\mathbb{1}, B) + 2d(\mathbb{1}, \varphi).$$

To estimate $d(\mathbb{1}, B)$ note that B is the time-one map of the path $g^{tT} \varphi g^{tT} \varphi^{-1}$, $t \in [0, 1]$. This path is generated by the Hamiltonian function $T\big(H(x) + H(\varphi^{-1} g^{-tT} x)\big)$ and thus, by definition, $d(\mathbb{1}, B)$ must be less than or equal to the length of this path:

$$d(\mathbb{1}, B) \leq T \int_0^1 \|H + H \circ \varphi^{-1} \circ g^{-tT}\| dt.$$

Note that

$$\|H + H \circ \varphi^{-1} \circ g^{-tT}\| = \|H \circ g^{tT} + H \circ \varphi^{-1}\| = \|H + H \circ \varphi^{-1}\|,$$

since the energy H is preserved by its Hamiltonian flow. Thus we have

$$d(\mathbb{1}, g^{2T}) \leq T\|H + H \circ \varphi^{-1}\| + 2d(\mathbb{1}, \varphi),$$

hence

$$\frac{d(\mathbb{1}, g^{2T})}{2T\|H\|} \leq \frac{\|H + H \circ \varphi^{-1}\|}{2\|H\|} + \frac{d(\mathbb{1}, \varphi)}{T\|H\|}.$$

This implies the estimate

$$\mu(H) \leq \inf_{\varphi \in D} \frac{\|H + H \circ \varphi^{-1}\|}{2\|H\|} = \delta(H)$$

and completes the proof. □

3.4. Proof of part (3) of Theorem 1.6.A. Combining the second statement of Theorem 3.2.A with Theorem 3.3.A we get $\mu(H) \leq \delta(H) < 1$. This completes the proof. □

3.5. Variational theory of geodesics. Let $H \in \mathfrak{d}$ be an optical Hamiltonian function of class $\mathcal{F}$. As we have shown above, the path g_H^t is not a minimal geodesic on the group D if $\Sigma(H)$ is not Lagrangian. Nevertheless, it can be shown that sufficiently short segments of g_H^t are minimal and hence g_H^t is a geodesic on D. In addition, the direct analysis of our proof gives nonminimality of the path g_H^t only for large scales of T.

On the other hand, one can expect that the variational theory of geodesics on D built in the papers **[U, L-M2]** works for manifolds with boundary. The results of this theory would imply that a segment $\{g_H^t\}_{t\in[0;T]}$ is a *local* minimum of the length functional with fixed end points provided there exists a point $x \in \Sigma(H)$ such that the linearized flow at this point has no nontrivial periodic orbits of period less than or equal to T. Note that if the tangent space at x to $\Sigma(H)$ is Lagrangian, then there are no such orbits for all T. Indeed, if a linear flow e^{tA} preserves a Lagrangian subspace, then the transformation A has only zero eigenvalues. In the dimension $2n = 4$ such a point x on $\Sigma(H)$ always exists since $\int_{\Sigma(H)} \omega = 0$. Therefore, if $\Sigma(H)$ is not Lagrangian, then g_H^t is a *locally* minimal but not minimal geodesic on D.

3.6. A generalization. The assertion of Theorem 3.3.A above can be slightly refined and extended to arbitrary symplectic manifolds.

Let $\mathfrak{d}$ be the Lie algebra of compactly supported Hamiltonian functions on a symplectic manifold M, and let D be the corresponding group of Hamiltonian diffeomorphisms. The L_∞-norm on $\mathfrak{d}$ defines a bi-invariant Finsler metric on D, which generates the distance function d.

Instead of $\mu(H)$ consider a finer characteristic $\widetilde{\mu}(H)$ of the asymptotic nonminimality of the path $\{g_H^t\}$, which takes into account the fundamental group of D. Namely, set $\alpha_H(T) = \inf_\gamma \operatorname{length} \gamma$, where the infimum is taken over all smooth paths γ on D that join the identity with g_H^T and are homotopic to the segment $\{g_H^t\}_{t\in[0;T]}$ with fixed endpoints. Again, $\alpha_H(T)$ is a subadditive function and hence the limit

$$\widetilde{\mu}(H) = \lim_{t\to+\infty} \frac{\alpha_H(T)}{T\|H\|}$$

is well defined. Obviously, $\mu(H) \leq \widetilde{\mu}(H)$.

Precisely as in 3.1, we set

$$\delta(H) = \inf_{\varphi \in D} \frac{\|H + H \circ \varphi\|}{\|H\|}.$$

THEOREM 3.6.A. *For all $H \in \mathfrak{d}$ the inequality*

$$\widetilde{\mu}(H) \leq \delta(H)$$

holds.

The proof is a minor modification of the one given in 3.3 above.

§4. Open problems

4.1. Homotopical dynamics in higher dimensions. Let H be a Hamiltonian function of class $\mathcal{F}$ on $T^*\mathbb{T}^n$ such that the minima set is Lagrangian. Let $L \subset T^*\mathbb{T}^n$ be an embedded Lagrangian torus that is homotopic to the zero section and lies in a regular level of H. *Is it true that the restriction of the Hamiltonian vector field to L is homotopically trivial?*

Theorem 1.5.B above gives a positive answer for $n = 2$. In higher dimensions the answer is known only for optical Hamiltonians (see 1.5.A above).

4.2. The critical energy $E(H)$ revisited. Let $H : B^*\mathbb{T}^n \to [-1; 0]$ be an optical function that belongs to $\mathcal{F} \cap \mathfrak{d}$ (see 1.6). Define $\widetilde{E}(H)$ as the supremum of those $E \in [0; 1]$ for which the energy level $\{H = -E\}$ contains an embedded Lagrangian torus homotopic to the zero section. Let us emphasize that in contrast to 1.6 such a torus is not assumed to be a section. Clearly, $\widetilde{E}(H) \geq E(H)$.

How to estimate $\widetilde{E}(H)$ from the above? This question is closely related to the following problem of the multi-dimensional Birkhoff's theory. Let L be an invariant torus of the Hamiltonian flow of an optical Hamiltonian function. Suppose that L lies in a regular energy level and is homotopic to the zero section.

Is it true that L is a section of the cotangent bundle? This was proved in **[P2, B-P1]** under the additional assumption that the dynamics on L has an invariant measure which is positive on open sets.

Notice that in all known examples the answer to this question is positive, and hence $E(H) = \widetilde{E}(H)$.

4.3. Ergodic sums and Hofer's geometry. Let D be the group of Hamiltonian diffeomorphisms of a symplectic manifold M. For a function H from the corresponding Lie algebra and a positive integer N define

$$\delta_N(H) = \inf_{\varphi \in D} \left\| \frac{1}{N} \sum_{j=0}^{N-1} H \circ \varphi^j \right\| \cdot \|H\|^{-1}.$$

Notice that $\delta_2(H)$ coincides with the invariant $\delta(H)$ introduced in 3.1.

How is the invariant δ_N related to Hofer's geometry? Theorem 3.3.A above gives an answer for $N = 2$.

References

[A] V. I. Arnold, *Sur une propriété topologique des applications globalement canoniques de la méchanique classique*, C. R. Acad. Sci. Paris **261** (1965), 3719–3722.

[Au-Laf] M. Audin and J. Lafontaine (eds.), *Holomorphic curves in symplectic geometry*, Birkhäuser, Boston, 1994.

[B-P1] M. Bialy and L. Polterovich, *Hamiltonian diffeomorphisms and Lagrangian distributions*, Geom. Funct. Anal. **2** (1992), 173–210.

[B-P2] ———, *Geodesics of Hofer's metric on the group of Hamiltonian diffeomorphisms*, Duke Math. J. **76** (1994), 273–292.

[C-Z] C. Conley and E. Zehnder, *The Birkhoff-Lewis fixed point theorem and a conjecture of V. I. Arnold*, Invent. Math. **73** (1983), 33–49.

[E-P1] Y. Eliashberg and L. Polterovich, *Unknottedness of Lagrangian surfaces in symplectic 4-manifolds*, Internat. Math. Res. Notices **11** (1993), 295–301.

[E-P2] ———, *New applications of Luttinger's surgery*, Comment. Math. Helv. **69** (1994), 512–522.

[E-P3] ———, *The problem of Lagrangian knots*, Proceedings of Georgia Int. Topology Conference (1994) (to appear).

[G] M. Gromov, *Pseudo-holomorphic curves in symplectic manifolds*, Invent. Math. **82** (1985), 307–347.

[Gi] E. Giroux, *Une structure de contact, même tendue, est plus ou moins tordue*, Ann. Sci. Ècole Norm. Sup. (4) **27** (1994), 697–705.

[H1] H. Hofer, *Lagrangian embeddings and critical point theory*, Ann. Inst. H. Poincaré Anal. Non Linèaire **2** (1985), 407–462.

[H2] ———, *On the topological properties of symplectic maps*, Proc. Roy. Soc. Edinburgh Sect. A **115** (1990), 25–38.

[H3] ———, *Estimates for the energy of symplectic map*, Comment. Math. Helv. **68** (1993), 48–72.

[H-Z] H. Hofer and E. Zehnder, *Symplectic invariants and Hamiltonian dynamics*, Birkhäuser, Boston, 1994.

[L-M1] F. Lalonde and D. McDuff, *The geometry of symplectic energy*, Ann. of Math. (2) **141** (1995), 349–371.

[L-M2] ———, *Hofer's L^∞-geometry: energy and stability of Hamiltonian flows. Parts* I, II, Preprint, 1994.

[La] F. Laudenbach, *Les formes différentielles de degré 1 non singulières: classes d'homotopie de leurs noyaux*, Comment. Math. Helv. **51** (1976), 447–464.

[La-S1] F. Laudenbach and J.-C. Sikorav, *Persistence d'intersections avec la section nulle au cours d'une isotopie Hamiltonienne dans un fibre cotangent*, Invent. Math. **82** (1985), 349–357.

[La-S2] ———, *Hamiltonian disjunction and limits of Lagrangian submanifolds*, Internat. Math. Res. Notices **4** (1994).

[M-Sa] D. McDuff and D. Salamon, *Introduction to symplectic topology*, Oxford Univ. Press, New York, 1994.

[P1] L. Polterovich, *New invariants of embedded totally real tori and one problem of Hamiltonian mechanics*, Methods of Qualitative Theory and Theory of Bifurcations, Gorkii, 1988, pp. 84–90. (Russian)

[P2] ———, *The second Birkhoff's theorem for optical Hamiltonian systems*, Proc. Amer. Math. Soc. **113** (1991), 513–516.

[P3] ———, *Symplectic displacement energy for Lagrangian submanifolds*, Ergodic Theory Dynamical Systems **13** (1993), 357–367.

[P4] ———, *An obstacle for non Lagrangian intersections*, A. Floer memorial volume, Birkhäser, Basel (to appear).

[R] B. Reinhart, *Line elements on the torus*, Amer. J. Math. 81 (1959), 617–631.

[S] J.-C. Sikorav, *Systèmes Hamiltoniens et topologie symplectique*, Dipartimento di Matematica dell' Università di Pisa, Ets Editrice, Pisa, 1990.

[U] I. Ustilovsky, *Conjugate points on geodesics of Hofer's metric*, Differential Geom. Appl. (to appear).

School of Mathematical Sciences, Tel Aviv University, Tel Aviv 69978, Israel

Amer. Math. Soc. Transl.
(2) Vol. **171**, 1996

On Morse-Smale Endomorphisms

M. Brin and Ya. Pesin

ABSTRACT. A C^1-map f of a compact manifold M is a Morse–Smale endomorphism if the nonwandering set of f is finite and hyperbolic and the local stable and global unstable manifolds of periodic points intersect transversally. Morse–Smale endomorphisms appear naturally in the dynamics of the evolution operator on the set of traveling wave solutions for lattice models of unbounded media. The main result of this paper is the openness of the set of Morse–Smale endomorphisms in the space $C^1(M, M)$ of C^1-maps of M into itself. The usual order relation on f (given by the intersections of local stable and global unstable manifolds) is used to describe the orbit structure of f and its small C^1-perturbations.

§1. Introduction

Morse–Smale endomorphisms arise naturally in lattice models of unbounded media with the evolution operator of diffusion type (see, e.g., [**AP**]). For such systems, the dynamics of the evolution operator on the set of traveling wave solutions is completely determined by the following multi-dimensional Hénon type map

$$F_\varepsilon(x_1, \dots, x_k, \dots, x_n) = (x_2, \dots, x_{k+1}, \dots, h(x_k) + \varepsilon g(x_1, \dots, x_n)),$$

where $x_i \in \mathbb{R}^d$, $h : \mathbb{R}^d \to \mathbb{R}^d$ is a C^r-diffeomorphism, $r \geq 1$, $g : \mathbb{R}^{dn} \to \mathbb{R}^d$ is a C^r-map, and ε is sufficiently small. If the map F_ε is *chaotic*, i.e., preserves an invariant mixing measure, then the lattice system displays a *spatial-temporal chaos*, i.e., there exists a measure on the set of traveling wave solutions, which is invariant and mixing with respect to both the evolution operator and the space translation operator. It is plausible that in several *physically interesting* situations the dynamics of the map F_ε is completely determined by the map h for all sufficiently small ε.

The first case is when the map h has a locally maximal hyperbolic set. One can easily see that the map F_0 also has a locally maximal hyperbolic set. Note that F_0 is not invertible, whereas F_ε may be a diffeomorphism (this is the case if, for example, one assumes that the matrix $\frac{\partial g}{\partial x_1}$ is nondegenerate). The stability of a locally maximal

1991 *Mathematics Subject Classification*. Primary 58F09.

Key words and phrases. Morse-Smale endomorphisms, stable and unstable manifolds, transversality, pseudo-orbits.

The second named author was partially supported by NSF Grant DMS 94-03723.

hyperbolic set for a C^1-endomorphism under small perturbations by endomorphisms (or diffeomorphisms) was established in [**AP**].

Another case is when the map h is a Morse–Smale diffeomorphism, i.e., its nonwandering set is finite and hyperbolic and the global stable and unstable manifolds of periodic points intersect transversally (since h acts on $\mathbb{R}^d$ one should also assume that infinity is a repelling fixed point for h).

From a physical point of view this situation often occurs when h is one-dimensional. The map F_0 is (see [**AP**]) a *Morse–Smale endomorphism*, i.e., its nonwandering set is finite and hyperbolic and the local stable and global unstable manifolds of periodic points intersect transversally (see §2).

In this context it is important to know whether Morse–Smale endomorphisms form an open set in the C^1-topology. The main result of this paper (see Theorem 4.1) provides a positive answer. A major still open question is whether Morse–Smale endomorphisms are structurally semi-stable, i.e., a small C^1-perturbation of a Morse–Smale endomorphism is topologically semiconjugate to it.

F. Przytycki [**Pr1, Pr2**] studied regular Axiom A endomorphisms (i.e., those that are locally invertible). He proved that an Axiom A endomorphism is structurally stable if and only if it is expanding or is a diffeomorphism.

In §2 we formulate the necessary properties of the stable and unstable manifolds.

In §3 we define Morse–Smale endomorphisms and consider the usual partial order relation $\geq$ on the set of nonwandering points of a Morse–Smale endomorphism f, i.e., $p \geq q$ if the unstable manifold of p intersects the local stable manifold of q. We prove that $\geq$ is a partial order without cycles, and that there are $\delta > 0$ and $\varepsilon > 0$ such that $p \geq q$ if and only if there is an ε-orbit of f from the δ-neighborhood of p to the δ-neighborhood of q. The last property is a major ingredient in the proof of the openness of Morse–Smale endomorphisms in §4.

§2. Stable and unstable manifolds for endomorphisms

We begin with a standard stable manifold theorem for a differentiable map (see [**Rob, Rue, Shu**]). Let p be a fixed point of a C^1-map $f : U \to \mathbb{R}^d$. Denote by $E^s(p)$, $E^u(p)$ the *stable* and *unstable subspaces* spanned by the generalized eigenvectors of $df(p)$ corresponding to the eigenvalues λ with $|\lambda| < 1$ and $|\lambda| > 1$, respectively. The point p is *hyperbolic* if no eigenvalue of $df(p)$ has absolute value 1, or equivalently, $E^s(p)$ and $E^u(p)$ span $\mathbb{R}^d$.

2.1. THEOREM (see [**Rob, Rue, Shu**]). *Let p be a hyperbolic fixed point of a C^1-map $f : U \to \mathbb{R}^d$. Then there exist local stable $W^s_{\text{loc}}(p)$ and unstable $W^u_{\text{loc}}(p)$ manifolds with the following properties*:

(1) *the manifolds $W^s_{\text{loc}}(p)$ and $W^u_{\text{loc}}(p)$ are of class C^1, pass through p, and are tangent at p to the subspaces $E^s(p)$ and $E^u(p)$, respectively*;

(2) *$W^s_{\text{loc}}(p)$ and $W^u_{\text{loc}}(p)$ are invariant under f, i.e.,*

$$f(W^s_{\text{loc}}(p)) \subset W^s_{\text{loc}}(p), \; f(W^u_{\text{loc}}(p)) \supset W^u_{\text{loc}}(p);$$

(3) *there are constants $C > 0$ and $\lambda \in (0, 1)$ such that for any $n > 0$,*

$$d(f^n x, f^n y) < C\lambda^n d(x, y)$$

if $x, y \in W^s_{\text{loc}}(p)$ and

$$d(f^n x, f^n y) > C\lambda^{-n} d(x, y)$$

if $f^k x, f^k y \in W^u_{\text{loc}}(p)$ for $k = 0, 1, \ldots, n$;

(4) *there is* $\delta > 0$ *such that*

$$W^s_{\text{loc}}(p) = \{x \in \mathbb{R}^d \ : \ d(f^n x, p) \le \delta \text{ for all } n \ge 0\},$$
$$W^u_{\text{loc}}(p) = \{x \in \mathbb{R}^d \ : \ \text{there exist points } x_n \in \mathbb{R}^d \text{ such that}$$
$$f^n x_n = x \text{ and } d(f^k x_n, p) \le \delta \text{ for all } n \ge 0 \text{ and } k = 0, 1, \ldots, n\}.$$

The existence of the local unstable manifold $W^u_{\text{loc}}(p)$ is shown in **[Shu]** (see Theorem 5.2). The existence of the local stable manifold $W^s_{\text{loc}}(p)$ is proved in **[Rob]** (see Theorem 10.1).

Denote by $C^1(U, \mathbb{R}^d)$ the space of C^1-maps of a neighborhood $U \subset \mathbb{R}^d$ into $\mathbb{R}^d$ with the C^1-topology.

2.2. THEOREM (see **[Shu, Rue]**). *Let p be a hyperbolic fixed point of a C^1-map $f\colon U \to \mathbb{R}^d$. Then for any $\varepsilon > 0$ there exists an open neighborhood $\mathcal{U} \ni f$ in $C^1(U, \mathbb{R}^d)$ such that every $g \in \mathcal{U}$ has a unique hyperbolic fixed point in the ε-neighborhood of p. The local stable and unstable manifolds of this point depend continuously on $g \in \mathcal{U}$.*

Let $f : U \to \mathbb{R}^d$ be a C^1-map with a hyperbolic fixed point $p \in U$. Define the *global* unstable manifold $W^u(p)$ of p by

$$W^u(p) = \bigcup_{n \ge 0} f^n(W^u_{\text{loc}}(p)).$$

We will need the following lemma which follows directly from the λ-lemma (or inclination lemma) of Palis (see **[PaMe]**).

2.3. LEMMA. *Let $p \in NW(f)$, $R, \varepsilon > 0$. Let G be a submanifold of dimension $k \ge u(p)$ which intersects $W^s_{\text{loc}}(p)$ transversally at a point x, i.e., the intersection of the tangent planes at x has dimension $\le \max(k + s(p) - d, 0)$.*

Then there is $n > 0$ such that $f^n G$ contains a submanifold $\widetilde{G}$ which is C^1 ε-close to the ball of radius R in $W^u(p)$ in the induced metric. □

§3. Morse–Smale endomorphisms and their orbit structure

Let $f : M \to M$ be a C^1-map of a compact d-dimensional Riemannian manifold M. Theorem 2.1 allows one to construct local stable and unstable manifolds $W^s_{\text{loc}}(p)$ and $W^u_{\text{loc}}(p)$ and the global unstable manifold $W^u(p)$ for any hyperbolic *periodic* point p of f.

3.1 DEFINITION. A C^1-map $f : M \to M$ of a compact d-dimensional manifold M is a *Morse–Smale endomorphism* if

(i) the nonwandering set $NW(f)$ is finite and hyperbolic, i.e., $NW(f)$ is the set $\text{Per}(f)$ of periodic points of f and all of them are hyperbolic;

(ii) the local stable and global unstable manifolds of periodic points intersect transversally, i.e., if $x \in W^s_{\text{loc}}(p) \cap W^u(q)$ with $p, q \in \text{Per}(f)$, then $T_x W^s_{\text{loc}}(p) \oplus T_x W^u(q) = T_x M$.

Note that if f is an invertible Morse–Smale endomorphism then it is a Morse–Smale diffeomorphism.

It follows immediately from the definition that any orbit of f eventually enters a small neighborhood of $NW(f)$ and stays in it forever. This implies the following important property of Morse–Smale endomorphisms.

3.2. PROPOSITION. *For any $x \in M$ there is $n > 0$ and $p \in NW(f)$ such that $f^n x \in W^s_{\text{loc}}(p)$.*

We assume now and for the remainder of this section that $f : M \to M$ is a Morse–Smale endomorphism. Following Smale's arguments in the proof of the spectral theorem for Axiom A diffeomorphisms, we define a partial order $\geq$ on $NW(f)$ by $p \geq q$ if $W^u(p) \cap W^s_{\text{loc}}(q) \neq 0$. A point x is called *heteroclinic* if $x \in W^u(p) \cap W^s_{\text{loc}}(q)$ and *transversal heteroclinic* if the intersection is transversal.

3.3. PROPOSITION. *The partial order $\geq$ is transitive and has no cycles, i.e., $p_1 \geq p_2 \geq \cdots \geq p_k = p_1$ implies that $p_i = p_1$, $i = 2, 3, \ldots, k$.*

PROOF. For $p \in NM(f)$ denote by $s(p)$ and $u(p)$ the dimensions of $W^s_{\text{loc}}(q)$ and $W^u(p)$, respectively. If $p \geq q$ then $u(p) \geq u(q)$ by the transversality of the intersections of local stable and global unstable manifolds. The transitivity of $\geq$ follows immediately from Lemma 2.3.

Assume now that $p_1 \geq p_2 \geq \cdots \geq p_k = p_1$. By Lemma 2.3 applied k times to $W^u(p_1)$, the submanifolds $W^u(p_1)$ and $W^s_{\text{loc}}(p_1)$ intersect transversally at a point $x \neq p_1$. It is easy to see that x is a nonwandering point of f with an infinite orbit. This contradicts the fact that f is a Morse–Smale endomorphism. □

For $\delta > 0$ denote by $U_\delta(A)$ the δ-neighborhood of A in M.

3.4. PROPOSITION. (1) *For every $\delta > 0$ there is $n(\delta)$ such that every finite orbit of length at least $n(\delta)$ must enter the δ-neighborhood of $NW(f)$, i.e., for every $x \in M$*

$$\bigcup_{i=0}^{n(\delta)} f^i x \bigcap U_\delta(NW(f)) \neq \emptyset.$$

(2) *For every $\delta > 0$ there is $N(\delta)$ such that the total time that an orbit can spend outside of the δ-neighborhood of $NW(f)$ does not exceed $N(\delta)$, i.e., for every $x \in M$,*

$$\sum_{i=0}^{\infty} \mathbf{1}_{U_\delta(NW(f))}(f^i x) \leq N(\delta).$$

PROOF. Assume that there is a number $\delta > 0$ and a sequence of points x_k such that $f^i x_k \notin U_\delta(NW(f))$ for all k and all $i \leq n_k$, where $n_k \to \infty$ as $k \to \infty$. Since M is compact, the sequence x_k has a limit point x whose positive semiorbit obviously stays out of $U_\delta(NW(f))$. An ω-limit point of x is a nonwandering point of f lying outside $U_\delta(NW(f))$. This is a contradiction, which proves the first statement. The second statement can be proved in a similar way. □

A sequence of points $z_k \in M$, $k = 1, \ldots, n$, is called an *ε-orbit* if $d(f z_k, z_{k+1}) \leq \varepsilon$. We formulate an analog of Proposition 3.4 for ε-orbits. The proof is quite similar to the proof of Proposition 3.4.

3.5. PROPOSITION. *For every $\delta > 0$ there is $n(\delta) > 0$ and $\varepsilon > 0$ such that for every ε-orbit $\{z_k\}$, $k = 1, \ldots, n$ with $n \geq n(\delta)$,*

$$\bigcup_{k=0}^{n} z_k \bigcap U_\delta(NW(f)) \neq \emptyset. \quad \square$$

We will characterize the partial order $\geq$ in terms of the behavior of the orbits of f. We do this under the additional assumption that the nonwandering set of f consists only of fixed points which is sufficient for the proof of the perturbation Theorem 4.1. However, the corresponding arguments in the proofs of Propositions 3.8, 3.9, and 3.11 below can be easily modified to work in the general case.

Assume that any point in $NW(f)$ is a fixed point of f. Denote by $U_\delta(x)$ the δ-neighborhood of $x \in M$.

Given $\delta > 0$ and $p, r \in NW(f)$ we say that r *δ-follows* p if there exist a sequence of points $x_n \in M$ and sequences of integers $a_n, b_n, c_n \to \infty$, $a_n < b_n < c_n$ such that

1. $x_n \to p$ and $f^{c_n} x_n \to r$;
2. $f^k x_n \in U_\delta(p)$ for $0 \leq k \leq a_n$ and $f^k x_n \in U_\delta(r)$ for $b_n < k \leq c_n$;
3. $f^k x_n \notin U_\delta(NW(f))$ for $a_n < k \leq b_n$.

We need the following two lemmas.

3.6. Lemma. *Let $p, q \in NW(f)$ and assume that there are sequences of points $x_n \to p$ and integers $t_n \to \infty$ such that $f^{t_n} x_n \to q$. Then there exists a point $r \in NW(f)$ such that r δ-follows p, a sequence of points $y_n \in M$ and a sequence of integers $\bar{k}_n$ such that $y_n \to r$, $f^{\bar{k}_n} y_n \to q$ and $\bar{k}_n \to \infty$ as $n \to \infty$.*

Proof. Given $n > 0$, we associate to each collection of points $f^i x_n$, $i = 0, \dots, t_n$, a word

$$w(n) = p_{i_1(n)}^{k_1(n)} p_{i_2(n)}^{k_2(n)} \cdots p_{i_{m(n)}(n)}^{k_{m(n)}(n)}$$

in the following way. The order of points $p_{i_j(n)}$, $j = 1, \dots, m(n)$, corresponds to the order in which the trajectory $f^i x_n, i = 1, \dots, k(n)$, enters the δ-neighborhoods of nonwandering points $p_1, \dots, p_s$ and the number $k_j(n)$, $j = 1, \dots, m(n)$, is the amount of time the trajectory spends in the corresponding neighborhood. Since $k_n \to \infty$ as $n \to \infty$, it follows from Proposition 3.5 that

1. there exist $M > 0$ such that $m(n) \leq M$ for any $n > 0$;
2. $\sum_{j=1}^{m(n)} k_j(n) \to \infty$ as $n \to \infty$.

We claim that there exists a point $p_s = r \in NW(f)$ and a subsequence of words $w(n_l)$ such that $k_s(n_l) \to \infty$ as $l \to \infty$ and $k_j(n_l) \leq$ const for all $j = 1, \dots, s-1$ and all l. It is easy to see that r δ-follows p. To prove the claim consider the smallest index j for which the sequence $k_j(n)$ is unbounded. Let $k_j(n_l) \to \infty$. Since there are finitely many possible values for the index $i_j(n_l)$, we can pass to a subsequence and assume that there is s such that $i_j(n_l) = s$ for all l, and the claim follows. □

3.7. Lemma. *If $p, r \in NW(f)$ are two points such that r δ-follows p, then $p \geq r$.*

Proof. Since the point r δ-follows p, we have the corresponding sequences x_n, a_n, b_n, c_n. Let $y_n = f^{a_n+1} x_n$. Let y be a limit point of the sequence $\{y_n\}$. Since $y_n \notin U_\delta(NW(f))$, we have that $y \notin U_\delta(NW(f))$. Clearly there is $K > 0$ such that $f^k y \in U_\delta(r)$ for all $k \geq K$. Hence for sufficiently small δ, by Statement 4 of Theorem 2.1, $f^K y \in W^s_{\text{loc}}(r)$. Similarly, one can show that $y \in W^u(p)$. □

The following proposition is an immediate corollary of Lemmas 3.6 and 3.7.

3.8. Proposition. *Let $p, q \in NW(f)$ and assume that there are sequences of points $x_n \to p$ and integers $t_n \to \infty$ such that $f^{t_n} x_n \to q$. Then $p \geq q$.* □

3.9. Proposition. *There exists* $\delta_0 > 0$ *such that for any* $\delta \le \delta_0$ *the following holds. Whenever* $p, q \in NW(f)$ *and there is a point* $x \in U_\delta(p)$ *for which* $f^k x \in U_\delta(q)$ *for some* $k > 0$, *we have* $p \ge q$.

Proof. Assume the contrary. Then there exist numbers $\delta_n \to 0$, $k_n \to \infty$ and points $p, q \in NW(f)$, $x_n \in U_{\delta_n}(p)$ such that $f^{k_n} x_n \in U_{\delta_n}(q)$ and it is not true that $p \ge q$. This contradicts Proposition 3.8. □

3.10 Remark. One can prove the following stronger version of Proposition 3.9. For any $\alpha > 0$ there is $\delta_0 > 0$ such that for any $\delta \le \delta_0$, whenever $p, q \in NW(f)$ and there is a point $x \in U_\delta(p)$ for which $f^k x \in U_\delta(q)$ for some $k > 0$, we have $p \ge q$ and there is a heteroclinic point $y \in W^u(p) \cap W^s_{\text{loc}}(q)$ with $d(x, y) < \alpha$.

An analog of Proposition 3.9 holds true for ε-orbits.

3.11. Proposition. *For any positive* $\delta \le \delta_0/4$ *there is* $\varepsilon > 0$ *such that, whenever* $p, q \in NW(f)$ *and there is an* ε*-orbit* $\{z_k\}$ *with* $z_1 \in U_\delta(p)$ *and* $z_n \in U_\delta(q)$, *we have* $p \ge q$.

Proof. Let p, q, z_n be as above. One can find a point $r \in NW(f)$ and numbers a, b, and c such that $z_k \in U_\delta(p)$ for $k = 1, \dots, a$, $z_k \in U_\delta(r)$ for $k = b, \dots, c$, and $z_k \notin U_\delta(NW(f))$ for $k = a+1, \dots, b$. By Proposition 3.4, $0 < b - a \le n(\delta/2)$ for a sufficiently small ε. Therefore, if ε is small enough, then there exists a point $x \in U_\delta(p)$ for which $f^k x \in U_\delta(r)$ for some $k > 0$. Thus by Proposition 3.9, $p \ge r$.

We repeatedly apply the above argument to the ε-orbit, and the proposition follows. □

§4. Perturbation theorem

Denote by $C^1(M, M)$ the space of C^1-maps of M.

4.1. Theorem. *Let* $f : M \to M$ *be a Morse-Smale endomorphism. Then there is* $\delta_0 > 0$ *such that for any positive* $\delta \le \delta_0$ *there exists an open neighborhood* $\mathcal{U} \ni f$ *in* $C^1(M, M)$ *with the property that any* $g \in \mathcal{U}$ *is a Morse-Smale endomorphism and*

1. *there is a bijection* $\chi : NW(f) \to NW(g)$ *with* $d(p, \chi(p)) < \delta$ *for any* $p \in NW(f)$;
2. *for* $p_1, p_2 \in NW(f)$ *we have* $p_1 \le p_2$ *if and only if* $\chi(p_1) \le \chi(p_2)$;
3. *for any* $q_1, q_2 \in NW(g)$ *we have that* $q_1 \le q_2$ *if and only if there is a point* $x \in U_\delta(q_2)$ *such that* $g^k x \in U_\delta(q_1)$ *for some* $k \ge 0$.

Proof. The following lemma allows us to reduce the theorem to the case when $NW(f)$ consists only of fixed points.

4.2. Lemma. *If* g *is a* C^1*-map of* M *such that* g^m *is a Morse-Smale endomorphism then* g *is a Morse-Smale endomorphism.*

Proof of Lemma 4.2. It is sufficient to show that any point $x \notin NW(g^m)$ is a wandering point for g. If x is such a point then, by Proposition 3.2, $g^{mn} x \in W^s_{\text{loc}}(p)$ for some $n > 0$ and $p \in NW(g^m)$. Hence, x is a wandering point under g. □

From now on, by switching to the corresponding power, we assume that $NW(f)$ consists only of fixed points. To show that any g close enough to f is a Morse-Smale endomorphism we have to prove that it satisfies properties (i) and (ii) of Definition 3.1.

Fix $\delta > 0$. By standard transversality arguments, if g is close enough to f, then for any $p \in NW(f)$ there is a unique hyperbolic fixed point $q = \chi(p)$ of g such that $d(p, q) < \delta$. Let x be a nonwandering point of g. Then arbitrarily close to x there is a point y and an arbitrarily large k such that the finite orbit $\mathcal{O} = \{y, gy, \dots, g^k y\}$ is a closed ε-orbit of f. If δ and $\mathcal{U}$ are small enough, Propositions 3.8 and 3.11 imply that $\mathcal{O}$ lies in a small neighborhood of a fixed point $p \in NW(f)$. It follows that $x = \chi(p)$. This completes the proof of property (i).

To prove (ii) we assume the contrary. Then there is a sequence of C^1-maps g_n that converges to f in the C^1-topology and each map g_n has a nontransversal heteroclinic point. To simplify the notation in the arguments below we use the following convention: p (possibly with an index) denotes a fixed point of f, $q(n)$ (possibly with an index), denotes a fixed point of g_n, $W^u(p)$, denotes the unstable manifold of f, $W^u(q(n))$, denotes the unstable manifold of g_n, and similarly for the local stable manifolds. By passing to a subsequence, if necessary, we can assume that for any sufficiently small $\delta > 0$

1. there are fixed points $p_0 \geq p_1 \geq \cdots \geq p_l$ of f and fixed points $q_j(n)$, $j = 0, 1, \dots, l$ of g_n such that $q_j(n) \to p_j$ as $n \to \infty$;
2. there are nontransversal heteroclinic points $y_n \in W^u(q_0(n)) \cap W^s_{\mathrm{loc}}(q_l(n))$ with the common unit vectors v_n of the tangent spaces such that $\kappa\delta \leq d(y_n, q_l(n)) \leq \delta$ for some $\kappa > 0$, the sequence $\{y_n\}$ converges to a point $y \in W^s_{\mathrm{loc}}(q_l(n))$ and $v_n \to v \in T_y W^s_{\mathrm{loc}}(p_l)$, $\|v\| = 1$;
3. there are points $x_n \in W^u(q_0(n))$ such that
$$d(x_n, q_0(n)) \leq \delta \leq d(g_n x_n, q_0(n)),$$
and the sequence $\{x_n\}$ converges to a point $x \in W^u(p_0)$;
4. there are sequences of integers $a_0(n) = 0 \leq b_0(n) < a_1(n) \leq b_1(n) < \cdots < a_l(n) \leq b_l(n)$ such that for $j = 1, \dots, l$ and $n = 1, 2, \dots$,
$$g_n^i x_n \in U_\delta(q_j(n)) \quad \text{if } a_j(n) \leq i \leq b_j(n)$$
and
$$g_n^i x_n \notin U_\delta(NW(g_n)) \quad \text{if } b_j(n) < i < a_{j+1}(n);$$
5. $g_n^{a_l(n)} x_n = y_n$.

Clearly $d(x, p_0) \leq \delta$, $d(fx, p_0) \geq \delta$ and $d(y, p_l) \leq \delta$. Hence, $d(x, p_0) \geq C\delta$, where $C = \max_{x \in M} \|df(x)\|$.

By Proposition 3.4,
$$\sum_{j=0}^{l-1} (a_{j+1}(n) - b_j(n)) < N(\delta).$$

Assume first that $a_l(n) - b_0(n)$ is bounded uniformly in n. Then $y = f^k x$ for some $k > 0$, and hence, $y \in W^u(p_0)$. Therefore y is a nontransversal heteroclinic point of f. This is a contradiction.

Suppose now that $a_l(n) - b_0(n)$ is not bounded in n. Then, by passing to a subsequence, decreasing δ, and deleting some of the fixed points if necessary, we may assume that for every $j = 1, \ldots, l-1$

6. $b_j(n) - a_j(n) \to \infty$ as $n \to \infty$ and the difference $a_{j+1}(n) - b_j(n)$ eventually becomes constant (which we denote by k_j);
7. the sequences of points $g_n^{a_j(n)} x_n$ and $g_n^{b_j(n)} x_n$ converge to points z_j and w_j respectively.

For convenience, we set $w_0 = x$ and $z_l = y$. Note that $z_j \in W^s_{\mathrm{loc}}(p_j) \cap W^u(p_{j-1})$ is a heteroclinic point of f.

In the argument below we need to compare two subspaces in the tangent spaces at two different points lying in the 2δ-neighborhood of p_j for $j = 0, \ldots, l$. For a sufficiently small δ, we identify the neighborhood with a ball in $\mathbb{R}^d$. We parallel translate any subspace at any point to 0 and calculate the distance between subspaces at 0 using, for example, the Grassmann metric.

Consider the image $E_n \subset T_{y_n}M$ of the tangent space to $W^u(q_0(n))$ at x_n under $dg_n^{b_l(n)}$. To obtain a contradiction we will show that for any $\varepsilon > 0$ and all sufficiently large n there is a subspace $V_n \subset E_n$ which is ε-close to $T_y W^u(p_{l-1})$ and not transversal to $W^s_{\mathrm{loc}}(q_l(n))$. This means that $W^u(p_{l-1})$ and $W^s_{\mathrm{loc}}(p_l)$ are not transversal at y which is impossible.

We need the following lemmas.

4.3. Lemma. *For any $\beta > 0$ there are $\alpha > 0$ and a neighborhood $\mathcal{V} \ni f$, $\mathcal{V} \subset \mathcal{U}$ such that for any $j = 0, \ldots, l$ the following holds true: if x is a point with $d(x, w_j) \le \alpha$, $E \subset T_x M$ is a subspace α-close to $T_{w_j} W^u(p_j)$, and $g \in \mathcal{V}$, then $d(g^{k_j}x, z_{j+1}) \le \beta$ and the subspace $dg^{k_j}E$ is β-close to $T_{z_{j+1}} W^u(p_j)$.*

Proof of Lemma 4.3. We have by Proposition 3.4(1) that $k_j \le n(\delta)$ for all j and the lemma follows. □

4.4. Lemma. *For any $\gamma > 0$ there are $\beta > 0$ and a neighborhood $\mathcal{W} \ni f$, $\mathcal{W} \subset \mathcal{U}$ such that for any $j = 0, \ldots, l$ the following holds true: if x is a point with $d(x, z_j) \le \beta$, $E \subset T_x M$ is a subspace β-close to $T_{z_j} W^u(p_{j-1})$, and $g \in \mathcal{V}$ is such that $d(g^k x, w_j) \le \beta$ for some integer $k > 0$, then the subspace $d_x g^k E$ contains a subspace E' which is γ-close to $T_{w_j} W^u(p_j)$.*

Proof of Lemma 4.4. Recall that z_j, $j = 1, \ldots, l$ are transversal heteroclinic points of f. As before, we identify the 2δ-neighborhoods of p_j's with balls in $\mathbb{R}^d$ and use parallel translation in $\mathbb{R}^d$ to identify subspaces at different points. By Theorem 2.2, for a sufficiently small $\delta > 0$, any g close enough to f has a unique hyperbolic fixed point $q_j = q_j(g)$ in $U_\delta(p_j)$, which depends continuously on g; the local stable and unstable manifolds of g at q_j depend continuously on g in the C^1-topology. Denote by F the orthogonal complement to $T_{z_j} W^s_{\mathrm{loc}}(p_j)$ in E and view it as a submanifold passing through x. It follows from the remarks above that if g is sufficiently close to f and β is small enough, then the submanifold F intersects $W^s_{\mathrm{loc}}(q_j)$ transversally at a point that is $C\beta$-close to x and z_j, where $C > 0$ does not depend on β and g. Note that $k \to \infty$ as $\beta \to 0$. Hence, by the λ-lemma of Palis (see **[PaMe]**) for a sufficiently small β we have that $d_x g^k F$ is γ-close to $T_{w_j} W^u(p_j)$. □

We now complete the proof of the theorem. Recall that $z_l = y$ and $W^u(p_{l-1})$ intersects $W^s_{\mathrm{loc}}(p_l)$ transversally at y. Therefore, the difference between any two unit vectors, one from $T_y W^s_{\text{loc}}(p_l)$ and another from $T_y W^u(p_{l-1})$, is uniformly bounded away from 0. We choose the first vector to be the accumulation vector v of the common vectors v_n for the nontransversal intersections above. By moving back from p_l to p_0 and applying repeatedly Lemmas 4.3 and 4.4, we construct vectors $\omega_n \in T_{x_n} W^u_{\text{loc}}(q_0(n))$ such that the vector $w_n = dg^{a_l(n)} v_n$ is arbitrarily close to the space $T_y W^u(p_{l-1})$ for a sufficiently large n. We multiply v_n by appropriate positive numbers to get w_n of unit length and obtain a contradiction. □

References

[AP] V. Afraimovich and Ya. Pesin, *Travelling waves in lattice models of multi-dimensional and multi-component media*: I. *General hyperbolic properties*, Nonlinearity **6** (1993), 429–455.

[PaMe] J. Palis and W. de Melo, *Geometric theory of dynamical systems: an introduction*, Springer-Verlag, Berlin and New York, 1982.

[Pr1] F. Przytycki, *On Ω-stability and structural stability of endomorphisms satisfying Axiom A*, Studia Math. **60** (1977), 61–77.

[Pr2] ———, *Anosov endomorphisms satisfying Axiom A*, Studia Math. **58** (1976), 249–285.

[Rob] C. Robinson, *Dynamical systems*: *stability, symbolic dynamics, and chaos*, CRC Press, Boca Raton, FL, 1995.

[Rue] D. Ruelle, *Elements of differentiable dynamics and bifurcation theory*, Academic Press, Boston, 1989.

[Shu] M. Shub, *Global stability of dynamical systems*, Springer-Verlag, Berlin and New York, 1987.

M. Brin, Department of Mathematics, University of Maryland, College Park, MD 20742, USA,
E-mail address: `mbrin@math.umd.edu`

Ya. Pesin, Department of Mathematics, Pennsylvania State University, University Park, PA 16802, USA,
E-mail address: `pesin@math.psu.edu`

Amer. Math. Soc. Transl.
(2) Vol. **171**, 1996

Continued Fractions and Geometrical Optics

L. A. Bunimovich

ABSTRACT. We develop a new approach to the study of continued fractions and prove some new results on their convergence. The crucial point of our approach is the partition of a continued fraction into segments with different "optical" properties.

§0. Introduction

For several centuries continued fractions have undergone extensive studies [**1**]. The reasons for this are their various applications in the number theory [**2, 3**], analysis [**4, 5**], statistics [**6**], etc., and, of course, the beauty of this mathematical object. In various investigations of continued fractions, not only numbers, but also functions, matrices, and operators, were considered as their elements. Obviously, the problem of convergence is the first one that was studied when dealing with infinite continuous fractions. However, this problem is far from being solved even for continued fractions whose elements are the real numbers.

The only exhaustive result is provided by the celebrated Seidel-Stern theorem, which gives the criterion for the convergence for the continued fractions with elements of the same sign (see e.g., [**2, 3**]).

On the other hand, there exist only some sufficient conditions for the convergence of continued fractions with elements of the varying signs (see e.g., [**3, 4**]). (It is worthwhile to mention that there are extensive results for some special classes of continued fractions, e.g., for the periodic ones [**4**].) All the results of this kind provide some sufficient conditions of the convergence that enables us to reduce the problem to the study of the convergence of some infinite series or infinite products. The important common feature of these results is their "locality," i.e., their conditions (usually in the form of some inequalities) always

1. relate in some way the neighboring elements of continued fractions;
2. involve *all* elements of a continued fraction

(see, e.g., the classical Scott-Wall theorem [**4**]).

1991 *Mathematics Subject Classification*. Primary 11A55; Secondary 78A05.
This research was partially supported by the NSF Grant #DMS-9303769.

In this paper we suggest the approach that is based on the partition of continued fractions into the segments with all the elements having the same sign or alternating signs. The corresponding (sufficient) convergence conditions involve all elements of segments with the alternating signs, but the boundary elements of the segments with constant signs. Thus, our conditions are, in a sense, global rather than local.

The idea of this approach was suggested by the studies of chaotic billiards with focusing boundaries **[7, 8]**.

It was Ya. G. Sinai who first used continued fractions in the analysis of billiards **[9]**. Formally, such continued fraction corresponds to any nonsingular trajectory of any billiard (see §1). Sinai showed that these continued fractions define the local stable and unstable foliations that play the central role in the investigations of hyperbolic dynamical systems. Moreover, there is a simple relation between the Kolmogorov-Sinai entropy of a hyperbolic billiard and the function whose value at any (nonsingular) point of the phase space equals to the continued fraction corresponding to its semitrajectory **[10]**. Sinai considered the billiards on a torus with smooth inward convex boundary and called such billiards "the dispersing ones". Now these billiards are called Sinai's billiards. For Sinai's billiards, as well as for the general class of dispersing billiards (that also includes billiards in $\mathbb{R}^n$ with inward convex smooth components of the boundary), the problem of convergence of the corresponding continued fractions is trivial, because all their elements have the same sign. Thus the Seidel-Stern theorem immediately implies that such continued fractions always converge **[9, 11]**. The same is true for continued fractions corresponding to trajectories of a billiard that reflect only from dispersing or neutral (plane) components of the boundary.

The presence of at least one focusing component immediately changes the character of the problem, because the continued fractions become that of a general type. It reflects the important fact that all dispersing billiards are the chaotic ones, while billiards with focusing components demonstrate the variety of behavior, from integrable to chaotic (see, e.g., **[12]**). Some results on the convergence of such continued fractions were obtained in **[7, 8]**.

So far, all the efforts were directed to the proof of the convergence of continued fractions corresponding to certain billiards with the goal to study ergodic properties of these billiards. In this paper we turn things around and suggest the approach, based on some ideas from the geometrical optics, to the study of the convergence of continued fractions of the general type.

A general continued fraction does not correspond to a trajectory of any billiard (even of a generalized one **[8]**). However, it always can be partitioned into some pieces that can be considered as (finite or infinite) trajectories of rays reflected from some mirrors.

The paper is structured as follows. We introduce the partition of any continued fraction into scattering, inverse scattering, and focusing segments (§1). In §2 we study absolutely focusing segments that form a subclass of focusing segments, and prove the results on convergence of continued fractions with focusing segments of this type (§3). In §4 we introduce another class of focusing segments for which we can also prove the convergence of continued fractions. We conclude the paper (§5) with some remarks on possible generalizations and on the application of our approach to numerical algorithms of approximation of continued fractions.

§1. Optical partitions of continued fractions

We shall study continued fractions of the form

(1) $$A = a_0 + \cfrac{1}{a_1 + \cfrac{1}{a_2 + \cfrac{1}{\ddots + \cfrac{1}{a_n + \cfrac{1}{\ddots}}}}}$$

where a_i, $i = 1, 2, \ldots$, are real numbers. The numbers a_i, $i = 1, 2, \ldots$, are called the elements of the continued fraction (1). It is more convenient to use instead of (1) the following notation

(2) $$A = a_0 + \frac{1}{a_1 +} \quad \frac{1}{a_2 + \cdots +} \quad \frac{1}{a_n + \ldots}.$$

Both expressions (1), (2) are of course the formal ones. We will be interested in the problem of the convergence of the series of *convergents*

$$S_n = a_0 + \frac{1}{a_1 +} \quad \frac{1}{a_2 + \cdots +} \quad \frac{1}{a_n} = [a_0; a_1, a_2, \ldots, a_n].$$

The limit $S = S_\infty = \lim_{n \to \infty} S_n$ (if it exists) defines the value of the continued fraction (1) (or (2)).

Consider any collection of consecutive elements $a_k, a_{k+1}, \ldots, a_{k+m}$. We call such a collection a *scattering* segment (S-segment) if $\operatorname{sgn} a_k = \operatorname{sgn} a_{k+1} = \cdots = \operatorname{sgn} a_{k+m} = 1$; an *inverse scattering* segment (IS-segment) if $\operatorname{sgn} a_k = \operatorname{sgn} a_{k+1} = \cdots = \operatorname{sgn} a_{k+m} = -1$, and a *focusing* segment (F-segment) if $\operatorname{sgn} a_k = -\operatorname{sgn} a_{k+1} = \operatorname{sgn} a_{k+2} = -\operatorname{sgn} a_{k+3} = \cdots = (-1)^m \operatorname{sgn} a_{k+m}$.

DEFINITION 1. A sequence of integers $n_i > 0$, $i = 1, 2, \ldots, k \le \infty$, defines an *optical partition* ξ of the continued fraction (1) with elements $C_i = \{a_{n_i - 1}, \ldots, a_{n_i}\}$ if

(i) $n_{i+1} > n_i$, for all $i = 1, 2, \ldots$;
(ii) $a_{n_i+1}, \ldots, a_{n_i+1}$ is an S-, IS- or F-segment;
(iii) (maximality of F-segments) the union of any two adjacent elements of ξ is not an F-segment

It is easy to see that a continued fraction can have several optical partitions. An optical partition can be finite or infinite. It is easy to see that, for a given continued fraction, either all optical partitions are finite or they are all infinite. If there exists a finite optical partition, then the problem of convergence reduces to the convergence of a continued fraction corresponding to the last element of ξ. Therefore, in view of the Seidel-Stern theorem, the problem of convergence is open only if the last element of ξ is an F-segment. We chose the partition $\hat{\xi}$ that has the longest F-segments among all other optical partitions for a given continued fraction.

Now we explain where all the names optical, scattering, etc., come from. Let $a_n, a_{n+1}, \ldots, a_{n+m}$, where $n, m > 0$ are integers, be an S-segment. Denote

(3) $$a_{n+m-2\ell} = \tau_\ell,$$

$\ell = 0, 1, \dots, m/2$, if m is even, or $\ell = 0, 1, \dots, \frac{m-1}{2}$ if m is odd, and

$$a_{n+m-2\ell-1} = \frac{2k_\ell}{\cos\varphi_\ell}, \tag{4}$$

$\ell = 0, 1, \dots, \frac{m}{2} - 1$ if m is even, or $\ell = 0, 1, \dots, \frac{m-1}{2}$ if m is odd, where $k_\ell > 0$ and $-\frac{\pi}{2} \le \varphi_\ell \le \frac{\pi}{2}$. (Of course such a representation of a_{n+r} is highly nonunique for all elements a_{n+r} with odd r. We take any of there representations.)

First we assume that m is odd. It is well known in the geometrical optics that the continued fraction

$$S_{n,n+m} = \frac{1}{a_n+} \quad \frac{1}{a_{n+1} + \cdots +} \quad \frac{1}{a_{n+m}} = [a_n, a_{n+1}, \dots, a_{n+m}] \tag{5}$$

gives a curvature of the front of an infinitesimal (initially) parallel beam of rays propagating with the unit velocity after it undergoes $\left(\frac{m+1}{2}\right)$ reflections (if m is odd) from some mirrors. We assume that at each reflection the angle of incidence equals the angle of reflection. Moreover, the curvature of a mirror at the point of reflection equals k_ℓ, and the corresponding angle of reflection equals φ_ℓ. The quantity τ_ℓ denotes a time of a free propagation (without reflections from the mirrors) between ℓth and $(\ell+1)$th reflections. If m is even, $S_{n,n+m}$ gives the curvature of a front at the time a_n after the $(m/2)$th reflection.

Now we recall two basic formulas from the geometrical optics (see, e.g., **[7, 8, 10, 12]**) to justify this "optical" meaning of a continuous fraction (5).

Let a narrow beam of rays have a curvature κ_0 at the moment $t = 0$. Suppose that this beam has no reflections from the mirrors before time t_0. Then the curvature κ_{t_0} of this beam at $t = t_0$ equals

$$\kappa_{t_0} = \frac{\kappa_0}{1 + t_0\kappa_0}. \tag{6}$$

The next formula shows the relation between the curvatures of a narrow beam of rays just before (κ_-) and just after (κ_+) the reflection from the mirror:

$$\kappa_+ = \kappa_- + \frac{2k}{\cos\varphi} \tag{7}$$

where k is the curvature of the mirror at the point of reflection and φ is the corresponding reflection angle.

The representation (3), (4) follows immediately from (6), (7). Thus the segment (5) of a continuous fraction corresponding to the series of reflection from dispersing mirrors (S-segment) should have all elements a_i, $i = n, n+1, \dots, n+m$, positive. One gets an IS-segment by changing the signs of all elements a_i, $i = n, n+1, \dots, n+m$. Finally, in a focusing segment every second element should be negative because the focusing mirrors have negative curvature. Therefore, the segment (5) with elements of alternating signs corresponds to a series of consecutive reflections from a focusing mirror. To get this representation, one should use the same change of variables as for S-segments, but with some negative curvatures k_ℓ.

Obviously, a continuous fraction (5) corresponds to a series of reflections of some beam of rays from some sequence of mirrors iff all its odd (even) elements are positive. Indeed each other element of such continued fraction should correspond to the time (> 0) between consecutive reflections.

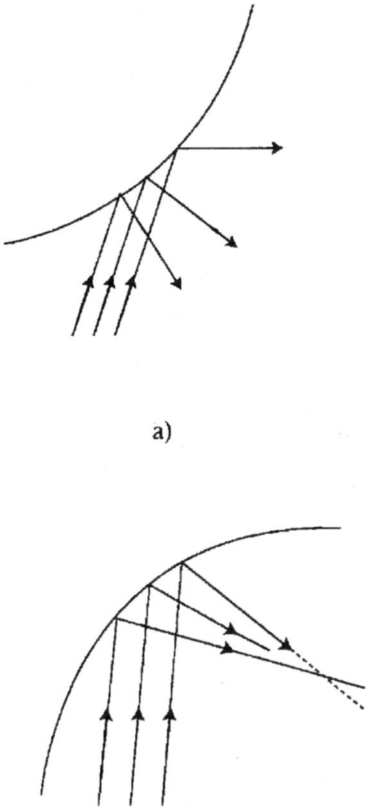

a)

b)

FIGURE 1. Dispersing (a) and focusing (b) mirrors

§2. Absolutely focusing segments

We introduce in this section some special subclass of focusing segments.

DEFINITION 2. A focusing segment of the form (5) of a continuous fraction is said to be an *absolutely focusing* (AF-segment) if there exist numbers $\delta_j \geq -1$, $j = 1, 2 \ldots, \frac{m}{2}$ (if m is even) or $\frac{m-1}{2}$ (if m is odd) such that

$$(8') \qquad |a_{n+2j-1}| \geq |a_{n+2j-2}|^{-1}\left(2 - \frac{\delta_j}{1+\delta_j}\right) + |a_{n+2j}|^{-1}(2 + \delta_{j+1})$$

or

$$(8'') \qquad |a_{n+2j}| \geq |a_{n+2j-1}|^{-1}\left(2 - \frac{\delta_j}{1+\delta_j}\right) + |a_{n+2j+1}|^{-1}(2 + \delta_{j+1}).$$

Elements a_{n+k} at the left-hand sides of (8′), (8″) are called time-elements (according to the optical analogy to be discussed later in this section).

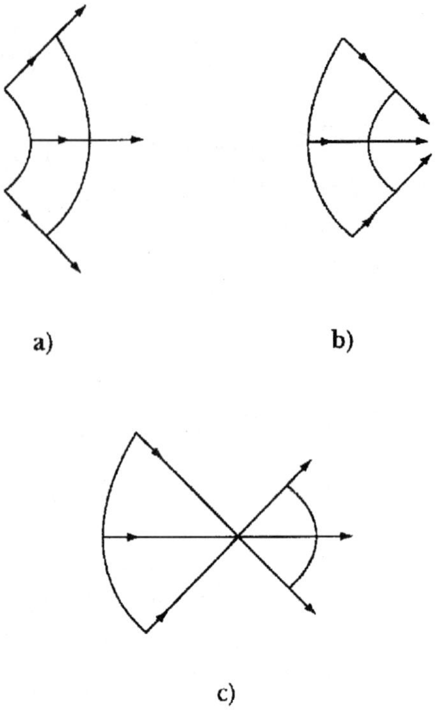

FIGURE 2. Types of beams of rays between two consecutive deflections from a mirror: (a) dispersing; (b) purely focusing; (c) defocusing

THEOREM 1. *If a continued fraction* (3) *is defocusing, then for any* ℓ, $0 \le \ell \le m$, *the sequence of continued fractions* $A_{n+\ell,n+k} = \frac{1}{a_{n+\ell}+\cdots+}\ \frac{1}{a_{n+k}}$, $k = \ell, \ell+1, \dots, m$, *is monotonous. Moreover,*

$$|S_{n+\ell,n+k}| \le |a_{n+\ell}|^{-1}\left(\frac{2+\delta_\ell}{1+\delta_\ell}\right), \tag{9}$$

where $\ell = 0, 2, 4, \dots, m$ (*if* m *is even*) *or* $m-1$ (*if* m *is odd*).

Proof of Theorem 1 can be obtained by the simple straightforward calculation (see **[8]**).

Now we discuss the geometric optics meaning of Definition 2. Consider a smooth focusing mirror (smooth convex curve). Let a narrow beam of rays have a series of consecutive reflections off this mirror. We assume that this beam was initially (before the first reflection) parallel if m is odd; otherwise we assume that after the first reflection its curvature equals $|a_{n+m}|\left(\frac{1+\delta_{m/2+1}}{2+\delta_{m/2+1}}\right)$.

Then as the direct consequence of (3), (4), (6), (7), (8′) (or (8″)) one gets (see **[8]**) that between any two consecutive reflections off the mirror this beam passes through a conjugate point and is focused after the last reflection off the mirror (Figure 3).

It is worth mentioning that in the process of reflections from a focusing (but not an absolutely focusing mirror), between consecutive reflections a beam can have any of the forms depicted in Figure 2(a)–(c).

The next statement follows immediately from Theorem 1 (see also [**8**]).

THEOREM 2. *If* $m = \infty$, *then a continued fraction* (5) *converges.*

§3. Convergence of continued fractions with absolutely focusing segments

In this section we prove some theorems on the convergence of continued fractions with focusing segments of the absolutely focusing type. These theorems generalize the classical Seidel-Stern theorem on the convergence of continued fractions with the elements of the same sign. Our results show that if the AF-segments of a continued fraction dominate in some sense the other segments, then the convergence criterion is basically the same as for the continued fractions with elements of the same sign. Contrary to the existing theorems on the convergence of continued fractions with the elements of different signs, our conditions are global rather than local (see, e.g., [**3, 4**]). More precisely, our conditions deal with the conjunctions of S-, IS-, and AF-segments. Hence, only the boundary elements of S- and IS-segments appear in these conditions. In the existing results, the conditions involve the relations between all neighboring elements (usually these conditions involve relations between a_n and a_{n+1} or between a_{n-1}, a_n, a_{n+1} for all integer n (see, e.g., [**3, 5**]). It is worth mentioning that the conditions that single out the absolutely focusing segments from the set of all focusing segments are local.

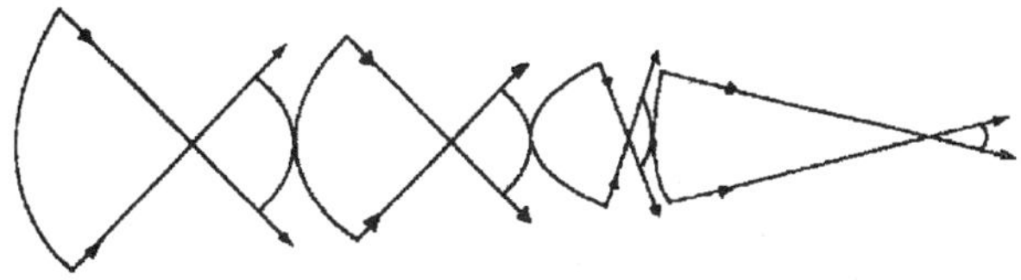

FIGURE 3. A series of reflections from an absolutely focusing mirror

THEOREM 3. *Let a continued fraction* (1) *contain S-, IS-, and AF-segments only and the total number of such segments is infinite. Suppose that for any AS-segment* $A_j = [a_{n_j+1}, \dots, a_{n_j+\ell_j}]$ *the following conditions hold*:

$$(10)\qquad \begin{aligned} |a_{n_j+1}| &> |a_{n_j+\ell_j}|^{-1} && \text{if } \ell_j = 2, \\ |a_{n_j}| &> |a_{n_j+1}|^{-1}(2+\delta_1^{(j)}) && \text{if } \ell_j > 2, \end{aligned}$$

and a_{n_j+2} *is the first time-element in* A_j. *Moreover, let* $\prod_{j=1}^{\infty}(2+\min_i \delta_i^{(j)}) = \infty$, *where* j *is the number of an AF-segment and* min *is taken over all* δ*'s in this segment* (*see* (8′), (8″)). *Then the series* (1) *converges.*

REMARK. If the total number of segments in a continued fraction (1) is finite then one of them is infinite (the "tail"). Then the problem reduces to the convergence of this tail. If such a tail is an AF-segment, then (1) always converges by Theorem 2. If there exists an infinite S- or IS-segment, then the Seidel-Stern theorem provides the condition for (1) to converge.

PROOF. We can assume that (1) contains infinitely many S-segments. If it is not the case, we change the signs of all elements in (1), i.e., consider the continued fraction $-A$ instead of A. If, again, $-A$ does not contain infinitely many S-segments then there exists n such that the continued fraction $A_{n,\infty}$ forms an infinite AF-segment. Thus, (1) converges according to Theorem 2.

Hence, without loss of generality, we can assume that $a_1 > 0$ and $a_2 > 0$. Hence

$$Q_1 = a_1 > 0 \quad \text{and} \quad Q_2 = a_2 a_1 + 1 > 0.$$

Denote by $S_n = \frac{P_n}{Q_n}$, $n = 1, 2 \ldots$, the nth convergent, where P_n, Q_n satisfy the standard recurrence relations (see, e.g., **[2–4]**)

$$P_k = a_k P_{k-1} + P_{k-2}, \qquad Q_k = a_k Q_{k-1} + Q_{k-2}. \tag{11}$$

LEMMA 1. *Let* $i > 0$. *If* $\operatorname{sgn} a_{i+1} = -\operatorname{sgn} a_{i+2}$, *then* $\operatorname{sgn}(S_{i+1} - S_i) = \operatorname{sgn}(S_{i+2} - S_{i+1})$.

PROOF. Let $[a_m, a_{m+1}, \ldots, a_{m+k}] = A_{m,m+k}$ be an S- or an IS-segment. It is well known **[2, 3]** that the sequences $S_{m,m+j}$, $j = 0, 2, \ldots$, and $S_{m,m+j}$, $j = 1, 3, \ldots$, are monotone decreasing and increasing (increasing and decreasing) respectively if $A_{m,m+n}$ is an S-segment (IS-segment).

Theorem 1 shows that the sequence $S_{m,m+j}$, $j = 0, 1, 2, \ldots, k$, is monotone if $A_{m,m+k}$ is an AF-segment.

Conditions (10) on the conjunction of these segments give the desired statement.

Lemma 1 shows that if $A_{m,m+k}$ is an AF-segment, then the sequence of convergents $S_m, S_{m+1}, \ldots, S_{m+k}$ is monotone. It is easy to see that this sequence is monotone increasing (decreasing) if $S_{m-1} < S_{m-2}$ $(S_{m-1} > S_{m-2})$.

LEMMA 2. *Let* $\operatorname{sgn} a_m = \operatorname{sgn} a_{m+1}$ *and* $\operatorname{sgn} a_{m+k} = \operatorname{sgn} a_{m+k+1}$, *where* $k > 1$. *Suppose also that the sequence* $\operatorname{sgn}(a_m), \operatorname{sgn}(a_{m+1}), \ldots, \operatorname{sgn}(a_{m+k})$ *contains at least one change of a sign. Then*

$$|S_{m+k+1} - S_m| < (1 - \gamma)|S_{m+1} - S_m| \tag{12}$$

for some positive $\gamma < 1$.

PROOF. Our conditions imply that $A_{m,m+k+1} = [a_m, a_{m+1}, \ldots, a_{m+k+1}]$ contains at least one AF-segment. Let this segment be $A_{m+\ell,m+r} = [a_{m+\ell}, a_{m+\ell+1}, \ldots, a_{m+r}]$, $r - \ell \geq 2$. Then it follows from Theorem 1 and (10) that the sequence $S_{m+\ell-1}, S_{m+\ell}, S_{m+\ell+1}, \ldots, S_{m+r}$ is monotone and moreover

$$|S_{m+r} - S_{m+\ell-1}| < |S_{m+\ell-1} - S_{m+\ell-2}| \tag{13}$$

if $1 \leq \ell \leq r$. Further, one gets by (10) that

$$|S_{m+r+1} - S_m| < \min_{\ell \leq s \leq r-1} |S_{m+s+1} - S_{m+s}|. \tag{14}$$

Lemma 1 together with (13), (14) imply Lemma 2.

The following statement gives an estimate for the quantity γ in (12).

LEMMA 3. *We have*

$$\gamma > 1 + \min_i \delta_i, \tag{15}$$

where δ_i*'s are the numbers that define the AF-segment* $A_{m+\ell,m+r}$ *(see* (8′), (8″)).

PROOF. We will use several times the following standard relation from the theory of continued fractions (see, e.g., **[1–3]**):

$$\frac{P_{n-1}}{Q_{n-1}} - \frac{P_n}{Q_n} = \frac{(-1)^n}{Q_{n-1}Q_N}. \tag{16}$$

First, we apply this relation with $n = m + \ell - 1$. From (11) and (16) we get that

$$\frac{P_{m+t}}{Q_{m+t}} - \frac{P_{m+\ell-1}}{Q_{m+\ell-1}} = \frac{(-1)^{m+\ell}}{Q_{m+\ell-1}(Q_{m+\ell-1}S_{m+\ell,m+t} + Q_{m+\ell-2})} \tag{17}$$

for $\ell \leq t \leq r$. Using the other famous relation (see, e.g., **[2, 3]**)

$$\frac{Q_n}{Q_{n-1}} = [a_n; a_{k-1}, \dots, a_1], \tag{18}$$

we obtain

$$\frac{|S_{m+r+1} - S_{m+r}|}{|S_{m+\ell-1} - S_{m+\ell-2}|} < (2 + \min_i \delta_i)^{-1}, \tag{19}$$

where min is taken over all δ_i that define the AF-segment $A_{m+\ell,m+r}$.

Now we can complete the proof of Theorem 3. Let a segment $[a_1, a_2, \dots, a_n]$ contain $k = k(n)$ AF-segments. Then

$$|S_{n+1} - S_n| < \prod_{j=1}^{k(n)} (2 + \min_i \delta_i^{(j)})^{-1} |S_2 - S_1|, \tag{20}$$

where min is taken over all δ's (see (8′), (8″)) that define the jth AF-segment. Theorem 3 is proved.

§4. Convergence of continued fractions with nonabsolutely focusing segments

The main property that characterized absolutely focusing segments is the monotonicity of the corresponding sequence of convergents. However, a general focusing segment of a continued fraction (i.e., a segment with elements of alternating signs) does not have any regularity in the distribution of its convergents.

Nevertheless, there is the other type of focusing segments that can demonstrate some regularity in the distribution of their convergents.

DEFINITION. A finite focusing segment of a continued fraction $[a_1, a_2, \dots, a_n]$ is called *purely focusing* (PF) if $\operatorname{sgn}[a_m, a_{m+k}] = \operatorname{sgn}[a_r, a_{r+s}]$ for any $1 \leq m < n$, $1 \leq r < n$, $m + k \leq n$, $r + s \leq n$, $k > 0$, $s > 0$.

LEMMA 4. *Let* $[a_1, a_2, \ldots, a_n]$ *be a purely focusing segment. Then the sequence* $|S_2|, |S_1|, |S_4|, |S_3|, |S_6|, \ldots, |S_n|$ *(if n is odd) or* $|S_2|, |S_1|, |S_4|, \ldots, |S_{n-1}|$ *(if n is even) is monotonically increasing.*

Proof of Lemma 4 easily follows from the definition of a purely focusing segment.

To explain the meaning of this definition from the point of view of the geometrical optics, we consider again a series of consecutive reflections of some narrow beam of rays off a focusing mirror Then purely focusing series arise when a beam is focused between any two consecutive reflections but never cross corresponding focusing points (Figure 4), contrary to the absolutely focusing case (Figure 3).

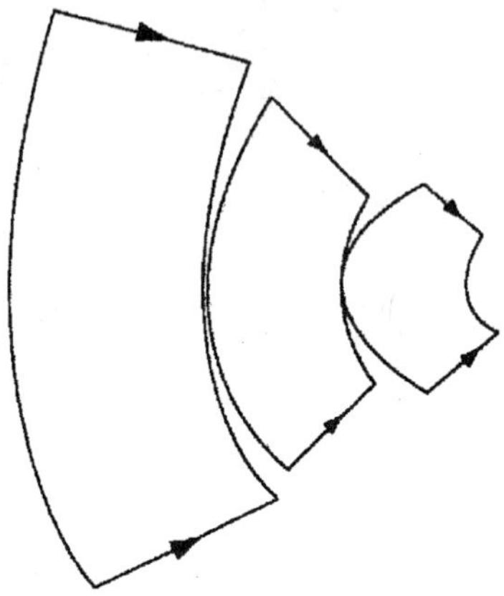

FIGURE 4. Purely focusing series

THEOREM 4. *Let a continued fraction* (1) *contains S-, IS-segments, and all its focusing segments are AF- and PF-segments only. Suppose that the conditions* (10) *hold and that for any PF-segment* $[a_m, a_{m+1}, \ldots, a_{m+k}] = A_{m,m+k}$ *we have*

$$|a_{m-1}| > |S_{m,m+k}|^{-1}(|a_{m-1}| > |S_{m,m+k-1}|^{-1}) \tag{21}$$

if k is even (odd), and $[a_m, a_{m+1}, \ldots, a_{m+k-1}, a_{m+k} + a_{m+k+1}^{-1}]$ *is also a PF-segment. Then* (1) *converges.*

Proof of Theorem 4 is completely analogous to the proof of Theorem 3.

REMARK. It is important to observe that to verify that a focusing segment $[a_1, a_2, \ldots, a_n]$ is a PF-segment, one should calculate all quantities $S_{m,k}$, $1 \le m < k \le n$. On the contrary, AF-segments are defined in local terms and usually can be localized inside a continued fraction without any calculations.

REMARK. Absolutely focusing and purely focusing segments are the only ones among all focusing segments that have some regularity property. Indeed, any other focusing segment corresponds to a sequence of reflections off a focusing mirror such that between consecutive reflections occurs either a defocusing (Figure 2(c)), or a pure focusing (Figure 2(b)), or a dispersing (Figure 2(a)).

§5. Concluding remarks

We have introduced the special class of continued fractions with alternating signs. These (absolutely focusing) fractions have some regularity property (Theorem 1) that allows us to prove certain sufficient conditions for the convergence of continued fractions with such segments. Contrary to the conditions considered earlier, our conditions are global. They involve the conditions on the conjunctions between the different types of segments of a continued fraction, rather than on all its elements. The simplest of these results is illustrated by Theorem 3, where all segments preserve their "optical" properties after combined together into a longer continued fraction. It is easy to formulate and prove the generalizations of this theorem that deal with the situations when such combinations sometimes produce the "finite perturbations" in the sense that only a finite number of segments gets destroyed.

Our "optical" approach provides also the natural algorithm to calculate the approximations for continued fractions. It starts with the optical partition of a continued fraction into S-, IS-, and F-segments. Then one should look for AF- and PF-segments and stop an approximation with such convergent that ends with a complete S-, IS-, PF-, or AF-segment.

On the other hand, this approach demonstrates that there is little hope for the existence of a convergence condition for continued fractions of a general type.

References

[1] C. Brezinski, *History of continued fractions and Padé approximants*, Springer-Verlag, Berlin, 1994.

[2] A. Ya. Khintchine, *Continued Fractions*, P. Noordhoff, Groningen, 1963.

[3] A. N. Khovanskii, *The application of continued fractions and their generalizations to problems in approximation theory*, Noordhoff, Groningen, 1963.

[4] H. S. Wall, *Analytic theory of continued fractions*, Van Nostrand, Toronto, 1948.

[5] W. B. Jones and W. J. Thron, *Continued fractions*, Addison-Wesley, London, 1980.

[6] K. O. Bowman and L. R. Shenton, *Continued fractions in statistical applications*, Marcel Dekker, New York, 1989.

[7] L. A. Bunimovich, *On billiards close to dispersing*, Mat. Sb. **94** (1974), no. 1, 49–73; English transl. in Math. USSR-Sb. **23** (1974).

[8] L. A. Bunimovich, *On absolutely focusing mirrors*, Lecture Notes in Math. (U. Krengel et al., eds.), vol. 1514, Springer-Verlag, Berlin and New York, 1991, pp. 62–82.

[9] Ya. G. Sinai, *Dynamical systems with elastic reflections*, Uspekhi Mat. Nauk **25** (1970), 137–189; English transl. in Russian Math. Surveys **25** (1970), 137–189.

[10] N. I. Chernov and R. Markarian, *Entropy of non-uniformly hyperbolic plane billiards*, Bol. Soc. Bras. Mat. **23** (1992), 121–135.

[11] J. Rehacek, *On the ergodicity of dispersing billiards*, Random and Computational Dynamics (to appear).

[12] L. A. Bunimovich, *Dynamical systems of hyperbolic type with singularities*, Encyclopedia of Mathematical Sciences (Ya. G. Sinai, ed.), vol. 2, Springer-Verlag, New York, 1989, pp. 151–178.

School of Mathematics and Center for Dynamical Systems and Nonlinear Studies, Georgia Institute of Technology, Atlanta, GA 30332

Institute of Oceanology, Russian Academy of Sciences, Moscow 117218
E-mail address: bunimovh@math.gatech.edu

Amer. Math. Soc. Transl.
(2) Vol. **171**, 1996

On Statistical Properties of Chaotic Dynamical Systems

N. I. Chernov

§1. Introduction

This paper is devoted to methods of investigating statistical properties of chaotic dynamical systems. By statistical properties we mean the rate of the decay of correlations, the central limit theorem, and other probabilistic limit theorems. The reader can find surveys of results in this area in [**3, 9, 8**].

An effective method of proving statistical properties is based on Markov approximations to dynamical systems. This approach is an alternative to the conventional Perron-Frobenius operator techniques. It was Ya. Sinai and his collaborators [**15, 16, 4, 5, 6**] who, starting in the mid-sixties, systematically developed the Markov partition method, Markov symbolic dynamics, and measure-theoretic Markov approximations to dynamical systems, such as Anosov maps and flows, geodesic flows, and billiards. Later Bowen and Ruelle [**2, 14**] extended these techniques to Axiom A diffeomorphisms and flows.

In [**4, 5**], Bunimovich and Sinai introduced a general construction of Markov chains approximating discrete-time dynamical systems and then studied it in [**6, 8**]. They showed that the chaotic behavior of the dynamical system ensures special conditions on transition probabilities of the approximating Markov chain. Under these conditions one can prove, in turn, the probabilistic limit theorems and establish strong bounds on correlation functions for the original dynamical system – some general results are presented in [**8**].

In this paper we further develop the technique of Markov approximations to dynamical systems. We explain how approximating Markov chains can be constructed for dynamical systems with discrete time and continuous time. We also introduce a new condition on the transition probabilities of Markov chains that implies strong bounds on the correlations. Our new condition is weaker than any of the conditions used in [**4, 6, 8**].

§2. Markov approximations for dynamical systems

A discrete-time dynamical system is a measurable transformation $T\colon M \to M$ of a measurable space M preserving a probability measure μ.

1991 *Mathematics Subject Classification*. Primary 58F15; Secondary 60F05.
This work was essentially done during my visit at the Georgia Institute of Technology (Atlanta), and I am grateful to L. A. Bunimovich, S.-N. Chow, and J. K. Hale for their kind hospitality.

Let $\mathcal{A} = \{A_i\}$ be a finite or countable measurable partition of M into subsets of positive measure. By Markov approximation for the map T we mean a probabilistic stationary Markov chain, whose transition probabilities are

$$\pi_{ij} = \mu(T^{-1}A_j/A_i) = \mu(T^{-1}A_j \cap A_i)/\mu(A_i), \tag{1}$$

and whose stationary distribution is

$$p_i = \mu(A_i). \tag{2}$$

This definition of Markov approximations for arbitrary measure-preserving transformations was introduced in **[8]**. It is one of possible implementations of the idea of "coarse-graining" of the phase space popular among physicists (see, for example, **[13]**). This definition is also very close to Ulam's construction **[17]** of Markov chains approximating interval maps.

The "discrepancy" of the Markov approximation generated by the Markov chain (1) and (2), within N iterates of the map T, is measured by the following quantity:

$$\begin{aligned} \nu_N := \sup_{n\le N} \sum_{i_0,\dots,i_n} &|\mu(T^{-n}A_{i_n}/T^{-(n-1)}A_{i_{n-1}} \cap \cdots \cap A_{i_0}) - \mu(T^{-1}A_{i_n}/A_{i_{n-1}})| \\ &\times \mu(T^{-(n-1)}A_{i_{n-1}} \cap \cdots \cap A_{i_0}). \end{aligned} \tag{3}$$

Here and further $\mu(A/B)$ means the conditional measure, $\mu(A/B) = \mu(A \cap B)/\mu(B)$. We always set it zero whenever $\mu(B) = 0$. The quantity ν_N measures how close (better to say, how distant!) the "long-memory" and "short-memory" conditional distributions are within the first N iterates of T.

Recall that given two probability distributions $P = \{p_i\}$ and $Q = \{q_i\}$ on the same index set $\{i\}$, the variational distance between P and Q is defined to be

$$\operatorname{Var}(P, Q) = \frac{1}{2}\sum_i |p_i - q_i|. \tag{4}$$

Now (3) estimates twice the mean variational distance between the long- and short-memory conditional distributions of $\{A_i\}$.

Using (3), one can estimate how close the finite dimensional distributions of the Markov chain,

$$p_{i_0 i_1 \cdots i_n} = p_{i_0}\pi_{i_0 i_1}\cdots\pi_{i_{n-1}i_n} \tag{5}$$

are to those of the dynamical system in the variational metric (4). It is shown in **[8]** that

$$\sum_{i_0,\dots,i_n} |\mu(T^{-n}A_{i_n} \cap \cdots \cap A_{i_0}) - p_{i_0 i_1 \cdots i_n}| \le (n-1)\nu_N \tag{6}$$

for any $n \le N$.

We now explain how Markov approximations with good properties can be constructed. Let (M, T, μ) be an Anosov diffeomorphism and μ a smooth invariant measure. Let $\mathcal{B}$ be a Markov partition of M into sufficiently small rectangles **[2, 15]**. Fix a large integer $K > 0$ and take $\mathcal{A} = \vee_{-K}^{K} T^k\mathcal{B}$. Then $\mathcal{A}$ is another Markov partition, whose atoms are exponentially small (in K). More precisely, there are constants $c_i, a_i > 0$ depending on the system (M, T, μ) such that for any $A \in \mathcal{A}$ we have $c_1e^{-a_1K} \le \operatorname{diam} A \le c_2e^{-a_2K}$ and $c_3e^{-a_3K} \le \mu(A) \le c_4e^{-a_4K}$. We do not go into details, but it is a standard argument that on any $A \in \mathcal{A}$ there is a product measure μ_A^p that approximates μ to the following degree of accuracy:

$$(7) \qquad \left| \frac{d\mu_A^p}{d\mu}(x) - 1 \right| \le c_5 e^{-a_5 K}$$

for every $x \in A$. It follows directly from the Markov property of the partition $\mathcal{A}$ that for any $n \ge 0$ and any atoms $A_{i_0}, \ldots, A_{i_n} \in \mathcal{A}$ we have

$$\mu^p_{A_{i_{n-1}}}(T^{-n}A_{i_n}/T^{-(n-1)}A_{i_{n-1}} \cap \cdots \cap A_{i_0}) = \mu^p_{A_{i_{n-1}}}(T^{-1}A_{i_n}/A_{i_{n-1}}).$$

It is now an immediate consequence of (7) that for the partition $\mathcal{A}$ we have $\nu_N \le 2c_5 e^{-a_5 K}$ for all $N > 0$. This approximation has the following advantage: ν_N is exponentially small (in K), so that, according to (6), the finite-dimensional distributions of the Markov chain and those of the dynamical system stay exponentially close (in K) on very long intervals of time $(0, N)$, at least for $N \approx e^{aK}$ with any $a < a_5$.

If the dynamical system (M, T, μ) is a smooth hyperbolic system with singularities and the measure μ is a Sinai-Bowen-Ruelle measure **[16]** (not necessarily absolutely continuous), the construction of partitions A with the above properties goes through, with some technical modifications, see **[6, 1]**.

Finally, Markov approximations can be constructed for dynamical systems with continuous time (flows). It is common to study flows using their special representations, which are called suspension flows or Kakutani flows, as defined below.

Let (M, T, μ) be a discrete time dynamical system and $l(x)$ a positive integrable function on M. A suspension flow build under the function $l(x)$ (this is called the ceiling function) is defined on the measurable space $\mathcal{M} = \{(x, s)\colon x \in M, 0 \le s < l(x)\}$ by the rule

$$(8) \qquad \Phi^t(x, s) = \begin{cases} (x, s + t) & \text{for } 0 \le t < l(x) - s, \\ (Tx, s + t - l(x)) & \text{for } l(x) - s \le t < l(Tx) + l(x) - s. \end{cases}$$

This flow is measurable and preserves the probability measure μ_f on $\mathcal{M}$ defined by $d\mu_f = c \cdot d\mu \times ds$, where $c^{-1} = \int_M l(x) d\mu(x)$ is the normalizing factor.

Let $\mathcal{A}$ be a partition of M generating a Markov approximation for T. We construct a Markov approximation to the flow Φ^t in the following way. First, let $\hat{l}(x)$ be the ceiling function l conditioned on the partition $\mathcal{A}$. Fix a small $\delta > 0$ (a "quantum of time") and set

$$(9) \qquad \hat{l}(x) = \hat{l}_{\delta,\mathcal{A}}(x) := ([\bar{l}(x)/\delta] + 2)\delta,$$

where $[a]$ stands for the integer part of a real number a. Now we consider another suspension flow $\hat{\Phi}^t$ over T, build under the function $\hat{l}$. Its phase space $\hat{\mathcal{M}} = \{(x, s)\colon x \in M, 0 \le s < \hat{l}(x)\}$ is naturally partitioned into the following blocks

$$X = A \times [k\delta, (k+1)\delta), \qquad A \in \mathcal{A}, \quad k = 0, 1, \ldots \hat{l}(A)/\delta - 1.$$

Denote this partition by $\hat{\mathcal{A}}$. Consider the map $\hat{T} = \hat{\Phi}^\delta$ on $\hat{\mathcal{M}}$. This map moves every atom of $\hat{\mathcal{A}}$ exactly onto another atom above it, breaks the top atoms into pieces, and transforms the pieces down to the bottom of $\hat{\mathcal{A}}$ according to the action of T on M.

The Markov chain approximating the map $\hat{T} = \hat{\Phi}^\delta$ is constructed by the same rules as above: its transition probabilities are $\hat{\pi}_{ij} = \hat{\mu}(X_j/\hat{T}X_i)$ and its stationary distribution is $\hat{p}_i = \hat{\mu}(X_i)$, where X_i, X_j are atoms of $\hat{\mathcal{A}}$ and $\hat{\mu}$ stands for the invariant measure of the suspension flow $\{\hat{\Phi}^t\}$. Since the map $\hat{T}$ acts very straightforwardly in the bulk of the partition $\hat{\mathcal{A}}$, this Markov chain provides very good approximation to the dynamical system $(\hat{\mathcal{M}}, \hat{T}, \hat{\mu})$. In fact, if the function l is bounded away from zero and infinity, i.e., $0 < l_{\min} \le l(x) \le l_{\max} < \infty$, then the quantity $\hat{v}_N$ defined by (3) for the Markov chain $||\hat{\pi}_{ij}||, ||\hat{p}_i||$ satisfies

$$\hat{v}_N \le \text{const} \cdot \delta v_{[bN\delta]}.$$

Here v is the quantity (3) generated by $\mathcal{A}$ for the Markov chain approximating T; $b > 0$ is a constant depending only on the original flow $\{\Phi^t\}$. We omit the proof of the above estimate.

§3. Mixing coefficients in Markov chains

In the previous section we showed how to construct Markov chains approximating dynamical systems.

Now we consider an abstract homogeneous Markov chain with a finite number of states that approximates a dynamical system. We denote the states by $1, 2, \dots, I$, the matrix of transition probabilities by $\Pi = ||\pi_{ij}||$ with $1 \le i, j \le I$, and the stationary distribution by $P = ||p_i||$. We also denote the m-step transition probabilities by $\pi_{ij}^{(m)}$, i.e., $||\pi_{ij}^{(m)}|| = \Pi^m$ and the set of indices $\{1, 2, \dots, I\}$ by $\mathcal{J}$.

To establish statistical properties for the underlying dynamical system, one usually must estimate the following quantities, called the mixing coefficients. For any $m \ge 1$, let

$$V_i^{(m)} = \frac{1}{2} \sum_{j=1}^{I} |\pi_{ij}^{(m)} - p_j|$$

and

$$V^{(m)} = \sum_{i=1}^{I} p_i V_i^{(m)}. \tag{10}$$

This last quantity is the mean variational distance between the m-step transition probabilities and the stationary distribution.

If the dynamical system is ergodic (mixing), then the approximating Markov chain is irreducible (aperiodic). For such chains, the mixing coefficient (10) monotonically decreases to zero as $m \to \infty$. The rate of the decay of this coefficient essentially represents the mixing rates of the original dynamical system. Moreover, it is possible to establish various statistical properties of the dynamical system based upon appropriate estimates of the mixing coefficients (10). Such an approach is developed in **[8]**.

A far more difficult problem is to establish estimates on the mixing coefficients (10) for Markov chains approximating dynamical systems. Dynamical properties of the system (e.g., hyperbolicity) can seldom provide such estimates directly. However, there are certain conditions on the transition probabilities of the Markov chain under which the coefficients (10) can be effectively bounded. It is also possible to verify those conditions by using the dynamical properties of the underlying system, such as hyperbolicity, existence of Markov partitions, etc. Several implementations of this strategy are described in **[8]**. The rest of this section is devoted to two conditions on the transition probabilities used in **[8]**.

One of them is the so-called Doeblin condition:

(11) $$d := 1 - \max_{i,j} \frac{1}{2} \sum_{k=1}^{I} |\pi_{ik}^{(s)} - \pi_{jk}^{(s)}| > 0$$

for some $s \geq 1$. The motivation to introduce this condition is based upon the classical D-condition **[11]** and Dobrushin's coefficient of ergodicity **[10]**. The Doeblin condition was explicitly introduced by Bunimovich and Sinai **[4, 5]** and then used in **[1]**. This condition implies the estimate

$$V_i^{(m)} \leq (1 - d/2)^{[m/s]}$$

for any $m \geq 1$ and all $i \in \mathcal{J}$, see proofs in Lemma 15, **[1]** and §4 below.

The second condition is

(12) $$r := \min_{i,j} \frac{\pi_{ij}^{(s)}}{p_j} > 0$$

for some $s \geq 1$. It was first explicitly introduced in **[6]** based on Ibragimov's regularity condition for stationary random processes **[12]**. It was also used in [7]. This condition implies the estimate

(13) $$V_i^{(m)} \leq (1 - r)^{[m/s]}$$

for any $m \geq 1$ and all $i \in \mathcal{J}$, see proofs in Lemma 4.3, **[6]**, and in §4 below.

The meaning of conditions (11) and (12) for the dynamics of the underlying system is the following. According to (12), the sth image of any atom $A_i \in \mathcal{A}$ intersects all the atoms of $\mathcal{A}$ and the conditional distribution on atoms $A \in \mathcal{A}$ (conditioned on $T^s A_i$) recovers a certain fraction of the invariant measure. This is a very stringent condition. However, it is possible to verify it for some uniformly hyperbolic maps with smooth invariant measure **[6, 8]**. Condition (11) means that the sth images of any two atoms $A_i, A_j \in \mathcal{A}$ are so close to each other that they intersect the same other atoms of $\mathcal{A}$ and the conditional distributions (conditioned on $T^s A_i$ and $T^s A_j$) are close to each other in the variational metric. This is a weaker condition than (12) (it does not require that $T^s A_i$ or $T^s A_j$ intersect all the atoms of $\mathcal{A}$), but it is still pretty stringent. It is verifiable for some hyperbolic attractors with Bowen-Ruelle-Sinai invariant measures **[1]**.

§4. A new condition on transition probabilities

In this section we introduce a new condition on the transition probabilities of the Markov chain, which is weaker than the two discussed above and provides good bounds for the mixing coefficients (10). Our new condition is designed to be verifiable in the case of Markov approximations to hyperbolic flows, in particular, for geodesic flows on surfaces of negative curvature. A verification of this condition is, however, beyond the scope of this paper.

We use the notations of the previous section. Denote $p_{\min} = \min_i p_i$. For every $i, j \in \mathcal{J}$ let

$$b_{i,j} = \sum_{k=1}^{I} \frac{\pi_{ik}\pi_{jk}}{p_k}.$$

THEOREM 1. *Suppose that the Markov chain satisfies the following condition*:

$$b = \min_{i,j} b_{i,j} > 0. \tag{14}$$

Then for any $m \geq 1$ and all $i \in \mathcal{J}$, we have

$$V_i^{(m)} \leq 50 b^{-1/2} p_{\min}^{-1} \cdot (1 - b/2)^{m/3} \tag{15}$$

and thus

$$V^{(m)} \leq 50 b^{-1/2} p_{\min}^{-1} \cdot (1 - b/2)^{m/3}.$$

REMARK. If the condition (14) is satisfied for the s-step transition probabilities $\pi_{ij}^{(s)}$ instead of π_{ij}, then Theorem 1 remains true with the exponent $m/3$ replaced by $[m/s]/3$ in the above bounds.

We will first compare our condition (14) with the two conditions described in the previous section.

LEMMA 2. *The regularity* (12) *implies the Doeblin condition* (11) *with $d \geq r$, and the latter* (*with $s = 1$*) *implies our condition* (14) *with $b \geq d^2$.*

PROOF. Without loss of generality we set $s = 1$. The Doeblin condition (11) is equivalent to

$$d = \min_{i,j} \left\{ \sum_{k=1}^{I} \min\{\pi_{ik}, \pi_{jk}\} \right\} > 0. \tag{16}$$

Clearly, (12) implies (16) with $d \geq r$. Using the Schwarz inequality one can show that (16) implies (14):

$$\begin{aligned} b_{i,j} &= \sum_{k=1}^{I} \frac{\pi_{ik}\pi_{jk}}{p_k} \geq \sum_{k=1}^{I} \left(\frac{\min\{\pi_{ik}, \pi_{jk}\}}{p_k} \right)^2 \cdot p_k \\ &\geq \left(\sum_{k=1}^{I} \frac{\min\{\pi_{ik}, \pi_{jk}\}}{p_k} \cdot p_k \right)^2 \geq d^2. \end{aligned}$$

The lemma is proved.

As Lemma 2 shows, our condition (14) is the weakest one among the three conditions.

Now we prove Theorem 1. First, we make an additional simplifying assumption that the stationary distribution P is uniform, i.e., $p_i = 1/I$ for all $1 \leq i \leq I$. In this case the matrix Π is doubly stochastic, i.e., its transpose Π^T is also a stochastic matrix. Consider the matrix $\tilde{\Pi} = \Pi\Pi^T$. It is a symmetric doubly stochastic matrix with the same uniform stationary distribution P. We denote its components by $||\tilde{\pi}_{ij}||$. We rewrite the condition (14) as follows:

$$b = \min_{i,j} \frac{\tilde{\pi}_{ij}}{p_j} > 0 \tag{17}$$

for any $i, j \in \mathcal{J}$. Note that this is exactly the regularity condition (12) applied to the stochastic matrix $\tilde{\Pi}$, with r replaced by b.

Due to (17), the operator $\tilde{\Pi}$ is a contraction on the simplex of probability distributions equipped with the variational distance (4), whose only "fixed point" is the distribution P, i.e.,

$$\text{Var}\,(P'\tilde{\Pi}, P) \leq (1-b)\text{Var}\,(P', P) \tag{18}$$

for any distribution $P' = ||p'_j||$. To show this, we denote by Σ_j^+ the summation over all j such that $(P'\tilde{\Pi})_j > p_j$. Then

$$\begin{aligned}\text{Var}\,(P'\tilde{\Pi}, P) &= \Sigma_j^+(\Sigma_i p'_i\tilde{\pi}_{ij} - \Sigma_i p_i\tilde{\pi}_{ij}) \\ &= \Sigma_i(p'_i - p_i)\Sigma_j^+\tilde{\pi}_{ij} = \Sigma_i(p'_i - p_i)\Sigma_j^+(\tilde{\pi}_{ij} - b/I).\end{aligned}$$

Now denote by Σ_i^+ the summation over those i for which $p'_i > p_i$. Since $\tilde{\pi}_{ij} - b/I \geq 0$ for all $i, j \in \mathcal{J}$ (this follows from (17)) the right-hand side of the last equation is bounded from above by

$$\Sigma_i^+(p'_i - p_i)\Sigma_j^+(\tilde{\pi}_{ij} - b/I) \leq \Sigma_i^+(p'_i - p_i)\Sigma_j(\tilde{\pi}_{ij} - b/I) \leq (1-b)\cdot\text{Var}\,(P', P).$$

The estimate (18) is proved.

Using the estimate (18) m times (in an obvious way) we get

$$\text{Var}\,(P'\tilde{\Pi}^m, P) \leq (1-b)^m\text{Var}\,(P', P). \tag{19}$$

In particular, the inequality (13) follows from (19) and the inequality $2^{-1}\sum_j |\tilde{\pi}_{ij} - p_j| \leq 1 - b$, which is a consequence of (17).

We now examine the spectrum of the matrix $\tilde{\Pi}$. Since it is a symmetric matrix, all its eigenvalues are real and its eigenvectors are mutually orthogonal. The vector P is an eigenvector with eigenvalue one. Let V be an eigenvector of $\tilde{\Pi}$ different from P, with an eigenvalue λ'. If $||V||$ is small enough, then $P' := P + V$ is a probability distribution. Then $P'\tilde{\Pi}^m = P + V\tilde{\Pi}^m = P + (\lambda')^m V$. Applying the estimate (19) we have

$$|\lambda'|^m \cdot \sum |v_i| \leq 2\text{Var}\,(P', P)(1-b)^m,$$

hence $|\lambda| \leq 1 - b$. Therefore, all but one of the eigenvalues of the matrix $\tilde{\Pi}$ lie in the interval $[-1+b, 1-b]$.

We denote by U the uniform distribution, i.e., $U = (1/I, \dots, 1/I)$. (We first assume that $P = U$, but later we remove this assumption.) We denote also by L_0 the hyperplane in $\mathbb{R}^I$ perpendicular to U. It is parallel to the simplex formed by probability distributions of which U is a center.

We now turn to the matrix Π. Fix an $i \in \mathcal{J}$. The ith row of the matrix Π^m is the vector $||\pi_{ij}^{(m)}||$, $1 \leq j \leq I$. It is equal to $E_i\Pi^m$, where $E_i = (0, \dots, 0, 1, 0, \dots, 0)$ is a unit row vector with its ith component equal to 1. The vector $E_i - P$ belongs to the subspace L_0, and this subspace is left invariant under both $\tilde{\Pi}$ and Π. As shown above, the restriction of $\tilde{\Pi}$ to L_0 is a contraction with all the eigenvalues lying in $[-1+b, 1-b]$. Hence, for any vector $V \in L_0$ we have $||V\tilde{\Pi}|| \leq (1-b)||V||$, where $||\cdot||$ is the Euclidean norm. Hence

$$||V\Pi||^2 = (V\Pi)\cdot(V\Pi) = V\Pi\Pi^T V^T = (V\tilde{\Pi})\cdot V \leq (1-b)||V||^2,$$

where the dot $(\cdot)$ stands for the scalar product of two row-vectors in $\mathbb{R}^I$. Therefore, $||(E_i - P)\Pi^m||^2 \leq (1-b)^m ||E_i - P|| < (1-b)^m$. This implies

$$(20)\quad V_i^{(m)} = \frac{1}{2}\sum_{j=1}^{I} |\pi_{ij}^{(m)} - p_j| \leq \frac{1}{2}\sqrt{I}\cdot ||(E_i - P)\Pi^m|| \leq 2^{-1} p_{\min}^{-1/2}\cdot (1-b)^{m/2}.$$

Hence, we obtain Theorem 1 under the additional assumption that P is uniform.

REMARK. In our calculations we have never used the fact that the components of the matrix Π are positive. So, the estimate (20) is still true if the matrix Π in Theorem 1 is a so-called quasistochastic matrix, i.e., a matrix whose components are not necessarily nonnegative but their sum in each row is one. (In other words, a matrix Π' is quasistochastic if and only if $\Pi' U^T = U^T$.) Of course, we have to assume, additionally, that the stationary distribution P is uniform: $P = U$.

We now prove Theorem 1 in full generality. The idea is to make the stationary distribution P uniform by an appropriate refinement and approximation and to use the above result.

By a refinement of a Markov chain we mean splitting each state $i \in \mathcal{J}$ into $I_i \geq 1$ of "equal" fragments. More precisely, we replace each state $i \in \mathcal{J}$ by a collection of I_i states labeled by i_r, $1 \leq r \leq I_i$. We then define a new Markov chain with the states i_r, $1 \leq r \leq I_i$, and $1 \leq i \leq I$. The total number of states is now $I' = I_1 + \cdots + I_I$. We denote the collection of new states by $\mathcal{J}'$. The transition probabilities are defined by $\pi'_{i_r j_s} = \pi_{ij}/I_j$ for every i_r and j_s; they form a matrix $\Pi' = ||\pi'_{i_r j_s}||$ of size $I' \times I'$. The stationary distribution is $P' = ||p'_{i_r}||$ with $p'_{i_r} = p_i/I_i$.

Clearly, for any i_r and j_s, we have

$$(21)\qquad b'_{i_r,j_s} := \sum_{k_t\in\mathcal{J}'} \frac{\pi'_{i_r k_t}\pi'_{j_s k_t}}{p'_{k_t}} = b_{i,j}$$

such that $b' := \min_{i_r,j_s} b'_{i_r,j_s} = b$. One can directly verify that for any $m \geq 1$ the components of the matrix $(\Pi')^m = ||\pi'^{(m)}_{i_r j_s}||$ satisfy the equation $\pi'^{(m)}_{i_r j_s} = \pi_{ij}^{(m)}/I_j$. Therefore,

$$(22)\qquad V'^{(m)}_{i_r} := \frac{1}{2}\sum_{j_s\in\mathcal{J}'} |\pi'^{(m)}_{i_r j_s} - p'_{j_s}| = V_i^{(m)}$$

for any $i_r \in \mathcal{J}'$.

Let the stationary probabilities p_i be rational numbers whose common denominator is D. In this case we can make a refinement as described above so that the stationary distribution will be uniform with probabilities $1/D$. Due to (21), we can use the version of Theorem 1 proved above. By (20) and (22), this gives the following upper bound:

$$(23)\qquad V_i^{(m)} \leq 2^{-1} D^{1/2}\cdot (1-b)^{m/2}.$$

We now return to the proof of Theorem 1. The idea is to change the matrix Π and the vector P slightly, so that the new stationary probabilities will be rational numbers with sufficiently small common denominator. Let $p > 0$ be a small parameter,

$p \ll p_{\min}$. We set $D = [p^{-1}] + 1$. There is a probability distribution $\bar{P} = (\bar{p}_1, \dots, \bar{p}_I)$ with rational components whose common denominator is D, such that

$$(24) \qquad |\bar{p}_i - p_i| \leq 1/D < p$$

for all $i \in \mathcal{J}$. Note that

$$(25) \qquad 2\mathrm{Var}\,(P, \bar{P}) \leq I/D < p/p_{\min}.$$

Let Γ be an $I \times I$ matrix defined by two conditions: (i) it is identical on the subspace L_0 defined above, i.e., $V\Gamma = V$ for every $V \in L_0$, and (ii) it maps $\bar{P}$ to P, i.e., $\bar{P}\Gamma = P$. Since both P and $\bar{P}$ are transversal to L_0, the matrix Γ is well defined and invertible. Moreover, both Γ and Γ^{-1} are quasistochastic (see Remark above). Since Γ and Γ^{-1} are identical on L_0, for any probability distribution Q we have

$$(26) \qquad Q\Gamma = (Q - \bar{P} + \bar{P})\Gamma = (Q - \bar{P})\Gamma + P = Q - \bar{P} + P.$$

Similarly,

$$(27) \qquad Q\Gamma^{-1} = (Q - P + P)\Gamma^{-1} = Q - P + \bar{P}.$$

We now consider the matrix $\bar{\Pi} = \Gamma\Pi\Gamma^{-1}$. It is quasistochastic, although not necessarily stochastic, and its stationary vector is $\bar{P}$, because $\bar{P}\bar{\Pi} = \bar{P}\Gamma\Pi\Gamma^{-1} = \bar{P}$.

When D is very large, the matrix Γ is very close to the identity matrix, and hence, the matrix $\bar{\Pi}$ is very close to Π. To make this precise, we employ the vectors E_i defined above and use (26) and (27) to obtain

$$E_i\bar{\Pi} = E_i\Gamma\Pi\Gamma^{-1} = E_i\Pi - \bar{P}\Pi + \bar{P}.$$

Then (25) gives

$$(28) \qquad \mathrm{Var}\,(E_i\bar{\Pi}, E_i\Pi) = \mathrm{Var}\,(\bar{P}, \bar{P}\Pi) \leq \mathrm{Var}\,(\bar{P}, P) + \mathrm{Var}\,(P\Pi, \bar{P}\Pi) \leq p/p_{\min}$$

(at the last step we used the classical estimate $\mathrm{Var}\,(P\Pi, Q\Pi) \leq \mathrm{Var}\,(P, Q)$, valid for any stochastic matrix Π and any probability distributions P, Q). We will need only the following consequence of (28):

$$(29) \qquad \delta_* := \max_{i,j} |\bar{\pi}_{ij} - \pi_{ij}| \leq p/p_{\min},$$

where $||\bar{\pi}_{ij}||$ are the components of $\bar{\Pi}$. It is now a simple calculation based on (14), (24), and (29) that for any $i, j \in \mathcal{J}$ we have

$$\sum_k \frac{\bar{\pi}_{ik}\bar{\pi}_{jk}}{\bar{p}_k} \geq \sum_k \frac{\pi_{ik}\pi_{jk} - \delta_*\pi_{ik} - \delta_*\pi_{jk} - \delta_*^2}{p_k(1 + p/p_{\min})}$$
$$\geq \left(b - 2p/p_{\min}^2 - p^2/p_{\min}^3\right) \cdot \left(1 + p/p_{\min}\right)^{-1}.$$

Assume now that $p/p_{\min}^2 < b/10$. Then

$$\sum_k \frac{\bar{\pi}_{ik}\bar{\pi}_{jk}}{\bar{p}_k} \geq b/2.$$

Therefore, the matrix $\bar{\Pi}$ is quasistochastic and satisfies condition (14) with b replaced by $b/2$. Its stationary vector has rational components with the common denominator D. The inequality (23) yields

$$(30)\quad \bar{V}_i^{(m)} := \frac{1}{2}\sum_j |\bar{\pi}_{ij}^{(m)} - \bar{p}_j| \le 2^{-1}D^{1/2}\cdot(1-b/2)^{m/2} \le p^{-1/2}\cdot(1-b/2)^{m/2}.$$

The estimate (28) admits the following generalization:

$$(31)\quad \operatorname{Var}(E_i\bar{\Pi}^m, E_i\Pi^m) \le \operatorname{Var}(\bar{P}, \bar{P}\Pi^m) \le \operatorname{Var}(\bar{P}, P) + \operatorname{Var}(P\Pi^m, \bar{P}\Pi^m) \le p/p_{\min}$$

for any $m \ge 1$. One can prove the above inequalities by arguments used in the proof of (28), and applied to the matrix $\bar{\Pi}^m = \Gamma\Pi^m\Gamma^{-1}$. By (30), (31), and (25) we have

$$(32)\quad \begin{aligned} V_i^{(m)} &= \frac{1}{2}\sum_j |\pi_{ij}^{(m)} - p_j| = \operatorname{Var}(E_i\Pi^m, P) \\ &\le \operatorname{Var}(E_i\Pi^m, E_i\bar{\Pi}^m) + \operatorname{Var}(E_i\bar{\Pi}^m, \bar{P}) + \operatorname{Var}(\bar{P}, P) \\ &\le p^{-1/2}\cdot(1-b/2)^{m/2} + 1.5p/p_{\min}. \end{aligned}$$

We now take

$$p = 10^{-1}bp_{\min}^2\cdot(1-b/2)^{m/3}$$

and obtain Theorem 1.

REMARK. Notice that the choice

$$p = 2^{-1}p_{\min}^{2/3}\cdot(1-b/2)^{m/3}$$

yields a slightly better estimate:

$$V_i^{(m)} \le 2p_{\min}^{-1/3}(1-b/2)^{m/3}.$$

But it is only valid when our assumption $p/p_{\min}^2 < b/10$ is satisfied, i.e., for

$$m \ge |\log p_{\min}^{4/3} + \log b - \log 10| \cdot |\log(1-b/2)|^{-1}.$$

§5. Relaxed condition on transition probabilities

In this section we relax condition (14). Our point is that in the case of dynamical systems with singularities some atoms of the partition $\mathcal{A}$ may be very "ugly" and their evolution may be totally out of control. In this case, it is enough to ensure a positive lower bound on $b_{i,j}$ for an "overwhelming majority" of pairs (i, j) rather than for every single pair (i, j). Based on this, we still can estimate $V^{(m)}$ as the following theorem shows.

THEOREM 3. *Suppose that there is a subset of pairs of indices, $\mathcal{R} \subset \mathcal{J}\times\mathcal{J}$, such that for every pair $(i, j) \in \mathcal{R}$ we have $b_{i,j} \ge b > 0$ and*

$$Q := \sum_{(i,j)\notin\mathcal{R}} p_i p_j < 1.$$

Then for any $m \ge 1$ we have

$$(33)\qquad V^{(m)} \le \text{const}\cdot\left[b^{-1/2}p_{\min}^{-2}(1-b/40)^{m/3} + m(p_{\min} + Q)\right],$$

where const *is an absolute constant* (*one can set* const= 50).

Note that Theorem 3 does not guarantee any convergence to the equilibrium. It is useful only when $m(p_{\min} + Q) \ll 1$, i.e., for relatively small values of m.

The last theorem in this section shows that the assumptions of Theorem 3 are stable under certain perturbations. Consider two Markov chains with matrices of transition probabilities $\Pi = ||\pi_{ij}||$ and $\Pi' = ||\pi'_{ij}||$ and with a common stationary distribution $P = ||p_i||$. Denote

$$b'_{i,j} = \sum_{k=1}^{I} \frac{\pi'_{ik}\pi'_{jk}}{p_k}.$$

THEOREM 4. *Let the Markov chain* (Π, P) *satisfy the assumptions of Theorem* 3, *and let*

$$v' := \frac{1}{2}\sum_{i,j=1}^{I} p_i|\pi_{ij} - \pi'_{ij}| < 1.$$

Then there is a subset $\mathcal{R}' \subset \mathcal{J} \times \mathcal{J}$ *such that for every pair* $(i, j) \in \mathcal{R}'$ *we have* $b'_{i,j} \geq b' = b/2$ *and*

$$Q' := \sum_{(i,j)\notin\mathcal{R}'} p_i p_j < Q + 50b^{-1}v'.$$

We now prove Theorem 3. The key idea of the proof is to add new states to the Markov chain considered in this theorem so that the new larger chain will meet the assumptions of Theorem 1 and, in a sense, will still be close enough to the original chain.

First, we notice that $b_{i,j} = b_{j,i}$, and so $(i, j) \in \mathcal{R}$ whenever $(j, i) \in \mathcal{R}$. Besides, $(i, i) \in \mathcal{R}$ for each $i \in \mathcal{J}$ since $b_{i,i} \geq 1$ for every i.

To implement our plan we need a "uniformization" of both the stationary distribution P and the set of "bad" pairs, $(i, j) \notin \mathcal{R}$. First we make the stationary vector P "fairly uniform'; this means that the ratio $\max_i p_i/p_{\min}$ does not exceed 2. To this end, we use the refinement techniques from the proof of Theorem 1 and simply break every state into two states whenever its probability is larger than $2p_{\min}$. After such a refinement, we define $\mathcal{R}'$ to be the union of all the pairs (i_r, j_s) (in the notations of §4) for which the pair of "predecessors" (i, j) was in $\mathcal{R}$ in the original chain. As it follows from (21), we have $b_{i_r,j_s} \geq b$ for every pair $(i_r, j_s) \in \mathcal{R}'$. It is also clear that

$$\sum_{(i_r,j_s)\notin\mathcal{R}'} p'_{i_r}p'_{j_s} = \sum_{(i,j)\notin\mathcal{R}} p_i p_j = Q.$$

The value of $V^{(m)}$ remains unchanged by virtue of (22), and, obviously, $p_{\min}$ is not altered by the above refinement.

Therefore, it is sufficient to prove Theorem 3 for the new, "refined" chain. In other words, we may assume, in addition to the assumptions of Theorem 3 that

$$\max_i p_i \leq 2p_{\min}. \tag{34}$$

Next, we make the set of "bad" pairs $\mathcal{J}^2 \setminus \mathcal{R}$ "fairly uniform" as follows. Denote $Q' = Q + p_{\min}$. For every $i \in \mathcal{J}$ let

$$q(i) := \sum_{j:\ (i,j)\notin\mathcal{R}} p_j.$$

For every $i \in \mathcal{J}$ such that $q(i) < 2Q'$ we remove from $\mathcal{R}$ one or more pairs (i, j) with some arbitrary $j \neq i$ in such a way that the value of $q(i)$ will increase and satisfy

$$2Q' \leq q(i) < 4Q.$$

This is possible, since $\max_i p_i \leq 2p_{\min} \leq 2Q'$. After that, we also remove the "transpose" (j, i) for every pair (i, j) that has been removed from $\mathcal{R}$. The remaining set $\mathcal{R}$ is still symmetric [i.e., $(i, j) \in \mathcal{R} \Leftrightarrow (j, i) \in \mathcal{R}$] and contains the diagonal $\{(i, i) : i \in \mathcal{J}\}$. It is then an easy calculation that the new value of Q, i.e.,

$$Q_* := \sum_{(i,j)\notin\mathcal{R}} p_i p_j = \sum_{i=1}^{I} p_i q(i)$$

satisfies $Q_* \leq Q + 8Q'$. Hence, $Q'_* = Q_* + p_{\min} \leq 9Q'$. In addition, for any $i \in \mathcal{J}$ we have

$$q(i) = \sum_{j:(i,j)\notin\mathcal{R}} p_j \geq 2Q' > \frac{1}{5} Q'_* \geq \frac{1}{5} Q_*. \tag{35}$$

We are now in a position to implement the plan mentioned at the beginning of the proof. For each unordered pair $(i, j) \notin \mathcal{R}$ we add a new state to our Markov chain, and we label it by (ij) [notice that $(ij) = (ji)$]. Next we specify a new Markov chain with the states $\{i : i \in \mathcal{J}\} \cup \{(ij) : (i, j) \notin \mathcal{R}\}$, which we call "old" and "new" states, respectively. The matrix of transition probabilities is defined by

$$\tilde{\pi}_{ij} = \frac{\pi_{ij}}{1 + q(i)}$$

for any pair $i, j \in \mathcal{J}$ of old states,

$$\tilde{\pi}_{i(ij)} = \frac{p_j}{1 + q(i)} \quad \text{and} \quad \tilde{\pi}_{(ij)k} = \frac{p_k q(k)}{Q_*}$$

for transitions between the old and new states and $\tilde{\pi}_{(ij)(kl)} = 0$ for any pair of new states (in particular, $\tilde{\pi}_{(ij)(ij)} = 0$). We also set $\tilde{\pi}_{k(ij)} = 0$ if k is different from i and j.

The new Markov chain has a stationary distribution with probabilities

$$\tilde{p}_i = p_i \frac{1 + q(i)}{1 + 2Q_*} \quad \text{and} \quad \tilde{p}_{(ij)} = \frac{2 p_i p_j}{1 + 2Q_*} \tag{36}$$

for the old and new states, respectively.

We now show that the new Markov chain meets the assumptions of Theorem 1. For any "good" pair $(i, j) \in \mathcal{R}$ of the old states we have $b_{i,j} \geq b$, and hence,

$$\sum_{k\in\mathcal{J}} \frac{\tilde{\pi}_{ik} \tilde{\pi}_{jk}}{\tilde{p}_k} \geq \frac{1}{8} b.$$

For any "bad" pair of the old states $(i, j) \notin \mathcal{R}$ we have

$$\frac{\tilde{\pi}_{i(ij)} \tilde{\pi}_{j(ij)}}{\tilde{p}_{(ij)}} \geq \frac{1}{8}.$$

For any pair of new states, (ij) and (lr), we have

$$\sum_{k\in\mathcal{J}} \frac{\tilde{\pi}_{(ij)k} \tilde{\pi}_{(lr)k}}{\tilde{p}_k} \geq 1.$$

The latter follows from the fact that $\tilde{\pi}_{(ij)k} = \tilde{\pi}_{(lr)k}$ for all $k = 1, \ldots, I$. Finally, for any old state i and any new state (lj) we have, by (35),

$$\sum_{k \in \mathcal{J}} \frac{\tilde{\pi}_{ik}\tilde{\pi}_{(lj)k}}{\tilde{p}_k} \geq \frac{1}{4} \sum_{k \in \mathcal{J}} \frac{\pi_{ik} q(k)}{Q_*} \geq \frac{1}{20}.$$

Therefore, the new Markov chain satisfies the hypotheses of Theorem 1 with b replaced by $b/20$. (Note that since $\sum_{i,j} p_i p_j b_{i,j} = 1$, we always have $b \leq 1$.) According to (36), the minimum of the stationary probabilities in the new Markov chain satisfies $p_{\min}^{\text{new}} \geq p_{\min}^2/2$.

By Theorem 1, for any $m \geq 1$ and every old state $i \in \mathcal{J}$ we have

$$\frac{1}{2} \sum_{j=1}^{I} |\tilde{\pi}_{ij}^{(m)} - \tilde{p}_j| \leq 50 b^{-1/2} p_{\min}^{-2} (1 - b/40)^{m/3}. \tag{37}$$

We now estimate the left-hand side of (33) as follows:

$$\sum_{i,j=1}^{I} |\pi_{ij}^{(m)} - p_j| p_i \leq \sum_{i,j=1}^{I} |\pi_{ij}^{(m)} - \tilde{\pi}_{ij}^{(m)}| p_i + \sum_{i,j=1}^{I} |\tilde{\pi}_{ij}^{(m)} - \tilde{p}_j| p_i + \sum_{i,j=1}^{I} |\tilde{p}_j - p_j| p_i. \tag{38}$$

The middle term at the right-hand side is readily bounded by (37) in view of (36). The last term at the right-hand side of (38) is bounded by

$$\sum_{i,j=1}^{I} |\tilde{p}_j - p_j| p_i \leq (1 + 2Q_*)^{-1} \sum_{i,j=1}^{I} (p_j q(j) + 2Q_* p_j) p_i \leq 3Q_* \leq 27Q'.$$

In order to estimate the first term at the right-hand side of (38), we expand the transition probability $\tilde{\pi}_{ij}^{(m)}$ as follows:

$$\tilde{\pi}_{ij}^{(m)} = \sum_{k_1, \ldots, k_{m-1}} \tilde{\pi}_{ik_1} \tilde{\pi}_{k_1 k_2} \cdots \tilde{\pi}_{k_{m-1} j},$$

where the variables $k_1, \ldots, k_{m-1}$ run over all the old and new states. We break this sum into two parts, $\tilde{\Sigma}_{ij}^{\text{old}}(m)$ and $\tilde{\Sigma}_{ij}^{\text{new}}(m)$, so that the former is taken over the old states only (i.e., over the states with $k_1, \ldots, k_{m-1} \in \mathcal{J}$) and the latter is taken over strings $k_1, \ldots, k_{m-1}$ that include at least one new state. Respectively,

$$\tilde{\pi}_{ij}^{(m)} = \tilde{\Sigma}_{ij}^{\text{old}}(m) + \tilde{\Sigma}_{ij}^{\text{new}}(m)$$

and

$$\sum_{i,j=1}^{I} |\pi_{ij}^{(m)} - \tilde{\pi}_{ij}^{(m)}| p_i \leq \sum_{i,j=1}^{I} |\pi_{ij}^{(m)} - \tilde{\Sigma}_{ij}^{\text{old}}(m)| p_i + \sum_{i,j=1}^{I} p_i \tilde{\Sigma}_{ij}^{\text{new}}(m). \tag{39}$$

Since the distribution (36) is stationary, rather straightforward calculations show the second term at the right-hand side of (39) can be estimated as follows:

$$\sum_{i,j=1}^{I} p_i \tilde{\Sigma}_{ij}^{\text{new}}(m) \leq (1 + 2Q_*) \sum_{i,j=1}^{I} \tilde{p}_i \tilde{\Sigma}_{ij}^{\text{new}}(m) \leq (1 + 2Q_*) m \sum_{(l,r) \notin \mathcal{R}} \tilde{p}_{(lr)}$$
$$\leq m Q_* \leq 9mQ'.$$

We then rewrite the first term at the right-hand side of (39) as follows

$$\begin{aligned}\sum_{i,j=1}^{I}|\pi_{ij}^{(m)}-\tilde{\Sigma}_{ij}^{\text{old}}(m)|p_i\\ &=\sum_{k_0,\dots,k_m=1}^{I}|p_{k_0}\pi_{k_0k_1}\pi_{k_1k_2}\cdots\pi_{k_{m-1}k_m}-p_{k_0}\tilde{\pi}_{k_0k_1}\tilde{\pi}_{k_1k_2}\cdots\tilde{\pi}_{k_{m-1}k_m}|\\ &=\sum_{k_0,\dots,k_m=1}^{I}\left|\sum_{t=0}^{m-1}p_{k_0}\pi_{k_0k_1}\cdots\pi_{k_{t-1}k_t}(\pi_{k_tk_{t+1}}-\tilde{\pi}_{k_tk_{t+1}})\tilde{\pi}_{k_{t+1}k_{t+2}}\cdots\tilde{\pi}_{k_{m-1}k_m}\right|.\end{aligned}$$

Since the distribution $\{p_k\}$ is stationary for the matrix $||\pi_{kl}||$, the last sum is bounded by

$$\sum_{t=0}^{m-1}\sum_{k_t,k_{t+1}=1}^{I}p_{k_t}|\pi_{k_tk_{t+1}}-\tilde{\pi}_{k_tk_{t+1}}|=m\sum_{i,j=1}^{I}p_i|\pi_{ij}-\tilde{\pi}_{ij}|.$$

Finally, since $|\pi_{ij}-\tilde{\pi}_{ij}|\le\pi_{ij}q(j)$ for any pair of old states $(i,j)\in\mathcal{J}^2$, the above sum is bounded by

$$m\sum_{i,j=1}^{I}p_i\pi_{ij}q(j)=mQ_*\le 9mQ'.$$

Combining the previous estimates of the three terms at the right-hand side of (38), we obtain

$$\frac{1}{2}\sum_{i,j=1}^{I}|\pi_{ij}^{(m)}-p_j|p_i\le 50b^{-1/2}p_{\min}^{-2}(1-b/40)^{m/3}+(9m+14)Q'.$$

Theorem 3 is proven.

We now prove Theorem 4. For any $i,j\in\mathcal{J}$ let $d_{ij}=\pi'_{ij}-\pi_{ij}$. Then,

$$\begin{aligned}b'_{i,j}&=\sum_k\frac{(\pi_{ik}+d_{ik})(\pi_{jk}+d_{jk})}{p_k}\\ &=\sum_k\frac{\pi_{ik}\pi_{jk}}{p_k}+\sum_k\frac{d_{ik}\pi_{jk}}{p_k}+\sum_k\frac{\pi_{ik}d_{jk}}{p_k}+\sum_k\frac{d_{ik}d_{jk}}{p_k}.\end{aligned}$$

Denote the last three sums by $D_{ij}^{(1)}$, $D_{ij}^{(2)}$, and $D_{ij}^{(3)}$, respectively. For each $s=1,2,3$ let $\mathcal{B}^{(s)}=\{(i,j)\colon |D_{ij}^{(s)}|>b/6\}$. Clearly, for any pair (i,j) in the set

$$\mathcal{R}'\colon=\mathcal{R}\setminus\bigcup_{s=1}^{3}\mathcal{B}^{(s)}$$

we have $b'_{i,j}\ge b/2$.

It remains to estimate the quantity Q'. First,

$$\sum_{(i,j)\in\mathcal{B}^{(1)}}p_ip_j<\frac{6}{b}\sum_{i,j}p_ip_j|D_{ij}^{(1)}|<\frac{6}{b}\sum_{i,j,k}p_k^{-1}p_ip_j|d_{ik}|\pi_{jk}=\frac{6}{b}\sum_{i,k}p_i|d_{ik}|=12b^{-1}\nu'.$$

A similar estimate holds for $\mathcal{B}^{(2)}$. Finally,

$$\sum_{(i,j)\in\mathcal{B}^{(3)}} p_i p_j < \frac{6}{b}\sum_{i,j} p_i p_j |D_{ij}^{(3)}| < \frac{6}{b}\sum_{i,j,k} p_k^{-1} p_i p_j |d_{ik}|\cdot|d_{jk}|$$

$$\leq \frac{12}{b}\sum_{i,k} p_i |d_{ik}| = 24b^{-1}\nu',$$

where we use the following upper bound

$$\sum_j p_j|d_{jk}| \leq \sum_j p_j\pi''_{jk} + \sum_j p_j\pi_{jk} = 2p_k.$$

Theorem 4 is proved.

References

[1] V. S. Afraimovich, N. I. Chernov, and E. A. Sataev, *Statistical properties of 2-D generalized hyperbolic attractors*, Chaos **4** (1994).

[2] R. Bowen, *Equilibrium states and the ergodic theory of Anosov diffeomorphisms*, Lecture Notes in Math., vol. 470, Springer-Verlag, Berlin, 1975.

[3] L. A. Bunimocvich, *Decay of correlations in dynamical systems with chaotic behavior*, Soviet Phys. JETP **62** (1985), 842–852.

[4] L. A. Bunimovich and Ya. G. Sinai, *Markov partitions for dispersed billiards*, Comm. Math. Phys. **78** (1980), 247–280.

[5] ———, *Statistical properties of Lorentz gas with periodic configuration of scatterers*, Comm. Math. Phys. **78** (1981), 479–497.

[6] L. A. Bunimovich, Ya. G. Sinai, and N. I. Chernov, *Statistical properties of two-dimensional hyperbolic billiards*, Russian Math. Surveys **46** (1991), 47–106.

[7] N. I. Chernov, *Statistical properties of the periodic Lorentz gas. Multidimensional case*, J. Statist. Phys. **74** (1994), 11–53.

[8] ———, *Limit theorems and Markov approximations for chaotic dynamical systems*, Probab. Theory Related Fields (1995) (to appear).

[9] M. Denker, *The central limit theorem for dynamical systems*, Dynamical Systems Ergodic Theory, Banach Center Publ., vol. 23, PWN–Polish Sci. Publ., Warsaw, 1989.

[10] R. L. Dobrushin, *Central limit theorem for nonstationary Markov chains* I, Theory Probab. Appl. **1** (1956), 65–79.

[11] J. L. Doob, *Stochastic processes*, Wiley, New York, 1990.

[12] I. A. Ibragimov and Y. V. Linnik, *Independent and stationary sequences of random variables*, Wolters-Noordhoff, Groningen, 1971.

[13] G. Nicolis, S. Martinez, and E. Tirapegui, *Finite coarse-graining and Chapman-Kolmogorov equation in conservative dynamical systems*, Chaos, Solutions and Fractals **1** (1991), 25–37.

[14] D. Ruelle, *Thermodynamic formalism*, Addison-Wesley, Reading, MA, 1978.

[15] Ya. G. Sinai, *Markov partitions and C-diffeomorphisms*, Functoinal. Anal. Appl. **2** (1968), 61–82.

[16] ———, *Gibbs measures in ergodic theory*, Russian Math. Surveys **27** (1972), 21–69.

[17] S. Ulam, *Problems in Modern Mathematics*, Independence Publishers, New York, 1960.

Department of Mathematics, University of Alabama at Birmingham, Birmingham, AL 35294
E-mail address: chernov@vorteb.math.uab.edu

Amer. Math. Soc. Transl.
(2) Vol. **171**, 1996

Lyapunov Exponents of the Schrödinger Equation with Certain Classes of Ergodic Potentials

Ilya Goldsheid and Eugene Sorets

ABSTRACT. We prove that all nonnegative Lyapunov exponents of difference Schrödinger equation

$$-\psi_{n+1} + Q_n\psi_n - \psi_{n-1} = E\psi_n, \qquad -\infty < n < +\infty,$$

are strictly positive. Here $\psi_n \in \mathbb{R}^m$ and Q_n is a symmetric $m \times m$ matrix whose off-diagonal elements do not depend on n. The diagonal elements of Q_n are given by the formula

$$q_{nj}(x) = \sigma f_j(T^n x) - E,$$

where all f_j are nonconstant "partly" analytic functions satisfying certain conditions, σ is sufficiently large, and T belongs to a class of ergodic transformations which, in particular, include irrational rotations and skew-shifts of a ν-dimensional torus.

§1. Introduction

We study asymptotic behavior of the solutions of the Schrödinger equation

$$-\psi_{n+1} + Q_n\psi_n - \psi_{n-1} = E\psi_n, \qquad -\infty < n < +\infty, \tag{1}$$

where ψ_n are m-vectors and Q_n are symmetric $m \times m$ matrices given by $Q_n(x) \overset{\text{def}}{=} Q(T^n x)$. Here $T\colon X \to X$ is an ergodic transformation. We prove that for certain dynamical systems T and classes of potentials Q the smallest nonnegative Lyapunov exponent of equation (1) will be strictly positive and obtain explicit lower bounds. Precise formulation of the results will be given in §2 for $m = 1$ and in §5 for $m > 1$.

This work is a continuation of **[GS]**, where we considered a class of potentials generated by the simplest ergodic dynamical system, irrational rotation of the circle. Here we extend the method of **[GS]** to wider classes of dynamical systems, which include, for example, mappings of multi-dimensional tori such as skew-shift and quasi-periodic rotation. From the physical point of view, we consider potentials that are

1991 *Mathematics Subject Classification.* Primary 35Q55; Secondary 58F15.

This paper was written during the author's stay at the IAS while supported by NSF grant DMS-860 1978.

given by different real-analytic functions at every point of the lattice. These functions must satisfy certain conditions, which we verify for the potentials generated by our dynamical systems (see Lemma 1).

The method of **[GS]** is an extension of the method from **[SS]**, which in turn grew out of the analysis of the work of M. Herman **[H]**. The difference between **[GS]** and the present paper is the explicit separation of the ergodic part of the proof from the integral estimates.

Other methods used by Ya. Sinai **[Si]**, J. Fröhlich, T. Spencer, and P. Wittwer **[FSW]** lead in the one-dimensional case to the proof of exponential localization and, in particular, to the positivity of the Lyapunov exponent for C^2 potentials generated by a Diophantine rotation of the circle.

Another paper dealing with the case of dynamical systems more general than the rotation of a circle (in the context of the spectral theory of Scrödinger operators) is that by V. Chulaevsky and Ya. Sinai **[CS]**, where the localization was proved for Diophantine rotation of the two-dimensional torus.

All the above results, as well as ours, hold for strong disorder and, in fact, at least some of them are not true for weak disorder **[BLT]**.

However, we would like to mention the recent nice nonpertubative approach **[J1, J2]** to the proof of the localization for a special class of potentials, namely the almost Mathieu equation.

It is useful to make it clear that the localisation problem remains unsolved for potentials generated by our dynamical systems (except of the just mentioned case of **[CS]**).

Our paper is organized as follows. We state the results for the one-dimensional case in §2. In §3 we describe the main examples, and in §4 we give the proof. In §5 we present the extension of the result to the case of arbitrary fixed $m > 1$. We would like to point out that the proof for the case $m = 1$ is self-contained and the simplicity of the proof is one of the advantages of this paper. We do not give a detailed proof for the case $m > 1$, since once the estimates of §4 are established, the proof of the extension is almost exactly the same as in **[GS]**.

§2. Formulation of results

In the next three sections we consider the case $m = 1$. To describe the class of equations (1) we are going to analyse, we introduce a class of dynamical systems and functions on the phase space of these dynamical systems which will generate the values of the coefficients of our equations.

The class of dynamical systems (X, μ, T) will have the following properties:

1. $X = S^1 \times Y$, where $S^1 = \{z \in \mathbb{C} : |z| = 1\}$ and Y is a compact metric space.
2. $\mu = ds \times d\kappa(y)$, where ds is the normalized Lebesque measure on S^1 and $\mu(X) = 1$.
3. T is ergodic with respect to μ and

$$T(z, y) = (\varphi_1(y)z, \varphi_2(y))$$

with $|\varphi_1(\cdot)| \equiv 1$ and $\varphi_2 \colon Y \to Y$.

The function $f \colon X \to [-1, 1]$ will have the following properties:

4. $f(z, y)$ is continuous in $(z, y) \in X$.
5. For every y, $f(\cdot, y)$ is a nonconstant analytic function on $\mathcal{A}(r, 1) \stackrel{\text{def}}{=} \{z : r \le |z| \le 1\}$ with $r < 1$.

6. For every real number t there exist a radius $r_t \in [r, 1]$ and $\delta_t > 0$ such that

$$|f(z, y) - t| \geq \delta_t$$

for all z with $|z| = r_t$. (In the other words, r_t and δ_t do not depend on y).

In the proof of the theorem we shall use the following consequence of our conditions. If

$$w(y_n) \stackrel{\text{def}}{=} \text{Wind}_{|z|=r}\left(f(T^n(z, y))\right),$$

then, since Y is compact,

$$\sup_{y_n} |w(y_n)| < \infty.$$

REMARK 1. Conditions 1–5 are quite natural and, in general, unavoidable. On the other hand, Condition 6 is purely technical and rather restrictive but, we believe, not necessary in many situations. Nevertheless, the theorem we prove here does provide new interesting classes of stationary potentials for Schrödinger operators such that the corresponding Lyapunov exponents are strictly positive.

Let us consider now the one-dimensional equation

$$-\psi_{n+1} + q_n \psi_n - \psi_{n-1} = 0, \qquad -\infty < n < +\infty, \tag{2}$$

where $q_n = q_n(\theta, y) = \sigma f \circ T^n(e^{2\pi i \theta}) - E$, with σ being a large parameter. Solutions of equation (2) can be rewritten in the form

$$\begin{pmatrix} \psi_{n+1} \\ \psi_n \end{pmatrix} = S(n) \begin{pmatrix} \psi_1 \\ \psi_0 \end{pmatrix},$$

where

$$S(n) \stackrel{\text{def}}{=} A_n \cdots A_1$$

and

$$A_j \stackrel{\text{def}}{=} \begin{pmatrix} q_j & -1 \\ 1 & 0 \end{pmatrix}.$$

The Lyapunov exponent γ of such a system is given by

$$\gamma(\theta, y) = \lim_n \frac{1}{n} \log \|S(n)(e^{2\pi i \theta}, y)\|. \tag{3}$$

Existence of the limit in (3) is guaranteed by the subadditive ergodic theorem **[Ki]** (see also **[O]** or **[GM]**). Since $\det S(n) = 1$ for all n, we have $\gamma \geq 0$. Our main result is the following theorem.

THEOREM 1. *If a dynamical system (X, μ, T) and a function f are as above, then for sufficiently large σ*

$$\gamma(\theta, y) \geq \log \sigma - C \quad \textit{for } \mu\textit{-a.e.} (\theta, y), \tag{4}$$

where the constant C is independent of σ.

We note that the estimate is fairly sharp. Indeed, it is very easy to show that

$$\gamma(\theta, y) \leq \log \sigma + O(\sigma^{-1}).$$

§3. Examples

We give now several examples which satisfy the conditions of Theorem 1.

EXAMPLE 1. Schrödinger operator with a potential and having multiple frequencies. Let X be a d-dimensional torus, $X = S^1 \times \cdots \times S^1$, μ the Lebesque measure on X, and

$$T(s_1, \ldots, s_d) = (s_1 + \alpha_1, \ldots, s_d + \alpha_d),$$

such that the coordinates of the vector $(1, \alpha_1, \ldots, \alpha_d)$ are rationally independent. The function f can be any continuous function which is a nonconstant analytic function in s_1 for every fixed value of $y \stackrel{\text{def}}{=} (s_2, \ldots, s_d)$, and thus, (taking $z = e^{2\pi i s_1}$) is analytic on $\mathcal{A}(r, 1)$ with $r \in (0, 1)$ independent of y. We also require that our function satisfy Condition 6.

EXAMPLE 2. Schrödinger operator with a potential generated by the skew-shift. Skew-shift on a torus is given by:

$$\begin{pmatrix} s_1 \\ s_2 \\ \vdots \\ s_d \end{pmatrix} \xrightarrow{T} \begin{pmatrix} s_1 + \alpha \\ s_2 + \varphi_1(s_1) \\ \vdots \\ s_d + \varphi_{d-1}(s_1, \ldots, s_{d-1}) \end{pmatrix},$$

where the functions φ_j have period 1 in each coordinate and are suitably restricted so that Condition 6 is satisfied, μ is the Lebesque measure, and T is ergodic with respect to it. We take $f(s_1, \ldots, s_d)$ to be a nonconstant analytic function of s_d for each fixed value of $y \stackrel{\text{def}}{=} (s_1, \ldots, s_{d-1})$. We note that f need only be continuous in y. Thus, for some function h, $f(T^n(s_d, y)) = f(y_n, s_d + h(y))$, where y_n does not depend on s_d, and this means that T satisfies Condition 3. Here we set $z = e^{2\pi i s_d}$.

REMARK 2. We belive that in the examples above Condition 6 is not necessary. But since we use this condition in the proof we want to indicate some cases when it can be verified. Suppose that our function f is of the form

$$f(s_1, \ldots, s_d) = p(s_1, \ldots, s_d) + \varepsilon q(s_1, \ldots, s_d),$$

where p satisfies Condition 6, q is an analytic function, and ε is a parameter. If the "radius of analyticity" of q is sufficiently large, it is easy to show that for sufficiently small ε Condition 6 is satisfied. In particular, Condition 6 is satisfied whenever $p(\cdot)$ is an analytic function in one variable.

§4. Proof of Theorem 1

Now we estimate the integral of the norm for a product of matrices of the form

$$A_n(z, y) = \begin{pmatrix} h_n(z, y) & -1 \\ 1 & 0 \end{pmatrix}.$$

Here $\{h_n(z, y)\}$ is a sequence of functions having a common domain of definition

$$\mathcal{D} = \{(z, y) : r \leq |z| \leq 1, y \in K\}.$$

We suppose that these functions are continuous on $\mathcal{D}$ and analytic in z on $\mathcal{A}(r,1)$ for each y. Introduce the product $S(n)(z,y) \stackrel{\text{def}}{=} A_n \cdots A_1$, and the vector

$$\begin{pmatrix} a_n(z,y) \\ b_n(z,y) \end{pmatrix} \stackrel{\text{def}}{=} S(n) \begin{pmatrix} 1 \\ 0 \end{pmatrix}.$$

LEMMA 1. *Suppose that*
1. $|h_n(re^{2\pi i\theta}, y)| \geq K_n \geq 2.5$ *for all* n, y, θ;
2. $\text{Wind}_{|z|=r}\, h_n(z,y) \leq C$ *for all* n, y.

Then

$$\int_X \log |a_n^2 + b_n^2|\, d\mu \geq \log \prod_{j=1}^{n} \left(K_j - \frac{1}{2}\right) - nC \log \frac{1}{r_0}.$$

REMARK 3. The lemma provides an a priori estimate for the norm of the product of matrices and does not depend on the way the functions $h_n(z,y)$ are constructed in our examples.

PROOF OF LEMMA 1. We use of the following inequality. If $g(z)$ is analytic on $\mathcal{A}(r,1)$, then

$$(5)\quad \int_0^1 \log |g(e^{2\pi i\theta})|\, d\theta \geq \int_0^1 \log |g(re^{2\pi i\theta})|\, d\theta + \left(\log \frac{1}{r}\right) \text{Wind}_{|z|=r}\, g(z).$$

For a proof, see **[SS]**. In our case, $g(z) = a_n^2(z,y) + b_n^2(z,y)$ and y is treated as a parameter. Let us note that the matrices $A_j(re^{2\pi i\theta}, y)$ preserve the cone

$$V \stackrel{\text{def}}{=} \{(v_1, v_2) \in \mathbb{C}^2 : |v_2| < (1/2)|v_1|\}.$$

The calculation which proves this claim is simple, but important:

$$A_j(re^{2\pi i\theta}, y) \begin{pmatrix} v_1 \\ v_2 \end{pmatrix} = \begin{pmatrix} h_j(re^{2\pi i\theta}, y)v_1 - v_2 \\ v_1 \end{pmatrix} = (h_j v_1 - v_2) \begin{pmatrix} 1 \\ \frac{v_1}{h_j v_1 - v_2} \end{pmatrix}.$$

Since

$$|h_j v_1 - v_2| > |h_j v_1| - (1/2)|v_1| > (K_j - 1/2)|v_1| > 2|v_1|,$$

the claim is true.

Repeating the above calculation n times, we see that

$$S(n)(re^{2\pi i\theta}, y) \begin{pmatrix} 1 \\ 0 \end{pmatrix} = \prod_1^n (h_j - \eta_j) \begin{pmatrix} 1 \\ \eta_{n+1} \end{pmatrix},$$

where $\eta_1 = 0$, $\eta_{j+1} = (h_j - \eta_j)^{-1}$, and $|\eta_j| \leq 1/2$ for all j. Therefore,

$$a_n^2 + b_n^2 = \prod_1^n (h_j - \eta_j)^2 \left(1 + \eta_{n+1}^2\right),$$

and

$$\begin{aligned}\int_X \log|a_n^2+b_n^2|\,d\mu &= \int_K dy \int_0^1 d\theta \log|a_n^2(e^{2\pi i\theta},y)+b_n^2(e^{2\pi i\theta},y)| \\ &\geq \int_K dy \int_0^1 d\theta \log|a_n^2(re^{2\pi i\theta},y)+b_n^2(re^{2\pi i\theta},y)| \\ &\quad + \int_K dy \left(\log\frac{1}{r}\right) \mathrm{Wind}_{|z|=r}(a_n^2+b_n^2) \\ &\geq 2\int_K dy \left[\log\prod_1^n\left(K_j-\frac{1}{2}\right)-\left(\log\frac{1}{r}\right)nC\right] \\ &\geq 2\log\prod_1^n\left(K_j-\frac{1}{2}\right)-2\left(\log\frac{1}{r}\right)nC.\end{aligned}$$

In the first inequality we use (5), and the other inequalities follow from the assumptions 1 and 2 of the lemma.

PROOF OF THEOREM 1. Since T is ergodic, γ defined in (3) is constant for μ-a.e. (θ, y) and

$$\gamma = \lim_n \frac{1}{n}\int_X \log\|S(n)\|\,d\mu \geq \lim_n \frac{1}{2n}\int_X \log|a_n^2+b_n^2|\,d\mu. \tag{6}$$

To estimate the last integral we use Lemma 1 with $h_n(z,y)=\sigma f(T^n(z,y))-E$, where $z=e^{2\pi i\theta}$. Let us represent the energy in the form $E=\sigma\lambda$. If E lies in the spectrum of the Schrödinger operator given by the left-hand side of equation (1), then $\lambda\in[-3,3]$.

Using Condition 6, for each fixed λ we choose $r_\lambda\in(r,1)$ such that

$$\min_{y,\theta}|f(r_\lambda e^{2\pi i\theta},y)-\lambda| \stackrel{\text{def}}{=} 2\delta_\lambda > 0. \tag{7}$$

Since f is continuous in (z,y,λ), there exists a neighborhood U of λ such that for $u\in U$

$$\min_{u\in U}\min_{y,\theta}|f(r_u e^{2\pi i\theta},y)-\lambda| > \delta_\lambda.$$

We choose a finite number of neighborhoods U that cover $[-3,3]$. Let $\delta=\min\delta_\lambda$, where the minimum is taken over δ_λ of this cover. From now on we assume that $\sigma>2.5/\delta$.

We are now ready to verify the assumptions of Lemma 1. First of all, the domain $\mathcal{D}$ depends on λ:

$$\mathcal{D}_\lambda \stackrel{\text{def}}{=} \{(z,y): r_\lambda\leq|z|<1, y\in K\}.$$

The numbers K_n are given by $K_n\stackrel{\text{def}}{=}\sigma\delta$ for all n, and

$$h_n(z,y)=\sigma\left(f(T^n(z,y))-\lambda\right).$$

Finally, the bound on the winding number (Condition 2 of Lemma 1) is satisfied because

$$\begin{aligned}\mathrm{Wind}_{|z|=r}h_n(z,y) &= \mathrm{Wind}_{|z|=r}\left(f(T^n(z,y))-\lambda\right) \\ &= \mathrm{Wind}_{|z|=r}\left(f(z,y)-\lambda\right) < \text{const}.\end{aligned}$$

In the last equality we used the fact that $T^n(z, y) = (c_n(y)z, \varphi_2^n(y))$. From (6) and Lemma 1 we have

$$\gamma \geq \frac{1}{2n} \log \left(\sigma\delta - \frac{1}{2} \right)^{2n} = \log \left(\sigma\delta - \frac{1}{2} \right).$$

§5. Extension to the strip

As we have already mentioned, we will only present the formulation of the theorem for the strip. We will not give a proof because it closely follows the arguments used for the proof of the corresponding theorem in **[GS]**. The matrices Q_n of equation (1) will have the following structure. The off-diagonal elements of Q_n are independent of n and the diagonal elements are

$$q_{nj} = \lambda f_j(T^n x),$$

with $x \in X$, where the functions $f_j(\cdot)$ satisfy Conditions 4–6 and the dynamical system (X, μ, T) satisfies Conditions 1–3. Equation (1) becomes equivalent to the finite-difference Schrödinger equation when the off-diagonal elements are chosen properly. For example, the case $q_{ij} = -1$ for $|i - j| = 1$ and $q_{ij} = 0$ for $|i - j| > 1$ corresponds to Schrödinger operator on the strip $\mathbb{Z} \times \{1, \dots, m\}$. Equation (1) can be written in the form

$$\begin{pmatrix} \psi_{n+1} \\ \psi_n \end{pmatrix} = \begin{pmatrix} Q_n & -I \\ I & 0 \end{pmatrix} \begin{pmatrix} \psi_n \\ \psi_{n-1} \end{pmatrix} \equiv A_n \begin{pmatrix} \psi_n \\ \psi_{n-1} \end{pmatrix} \tag{8}$$

so that the asymptotic behavior of the solutions of equation (1) is determined by the asymptotic behavior of the product $S(n) \stackrel{\text{def}}{=} A_n \cdots A_1$.

To describe our result let us consider the following decomposition of $S(n)$:

$$S(n) = U(n)D(n)V(n),$$

where $U, V \in O(n)$, and $D(n) = \operatorname{diag}(d_1^{(n)}, \dots, d_{2m}^{(n)})$ with $d_1^{(n)} \geq d_2^{(n)} \geq \cdots \geq d_{2m}^{(n)} > 0$. Since $A_n \in Sp(m, \mathbb{R})$ for all n, we have $d_k = d_{2m-k+1}^{-1}$. The kth Lyapunov exponent γ_k is defined by

$$\gamma_k \stackrel{\text{def}}{=} \lim_{n \to \infty} \frac{1}{n} \log d_k(n). \tag{9}$$

Clearly, $\gamma_k = -\gamma_{2m-k+1}$ and $\gamma_1 \geq \gamma_2 \geq \cdots \geq \gamma_m \geq 0$. Existence of the limit in (9) for almost every θ and its independence of θ are guaranteed by the Subadditive Ergodic Theorem **[Ki]** and the ergodicity of the underlying dynamical system $\theta \mapsto \theta + \alpha$.

For the strip the main result is the following theorem.

THEOREM 2. *There exists λ_0 such that for all $\lambda > \lambda_0$ and all E there exists a set $\Omega(E) \subset X$ of measure* 1 *with*

$$\gamma_m(E) = \gamma_m(E, x) > 0 \quad \textit{for all } x \in \Omega(E). \tag{10}$$

Together with the Oseledets Multiplicative Ergodic Theorem **[O, GM]**, this implies the existence of m solutions of equation (1) that decrease exponentially as $n \to +\infty$ and grow exponentially as $n \to -\infty$, and m other solutions of equation (1) that decrease exponentially as $n \to -\infty$ and grow exponentially as $n \to +\infty$. These $2m$ solutions are linearly independent and form a basis of all solutions of equation (1).

Let us mention two important consequences of this fact.

1. The spectrum of the operator H defined by the left-hand side of equation (1) is singular.
2. The Green function of H decays exponentially for almost every energy E.

References

[BLT] J. Bellisard, R. Lima, and D. Testard, *A Metal-insulator transition for the almost Mathieu model*, Comm. Math. Phys. **88** (1983), 207–234.

[CS] V. Chulaevsky and Ya. Sinai, *Anderson localization for multi-frequency quasi-periodic potentials in one dimension*, Comm. Math. Phys. **125** (1989), 91–112.

[FSW] J. Fröhlich, T. Spencer, and P. Wittwer, *Localization for a class of one-dimensional quasi-periodic potentials*, Comm. Math. Phys. **132** (1990), 5–25.

[GM] I. Ya. Goldsheid and G. A. Margulis, *Lyapunov indices of products of random matrices*, Uspekhi Mat. Nauk **44** (1989), no. 5, 13–60; English transl., Russian Math. Surveys **44** (1989), no. 5, 11–71.

[GS] I. Ya. Goldsheid and E. Sorets, *Lyapunov exponents of the Schrödinger equation with quasi-periodic potential on a strip*, Comm. Math. Phys. **145** (1992), 507–513.

[H] M. Herman, *Une méthode pour minorer les exposants de Lyapunov et quelques exemples montrant le charactère local d'un théorème d'Arnold et de Moser sur le tore en dimension* 2, Commun. Math. Helv. **58** (1983), 453–502.

[J1] S. Jitomirskaya, *Anderson localization for the almost Mathieu equation*: *A Nonpertubative Proof*, Preprint (1993).

[J2] ———, *Anderson localization for the almost Mathieu equation*: II. *Point Spectrum for* $\lambda > 2$, Preprint (1994).

[Ki] J. F. C. Kingman, *Subadditive processes*, Lecture Notes Math., vol. 539, Springer-Verlag, Heidelberg, 1976.

[O] V. I. Oseledets, *A multiplicative ergodic theorem, characteristic Lyapunov exponents of dynamical systems*, Trudy Moskov. Mat. Obshch. **19** (1968), 179–210; English transl., Trans. Moscow Math. Soc. **19** (1968), 197–231.

[Si] Ya. Sinai, *Anderson localization for one-dimensional difference Schrödinger operator with quasi-periodic potential*, J. Stat. Phys. **46** (1987), 861–918.

[SS] E. Sorets and T. Spencer, *Positive Lyapunov exponents for Schrödinger operator with quasi-periodic potentials*, Comm. Math. Phys. **142** (1992), 543–566.

School of Mathematics, Queen Mary and Westfiel College, University of London, London, UK

Computer Science Department, Yale University, New Haven, USA

Amer. Math. Soc. Transl.
(2) Vol. 171, 1996

Geometric Interpretation of Entropy for Random Processes

B. M. Gurevich

§1. Introduction

One of the earliest and the most famous Sinai's mathematical achievements is his contribution to the entropy theory of dynamical systems. It is well known that the measure-theoretic entropy characterizes the greatest rate of refinement for finite partitions of the phase space. There are systems for which the entropy admits another, somewhat more geometric, interpretation. We outline this for symbolic dynamical systems, i.e., discrete random processes. The general idea is that the entropy measures the boundary distortion rate for some simple sets in the phase space. For mechanical systems it was first observed by Zaslavsky **[Z]** who, however, gave no adequate mathematical description.

Let A be a finite set and $X = A^{\mathbb{Z}}$ the corresponding sequence space endowed with the metric

$$d^\theta(x, x') = \sum_{n\in\mathbb{Z}} \theta^{|n|}(1 - \delta(x_i, x_i')), \qquad x = (x_i, i \in \mathbb{Z}), \quad x' = (x_i', i \in \mathbb{Z}),$$

where θ is an arbitrary number in $(0,1)$ and $\delta(\cdot,\cdot)$ is the Kronecker symbol. The topology in X induced by d^θ does not depend on θ. The space (X, d^θ) is compact for each θ and the shift transformation S defined by $(Sx)_i = x_{i+1}$, $i \in \mathbb{Z}$, is a homeomorphism.

A subset $Y \subset X$ is called a Markov set of order $m \geq 1$ if there exists a family $\mathcal{F}$ of sequences from A^{m+1} such that $y = (y_i,\ i \in \mathbb{Z}) \in Y$ if and only if $(y_j, \dots, y_{j+m}) \in \mathcal{F}$ for all $j \in \mathbb{Z}$. Clearly, every Markov set is closed and S-invariant. Also, a Markov set of order m is a Markov set of order $m+1$ as well.

For any $\varepsilon > 0$, $\theta \in (0,1)$, and $x \in X$ we denote by $B^\theta(x, \varepsilon)$ the ball in (X, d^θ) of radius ε centered at x; for any set $D \subset (X, d^\theta)$ denote by $\mathcal{O}_\varepsilon^\theta(D)$ its open ε-neighborhood.

We now state the main result of this paper.

THEOREM. *Let μ be an S-invariant Borel probability measure on X concentrated on a Markov set $Y \subset X$ (i.e., $\mu(Y) = 1$) and ergodic (more precisely, the dynamical system (X, μ, S) is ergodic). Then*

1991 *Mathematics Subject Classification*. Primary 28D20; Secondary 60G10.

(i) *for any function* $t : \mathbb{R}^+ \to \mathbb{Z}^+$ *such that*

$$\lim_{\varepsilon\to 0} t(\varepsilon)/\log\varepsilon = 0, \tag{1}$$

and for any $\theta \in (0,1)$ *we have*

$$\lim_{\varepsilon\to 0}\frac{1}{t(\varepsilon)}\log\frac{\mu(O_\varepsilon^\theta(S^{t(\varepsilon)}(Y\cap B^\theta(x,\varepsilon))))}{\mu(B^\theta(x,\varepsilon))} = h_\mu, \tag{2}$$

where h_μ *is the measure-theoretic entropy of the dynamical system* (X,μ,S) *and the convergence is meant in the topology of the space* $L_1(X,\mu)$;

(ii) *for any function* t *satisfying* (1) *there exists a sequence* $\{\varepsilon_n\}$ *such that* (2) *holds for all* $\theta \in (0,1)$ *and* μ*-almost all* $x \in X$, *provided that* ε *is replaced by* ε_n *and* $n \to \infty$.

REMARK 1. The L_1-convergence as $\varepsilon \to 0$ certainly implies the a.e. convergence for some sequence $\{\varepsilon_n\}$ as $n \to \infty$. Arguing this way, however, we would have to come to a sequence depending on θ.

§2. Proof of the theorem

(i) Assume that to each point $x \in X$ one assigns a set $V(x) \subset X$. For any $U \subset X$ we denote by $\bigcup\{V(x), x \in U\}$ the union of sets $V(x)$ over all $x \in U$. For any $x = (x_i,\ i \in \mathbb{Z}) \in X$ and any $k,l \in \mathbb{Z}$ we denote by $C_k^l(x)$ the cylinder set with support $[k,l]$ containing x, i.e., $C_k^l(x) = \{y = (y_i, i \in \mathbb{Z}) : y_i = x_i, k \le i \le l\}$.

It is easy to see that given $\theta \in (0,1)$ there exists a positive integer $\kappa = \kappa(\theta)$ such that

$$C_{-k-\kappa}^{k+\kappa}(x) \subseteq B^\theta(x,\theta^k) \subseteq C_{-k}^k(x)$$

for all $x \in X,\ k \in \mathbb{Z}^+$. It follows that if $\theta^{k+1} < \varepsilon \le \theta^k$, i.e.,

$$k = k(\theta,\varepsilon) = [\,|\log\varepsilon|/|\log\theta|\,], \tag{3}$$

(here $[c]$ is the integer part of c), then

$$C_{-k-\kappa-1}^{k+\kappa+1}(x) \subseteq B^\theta(x,\varepsilon) \subseteq C_{-k}^k(x), \tag{4}$$

hence

$$\mu(C_{-k-\kappa-1}^{k+\kappa+1}(x)) \le \mu(B^\theta(x,\varepsilon)) \le \mu(C_{-k}^k(x)). \tag{5}$$

Using (4), we will obtain two-sided bounds for the set

$$D(x,\theta,\varepsilon) = Y \cap O_\varepsilon^\theta(Y \cap S^{t(\varepsilon)}B^\theta(x,\varepsilon)) = \cup\{B^\theta(y,\varepsilon), y \in S^{t(\varepsilon)}(Y \cap B^\theta(x,\varepsilon))\}.$$

Indeed, for $k = k(\theta,\varepsilon)$ (see (3)) and $t = t(\varepsilon)$, we obtain the following inclusions using (4) and the S-invariance of Y

$$\begin{aligned} D(X,\theta,\varepsilon) &\subseteq \cup\{Y \cap C_{-k}^k(y),\ y \in Y \cap S^t C_{-k}^k(x)\} \\ &= \cup\{Y \cap C_{-k}^k(y),\ y \in Y \cap C_{-k-t}^{k-t}(S^t x)\}, \end{aligned} \tag{6}$$

$$\begin{aligned} D(X,\theta,\varepsilon) &\supseteq \cup\{Y \cap C_{-k-\kappa-1}^{k+\kappa+1}(y),\ y \in Y \cap S^t C_{-k-\kappa-1}^{k+\kappa+1}(x)\} \\ &= \cup\{Y \cap C_{-k-\kappa-1}^{k+\kappa+1}(y),\ y \in Y \cap C_{-k-\kappa-1-t}^{k+\kappa+1-t}(S^t x)\}. \end{aligned} \tag{7}$$

Since Y is a Markov set, one can simplify the right-hand sides of (6) and (7) in the way described below.

In the following lemma we will use the notation $C_{[i,j]}(x)$ or $C_\Delta(x)$ instead of $C_i^j(x)$, if $x \in X$, $\Delta = [i,j]$, $i \le j$, $i, j \in \mathbb{Z}$. We denote by $|\Delta|$ the length of the interval $\Delta \subset \mathbb{Z}$, i.e., the number of points in it. For each sequence $y = (y_j, j \in \mathbb{Z}) \in X$ denote by $y_{[i,j]}$ (or y_Δ, when $\Delta = [i,j]$) the block $(y_i, \dots, y_j) \in A^{|\Delta|}$. We will also write $y_{(-\infty,j]}$ and $y_{[i,\infty)}$ when $i = -\infty$ and $j = \infty$, respectively.

LEMMA 1. *If $Y \subset X$ is a Markov set of order m, then for any $x \in Y$ and any two intervals $\Delta_i \subset \mathbb{Z}$, $i = 1,2$, such that $|\Delta_1 \cap \Delta_2| \ge m+1$, the following equality holds*:

$$\cup\{Y \cap C_{\Delta_1}(y),\ y \in Y \cup C_{\Delta_2}(x)\} = Y \cap C_{\Delta_1 \cap \Delta_2}(x). \tag{8}$$

PROOF. Clearly, the set at the left-hand side of (8) is contained in the set at the right-hand side regardless of both the Markov property of Y and the restrictions on Δ_1 and Δ_2. In order to prove the opposite inclusion, consider Y, Δ_1, Δ_2 as in the statement and let $z \in Y \cap C_{\Delta_1 \cap \Delta_2}(x)$, where $\Delta_1 = [i_1, j_1]$, $\Delta_2 = [i_2, j_2]$. Define a new sequence $y \in X$ as follows:

$$\begin{aligned}
&y = z, && \text{if } \Delta_1 \supseteq \Delta_2,\\
&y = x, && \text{if } \Delta_1 \subseteq \Delta_2,\\
&y_{(-\infty,j_1]} = z_{(-\infty,j_1]},\ y_{[j_1+1,\infty)} = x_{[j_1+1,\infty)}, && \text{if } i_1 < i_2 < j_1 < j_2,\\
&y_{(-\infty,j_2]} = x_{(-\infty,j_2]},\ y_{[j_2+1,\infty)} = z_{[j_2+1,\infty)}, && \text{if } i_2 < i_1 < j_2 < j_1.
\end{aligned} \tag{9}$$

The inequality $|\Delta_1 \cap \Delta_2| \ge m+1$ eliminates all the other mutual dispositions of Δ_1 and Δ_2, and also implies that any block of length $m+1$ in y is at the same time a block in either x or z. By our assumption, $x, z \in Y$. Since Y is a Markov set of order m, we get $y \in Y$. Finally, the equality $z_{\Delta_1 \cap \Delta_2} = x_{\Delta_1 \cap \Delta_2}$ guarantees that in all cases listed in (9) we have $y_{\Delta_1} = z_{\Delta_1}$, $y_{\Delta_2} = x_{\Delta_2}$, i.e., $y \in C_{\Delta_2}(x)$, $z \in C_{\Delta_1}(y)$. Therefore, $z \in Y \cap C_{\Delta_1}(y)$, where $y \in Y \cap C_{\Delta_2}(x)$. It means that z belongs to the left-hand side of (8), which completes the proof of Lemma 1.

From (3) and the assumption $t(\varepsilon) = o(\log \varepsilon)$ (see (1)) it follows that $t(\varepsilon) = o(k(\theta, \varepsilon))$ as $\varepsilon \to 0$ and θ is fixed. Hence, if ε is sufficiently small, the two pairs of intervals, $\Delta_1 = [-k, k]$, $\Delta_2 = [-k-t, k+t]$ and $\Delta_1 = [-k-\kappa-1, k+\kappa+1]$, $\Delta_2 = [-k-\kappa-1-t, k+\kappa+1-t]$, where $k = k(\theta,\varepsilon)$, $t = t(\varepsilon)$ satisfy the assumptions of Lemma 1. From (6)–(9), Lemma 1, and the equality $\mu(Y) = 1$, we obtain

$$\begin{gathered}
Y \cap C_{-k-\kappa-1}^{k+\kappa+1-t}(S^t x) \subseteq Y \cap O_\varepsilon^\theta(Y \cap S^t B^\theta(x,\varepsilon)) \subseteq Y \cap C_{-k}^{-k-t}(x),\\
\frac{\mu(C_{-k-\kappa-1}^{k+\kappa+1-t}(S^t x))}{\mu(C_{-k}^{k}(x))} \le \frac{\mu(O_\varepsilon^\theta(Y \cap S^t B^\theta(x,\varepsilon)))}{\mu(Y \cap B^\theta(x,\varepsilon))} \le \frac{\mu(C_{-k}^{-k-t}(S^t x))}{\mu(C_{-k-\kappa-1}^{k+\kappa+1}(x))}.
\end{gathered} \tag{10}$$

The rest of the proof can be conducted using well-known arguments in the proof of Shannon–McMillan–Breiman Theorem (see, for instance, [**B**]). Consider the set

$$X^0 = \{x \in X : \mu(C_i^j(S^m x)) > 0 \text{ for all } i, j, m \in \mathbb{Z},\ i \le j\}$$

and the functions

$$g_0^+(x) = g_0^-(x) = -\log \mu(C_0^0(x)), \quad g_r^+(x) = -\log\big(\mu(C_{-r}^0(x))/\mu(C_{-r}^{-1}(x))\big),$$

$$g_r^-(x) = -\log\big(\mu(C_0^r(x))/\mu(C_1^r(x))\big), \quad r \ge 1,\ x \in X^0.$$

Since $\mu(X^0)=1$, these functions are well defined a.e. Moreover, they are nonnegative and satisfy the identities

$$\begin{aligned}-\log\mu(C_a^b(x)) &= \sum_{r=0}^{b-a} g_r^+(S^{r+a}x)\\ &= \sum_{r=0}^{b-a} g_r^-(S^{-r+b}x),\quad x\in X^0,\quad a,b\in\mathbb{Z},\quad a\le b.\end{aligned}\tag{11}$$

If ε is sufficiently small, using (11), one can obtain the following relations:

$$\begin{aligned}\frac{1}{t}\log\frac{\mu(C_{-k-\kappa-1}^{k+\kappa+1-t}(S^tx))}{\mu(C_{-k}^k(x))} &= \frac{1}{t}\log\frac{\mu(C_{-k-\kappa-1+t}^{k+\kappa+1}(x))}{\mu(C_{-k}^k(x))}\\ &= \frac{1}{t}\log\left(\frac{\mu(C_{-k}^{k+\kappa+1}(x))}{\mu(C_{-k}^k(x))}\frac{\mu(C_{-k-\kappa-1+t}^{k+\kappa+1}(x))}{\mu(C_{-k}^{k+\kappa+1}(x))}\right)\\ &= -\frac{1}{t}\sum_{r=2k+1}^{2k+\kappa+1} g_r^+(S^{r-k}x)+\frac{1}{t}\sum_{r=2(k+\kappa+1)+1-t}^{2k+\kappa+1} g_r^-(S^{-r+k+\kappa+1}x),\quad x\in X^0.\end{aligned}\tag{12}$$

Similarly,

$$\begin{aligned}\frac{1}{t}\log\frac{\mu(C_{-k}^{k-t}(S^tx))}{\mu(C_{-k-\kappa-1}^{k+\kappa+1}(x))} &= \frac{1}{t}\sum_{r=2k+\kappa+2}^{2(k+\kappa+1)} g_r^+(S^{r-k-\kappa-1}x)\\ &\quad+\frac{1}{t}\sum_{r=2k+1-t}^{2k+\kappa+1} g_r^-(S^{-r+k}x),\qquad x\in X^0.\end{aligned}\tag{13}$$

As in the proof of Shannon–McMillan–Breiman Theorem ([**B**, Sect. 13]) one shows that the function $G^+(x)=\sup_r g_r^+(x)$ belongs to $L_1(X,\mu)$ (actually, to $L_p(X,\mu)$ for every $p>0$) and the following limit exists μ-a.e.:

$$\lim_{r\to\infty} g_r^+(x)=g^+(x)$$

with $Eg_r^+=h_\mu$ (here E denotes the integral with respect to μ). Similarly,

$$\int_X \sup_r g_r^-(x)\mu(dx)<\infty,\tag{14}$$

and there exists the limit

$$\lim_{r\to\infty} g_r^-(x)=g^-(x)\tag{15}$$

such that

$$Eg^-=h_\mu.\tag{16}$$

From (14) it readily follows that the first of the two terms at the right-hand side of (12) tends to zero in $L_1(X,\mu)$ as $\varepsilon\to 0$. The same holds true for (13). The second term in (12) can be represented in the form

$$(17)\qquad \begin{aligned}&\frac{1}{t}\sum_{r=2(k+\kappa+1)+1-t}^{2k+\kappa+1} g^{-}(S_1^{r-k-\kappa-1}x)\\ &\quad+\frac{1}{t}\sum_{r=2(k+\kappa+1)+1-t}^{2k+\kappa+1}\left(g_r^{-}(S_1^{r-k-\kappa-1}x)-g^{-}(S_1^{r-k-\kappa-1}x)\right),\end{aligned}$$

where $S_1 = S^{-1}$. By the ergodic theorem and ergodicity of the dynamical system (X, μ, S_1), the first term in (17) tends to h_μ in $L_1(X, \mu)$ as $\varepsilon \to 0$ (see (16)). The L_1-norm (denoted below by $\|\cdot\|$) of the second term does not exceed $(t-\kappa-1)\|G^{-}_{2(k+\kappa+1)+1-t}\|/t$, where

$$(18)\qquad G_n^{-}(x) = \sup_{r\geq n}|g_r^{-}(x) - g(x)|.$$

From (14), (15), and (18) it follows that

$$(19)\qquad \lim_{n\to\infty}\|G_n^{-}\| = 0.$$

This implies the L_1-convergence of the right-hand side of (12) to h_μ as $\varepsilon \to 0$. The same holds true for (13). Now (10), (12), and (13) together imply statement (i) of Theorem.

(ii) We begin with the following lemma. Its simple proof is omited.

LEMMA 2. *Let $f_0, f_1, \ldots$ be a sequence of functions from $L_1(X, \mu)$ such that $|f_i(x)| \leq f_0(x)$ a.e. for all $i \geq 1$. Let also $a_1, a_2, \ldots$ be a sequence of positive integers satisfying $\sum_n a_n^{-1} < \infty$. Then for any $b_{n,i} \in \mathbb{Z}$, $n, i = 1, 2, \ldots$, and any $c_n \in \mathbb{Z}^+$, $n = 1, 2, \ldots$, such that* $\max_n c_n < \infty$, *we have the a.e. convergence*

$$\lim_{n\to\infty}\frac{1}{a_n}\sum_{i=0}^{c_n} f_i(S^{b_{n,i}}x) = 0.$$

We shall choose the sequence $\{\varepsilon_n\}$ such that

$$(20)\qquad \sum_{n=1}^{\infty}\left(t(\varepsilon_n)\right)^{-1} < \infty.$$

By Lemma 2, the first terms at the right-hand sides of both (12) and (13) converge a.e. to zero if we replace ε_n by ε and let $n \to \infty$. One can rewrite the second term at the right-hand side of (12) in the form (17). Letting again $\varepsilon = \varepsilon_n$ and denoting the result by $F(x, \kappa, n)$, we can continue transforming it as follows:

$$\begin{aligned}F(x,\kappa,n) &= \frac{1}{t}\sum_{j=0}^{t-\kappa-2} g^{-}(S_1^{k+\kappa+2-t+j}x)\\ &\quad+\frac{1}{t}\sum_{j=0}^{t-\kappa-2}\left[g^{-}_{j+2k+2\kappa+3-t}(S_1^{k+\kappa+2-t+j}x) - g^{-}(S_1^{k+\kappa+2-t+j}x)\right]\\ &= \frac{1}{t}\sum_{j=0}^{t-1} g^{-}(S_1^{k+\kappa+2-t+j}x)\end{aligned}$$

$$+\frac{1}{t}\sum_{j=0}^{t-1}[g^-_{j+2k+2\kappa+3-t}(S_1^{k+\kappa+2-t+j}x) - g^-(S_1^{k+\kappa+2-t+j}x)] \tag{21}$$
$$-\frac{1}{t}\sum_{j=0}^{\kappa} g^-(S_1^{k+1+j}x)$$
$$-\frac{1}{t}\sum_{j=0}^{\kappa}[g^-_{j+2k+\kappa+2}(S_1^{k+1+j}x) - g^-(S_1^{k+1+j}x)].$$

Let us look at the four terms constituting the right-hand side of (21). Applying once again Lemma 2 and taking into account (15) and (18), one can see that the last two of them tend to zero a.e. as $n \to \infty$. Furthermore, for any $N > 0$ and any n large enough to guarantee the inequality $2k + 2\kappa + 3 - t \geq N$, the second term does not exceed

$$\frac{1}{t}\sum_{j=0}^{t-1} G_N^-(S_1^{k+\kappa+2-t+j}x) \tag{22}$$

(see (18)).

To study the asymptotic behavior of both (22) and the first term on the right-hand side of (21), we use results from **[BJR]**. We need the following statement.

LEMMA 3. *There exists a sequence of positive numbers* ε_n *such that* $\lim_{n\to\infty}\varepsilon_n = 0$ *and for any* $\theta \in (0,1)$ *one can find a number* $n(\theta) > 0$ *such that*

$$k(\theta,\varepsilon_{n+1}) - t(\varepsilon_{n+1}) > k(\theta,\varepsilon_n), \quad t(\varepsilon_{n+1}) > k(\theta,\varepsilon_n) - t(\varepsilon_n), \quad t(\varepsilon_n) > n^2, \tag{23}$$

whenever $n > n(\theta)$.

PROOF. Let $l(\varepsilon) = |\log\varepsilon|$ and $\varphi(\varepsilon) = l(\varepsilon)/3t(\varepsilon)$. We choose ε_0 to be an arbitrary positive number ε for which $t(\varepsilon) > 1$ and then proceed by induction. Suppose that for all $k \leq n-1$, with $n \geq 1$, we have ε_k satisfying

$$l(\varepsilon_{k+1}) - l(\varepsilon_k) > \varphi(\varepsilon_{k+1})t(\varepsilon_{k+1}), \quad t(\varepsilon_{k+1}) + t(\varepsilon_k) > \varphi(\varepsilon_k)l(\varepsilon_k), \quad t(\varepsilon_k) > k^2.$$

Now we can find a number $\varepsilon_{n+1} > 0$ such that

$$t(\varepsilon_{n+1}) > \max\{(n+1)^2,\ \varphi(\varepsilon_n)l(\varepsilon_n) - t(\varepsilon_n)\}, \quad l(\varepsilon_{n+1}) > 2l(\varepsilon_n).$$

The last inequality implies that

$$l(\varepsilon_{n+1}) - l(\varepsilon_n) > (1/2)l(\varepsilon_{n+1}) = (3/2)\varphi(\varepsilon_{n+1})t(\varepsilon_{n+1}) > \varphi(\varepsilon_{n+1})t(\varepsilon_{n+1}).$$

Hence, for all n we have

$$l(\varepsilon_{n+1}) - l(\varepsilon_n) > \varphi(\varepsilon_{n+1})t(\varepsilon_{n+1}), \quad t(\varepsilon_{n+1}) + t(\varepsilon_n) > \varphi(\varepsilon_n)l(\varepsilon_n), \quad t(\varepsilon_n) > n^2. \tag{24}$$

Taking into account the definition of $k(\theta,\varepsilon)$ (see (3)) and the equality $\lim_{\varepsilon\to 0}\varphi(\varepsilon) = \infty$ (see (1)), we come to the conclusion that (24) implies (23) for all sufficiently large n. This completes the proof of Lemma 3.

LEMMA 4. *Let* $\{\varepsilon_n\}$ *be any sequence satisfying the first two inequalities in* (23). *Then for any* $\theta \in (0,1)$, *any* $f \in L_1(X,\mu)$, *and* μ*-almost all* $x \in X$ *we have*

$$\lim_{n\to\infty} \frac{1}{t(\varepsilon_n)} \sum_{j=0}^{t(\varepsilon_n)-1} f(S^{k(\theta,\varepsilon_n)-t(\varepsilon_n)+j}x) = Ef.$$

This is an immediate consequence of Corollary 3 in **[BJR]**.

To complete the proof of the theorem we notice that by Lemma 4 for μ-almost all $x \in X$ the first term at the right-hand side of (21) tends to $Eg^- = h_\mu$ as $n \to \infty$ (see (16)), whereas the second term tends to zero (see (22) and (19)). Moreover, the latter is true for all other terms, because $\sum_n (t(\varepsilon_n))^{-1} < \infty$ (see (20), (23)). Hence,

$$\lim_{n\to\infty} F(x,\kappa,n) = h_\mu.$$

Using same arguments we can prove that the right-hand side of (13) (with ε replaced by ε_n) tends to h_μ a.e. as $n \to \infty$. The theorem is proved.

REMARK 2. One can show that the theorem remains true for some non-Markovian sets $Y \subset X$. Moreover, we believe that it is also true for many non-symbolic systems, e.g. for hyperbolic and some one-dimensional maps. On the other hand, we conjecture that there exists a closed shift-invariant set $Y \subset X$ and a probability measure μ concentrated on Y for which the theorem is false, and hence the left-hand side of (2) provides only a lower bounds for h_μ.

Acknowledgements. The first version of this paper was written during the author's stay at the University of Erlangen-Nurnberg in June, 1994. I am deeply indebted to this university, especially to Professor G. Keller, for their kind hospitality and to Volkswagen Stiftung for financial support. I am also grateful to I. Kornfeld for drawing my attention to the work **[BJR]**.

References

[B] P. Billingsley, *Ergodic theory and information*, Wiley, New York, 1965.

[BJR] A. Bellow, R. Jones, and J. Rosenblatt, *Convergence for moving averages*, Ergodic Theory Dynamical Systems **10** (1990), no. 1, 43–62.

[Z] G. M. Zaslavsky, *Chaos in dynamic systems*, Harwood, Chur, 1985.

DEPARTMENT OF MATHEMATICS, MOSCOW STATE UNIVERSITY, MOSCOW 119899, RUSSIA

Amer. Math. Soc. Transl.
(2) Vol. **171**, 1996

A Two-Dimensional Version of the Folklore Theorem

Michael Jakobson* and Sheldon Newhouse†

ABSTRACT. We formulate some sufficient conditions for the existence of Sinai-Ruelle-Bowen measures for piecewise C^2 diffeomorphisms with unbounded derivatives. The result can be viewed as a two-dimensional version of the well known one-dimensional Folklore Theorem on the existence of absolutely continuous invariant measures. Here we formulate the results and outline the main ideas and tools of our approach. The detailed version will appear elsewhere.

§1. The Folklore Theorem and SRB measures

The well-known Folklore Theorem in one-dimensional dynamics can be formulated as follows.

FOLKLORE THEOREM. *Let $I = [0, 1]$ be the unit interval and $\{I_1, I_2, \dots\}$ a countable collection of disjoint open subintervals of I such that $\bigcup_i I_i$ has the full Lebesgue measure in I. Suppose there are constants $K_0 > 1$ and $K_1 > 0$ and mappings $f_i : I_i \to I$ satisfying the following conditions:*

1. *f_i extends to a C^2 diffeomorphism from* $\mathrm{Closure}(I_i)$ *onto* $[0, 1]$, *and* $\inf_{z \in I_i} |Df_i(z)| > K_0$ *for all i.*
2. $\sup_{z \in I_i} \dfrac{|D^2 f_i(z)|}{|Df_i(z)|} |I_i| < K_1$ *for all i,*

where $|I_i|$ denotes the length of I_i. Then the mapping $F(z)$ defined by $F(z) = f_i(z)$ for $z \in I_i$ has a unique invariant ergodic probability measure μ equivalent to Lebesgue measure on I.

For a proof of the Folklore Theorem and the ergodic properties of μ, see, for example, [**1**] and [**17**].

Here we formulate a theorem that can be considered as a two-dimensional version of this Folklore Theorem.

Let $\tilde{Q}$ be a Borel subset of the unit square Q in the plane $\mathbf{R}^2$ with positive Lebesgue measure, and let $F : \tilde{Q} \to \tilde{Q}$ be a Borel measurable map. An F-invariant Borel probablility measure μ on Q is called a *Sinai-Ruelle-Bowen* measure (or SRB-measure)

1991 *Mathematics Subject Classification*. Primary 58F12; Secondary 58F15.
*Partially supported by NSF Grant 9303369.
†Partially supported by ARPA.

for F if μ is an ergodic measure with non-zero Lyapunov exponents and there is a set $A \subset \tilde{Q}$ of positive Lebesgue measure such that for $x \in A$ and any continuous real-valued function $\varphi : Q \to \mathbf{R}$, we have

$$\lim_{n\to\infty} \frac{1}{n} \sum_{k=0}^{n-1} \varphi(F^k x) = \int \varphi \, d\mu. \tag{1}$$

The set of all points x for which (1) holds is called the *basin* of μ.

We are interested in giving conditions under which certain two-dimensional maps F which piecewise coincide with hyperbolic diffeomorphisms f_i have SRB measures. As in the one-dimensional situation, there is an essential difference between a finite and an infinite number of f_i. In the case of an infinite number of f_i, their derivatives grow with i and relations between first and second derivatives become crucial.

A different class of piecewise hyperbolic maps $F = f_i \mid E_i$, in which the number of domains E_i is finite but derivatives $Df_i(z)$ are allowed to grow when z approaches the boundary of E_i, was considered in **[10]** and **[13]**. In these works it was assumed that the derivatives $DF(z)$ grow at most exponentially depending on the distance between the point z and the singular set where F is not defined. In our work we allow the domains E_i to accumulate towards the limit set in an arbitrary way. Then the conditions of **[10]**, **[13]** are typically violated even for linear maps f_i.

In **[12]** a method was developed that reduces the existence of an SRB measure for F to the existence of an absolutely continuous invariant measure for the one-dimensional map G obtained from F by factorizing along the stable manifolds. The technique of distortion estimates that we develop here might be used to check the conditions of **[12]**. In general, for the systems under consideration, checking the conditions of **[12]** looks similar to the straightforward generalization of **[3]**.

§2. Hyperbolicity and geometric conditions

Consider a countable collection $\xi = \{E_1, E_2, \dots,\}$ of full height closed curvilinear rectangles in Q. Assume that each E_i lies inside a domain of definition $\mathcal{E}_i$ of a C^2-diffeomorphism f_i which maps E_i onto its image $S_i \subset Q$. We assume each E_i connects the top and the bottom of Q. Thus each E_i is bounded from above and from below by two subintervals of the line segments $\{(x, y) : y = 1,\ 0 \le x \le 1\}$ and $\{(x, y) : y = 0,\ 0 \le x \le 1\}$. We assume that the left and right boundaries of E_i are graphs of smooth functions $x^{(i)}(y)$ with $\left|\dfrac{dx^{(i)}}{dy}\right| < \alpha$. We further assume that the images $f_i(E_i) = S_i$ are strips connecting the left and right sides of Q and that they are bounded on the left and right by the two subintervals of the line segments $\{(x, y) : x = 0,\ 0 \le y \le 1\}$ and $\{(x, y) : x = 1,\ 0 \le y \le 1\}$ and above and below by the graphs of smooth functions $Y^i(X)$, $\left|\dfrac{dY^{(i)}}{dX}\right| < \beta$. The bounds on derivatives follow from hyperbolicity conditions H1, H2 that we formulate below.

The sets E_i are called *posts*, the sets S_i are called *strips*. We say that the E_i's are of *full height* in Q, while the S_i's are of *full width* in Q.

We shall assume that the margins between the posts E_i and the extended domains $\mathcal{E}_i$ containing these posts are comparable to E_i in the following sense. For a point $z \in Q$, let l_z denote the horizontal line through z. If $E \subseteq Q$ is a subset of Q, let $\delta_z(E)$ denote the diameter of the horizontal section $l_z \bigcap E$. We call $\delta_z(E)$ the *z-width* of E.

For a given $k > 0$ and any $z \in \operatorname{int} E_i$, let $B_{(k)}(z)$ be the ball of radius $k\delta_z(E_i)$ centered at z. We assume there exists $k > 0$ independent of i and $z \in \operatorname{int} E_i$ such that $\mathcal{E}_i$ contains the set $\bigcup_{z \in E_i} B_{(k)}(z)$.

We assume the following *geometric conditions*:

G1. $\operatorname{int} E_i \cap \operatorname{int} E_j = \emptyset$ for $i \neq j$;

G2. $\operatorname{mes}(Q \setminus \cup_i \operatorname{int} E_i) = 0$ where mes is the Lebesgue measure;

G3. $S_i \cap S_j = \emptyset$ for $i \neq j$ and S_i are disjoint from the top and from the bottom of Q;

and the condition G4 will be formulated below.

In the standard coordinate system for a map $f : (x, y) \to (f_1(x, y), f_2(x, y))$ we use $Df(x, y)$ to denote the differential of f at some point (x, y) and $f_{jx}, f_{jy}, f_{jxx}, f_{jxy}$, etc., for partial derivatives of f_j, $j = 1, 2$.

We use a version of hyperbolicity conditions introduced by Alekseev ([**2**]) who generalized the conditions of Smale ([**14**]). See [**6**] for another version of such conditions.

Hyperbolicity conditions. There exist positive constants $\mu_1, \mu_2, \varepsilon_{12}, \varepsilon_{21}$, such that for all i the map

$$F(z) = f_i(z) \quad \text{for } z \in \mathcal{E}_i$$

satisfies the conditions

H1. $|F_{2y} - F_{1y}F_{2x}F_{1x}^{-1}| \leq \mu_1, |F_{1x}^{-1}| \leq \mu_2, \left|\dfrac{F_{1y}}{F_{1x}}\right| \leq \varepsilon_{12}, \left|\dfrac{F_{2x}}{F_{1x}}\right| \leq \varepsilon_{21}$;

H2. $\mu_1\mu_2 < 1$;

H3. $\mu_1 + \mu_2 - \mu_1\mu_2 + \varepsilon_{12}\varepsilon_{21} \leq 1$.

For a positive real number $\alpha > 0$, we define the cones

$$K_\alpha^u = \{(v_1, v_2) : |v_2| \leq \alpha|v_1|\},$$
$$K_\alpha^s = \{(v_1, v_2) : |v_1| \leq \alpha|v_2|\}.$$

As is proved in [**2**], the hyperbolicity conditions H1–H3 imply that there exist two disjoint families of cones K_α^u and K_β^s independent of the point z such that DF maps K_α^u into the interior of K_α^u and DF^{-1} maps K_β^s into the interior of K_β^s. Also there exists a constant $K_0 > 1$ and a constant $c_0 > 0$ such that for any $v^{(u)} \in K_\alpha^u$ and for any $v^{(s)} \in K_\beta^s$ we have

$$|DF^n v^{(u)}| \geq c_0 K_0^n |v^{(u)}|, \tag{2}$$

$$|DF^{-n} v^{(s)}| \geq c_0 K_0^n |v^{(s)}|. \tag{3}$$

The cone conditions imply that any intersection $FE_i \cap E_j$ is full width in E_j. Also $E_{ij} = E_i \cap F^{-1}E_j$ is a full height subpost of E_i and F^2E_{ij} is a full width substrip of Q.

Proceeding with finite strings, we get that each set

$$E_{i_{-n}} \cap F^{-1}(E_{i_{-n+1}}) \ldots F^{-n}(E_{i_0}) = P_{i_{-n}i_{-n+1}\ldots i_0}$$

is a full height subpost of $E_{i_{-n}}$, and each set

$$FE_{i_0} \cap F^2 E_{i_{-1}} \ldots F^{n+1} E_{i_{-n}} = F^{n+1}(P_{i_{-n}i_{-n+1}\ldots i_0})$$

is a full width strip in Q. Then it follows from Theorem 4 in [**2**] that the corresponding infinite intersections are C^1 curves.

Namely, the following proposition holds.

PROPOSITION 1. *Any C^1 map F satisfying the above geometric conditions* G1–G3 *and hyperbolicity conditions* H1–H3 *has a "topological attractor"*

$$\Lambda = \bigcup_{\dots i_{-n}\dots i_{-1} i_0} \bigcap_{k=0}^{\infty} F^{k+1} E_{i_{-k}}.$$

The infinite intersections $\bigcap_{k=0}^{\infty} F^{k+1} E_{i_{-k}}$ define C^1 curves $\gamma = y(x)$, that are the unstable manifolds for the points of the attractor. The infinite intersections $\bigcap_{n=0}^{\infty} P_{i_{-n} i_{-n+1} \dots i_0}$ define C^1 curves $x(y)$, that are the stable manifolds for the points of the attractor.

REMARK 1. Let $\tilde{Q}$ be the set of points whose forward orbits always stay in $\bigcup_i \operatorname{int} E_i$. Then $\tilde{Q}$ has full Lebesgue measure in Q and F maps $\tilde{Q}$ into itself. The union of the stable manifolds contains $\tilde{Q}$, which has full measure in Q. The trajectories of all points in $\tilde{Q}$ converge to Λ. This is the reason to call Λ a topological attractor. However, the convergence of Birkhoff averages to the unique SRB measure is a much stronger property.

REMARK 2. The distortion condition D1 and *distortion estimates* below imply that if our maps f_i are C^2 smooth, then the unstable manifolds are actually C^2 smooth. Similar conditions on the inverses of f_i imply that the stable manifolds are C^2 smooth, see **[9]**.

§3. Distortion conditions and the main theorem

Since we have a countable number of domains, the derivatives of f_i grow. We will need certain assumptions on the second derivatives.

In a given coordinate system, we write $f_i(x, y) = (f_{i1}(x, y), f_{i2}(x, y))$. We use $f_{ijx}, f_{ijy}, f_{ijxx}, f_{ijxy}$, etc. for partial derivatives of f_{ij}, $j = 1, 2$. Next we formulate distortion conditions. These will be used to control the fluctuation of the derivatives of iterates of F as in Lemma 1 below.

Suppose there is a constant $C_0 > 0$ such that the following *distortion conditions* hold:

D1. $\sup_{z \in \mathcal{E}_i, i \geq 1} \dfrac{|f_{ijkl}(z)|}{|f_{i1x}(z)|} \delta_z(E_i) < C_0$, where $j = 1, 2$, $k, l = x, y$.

REMARK 1. The widths $\delta_z(E_i)$ vary continuously as z varies in E_i, but as z moves from the top to the bottom of Q, the widths can become arbitrarily small.

The following geometric condition is sufficient to control the fluctuations of the widths of E_i.

Let $\delta_{i,\min} = \min_{z \in E_i} \delta_z(E_i)$, $\delta_{i,\max} = \max_{z \in E_i} \delta_z(E_i)$. Let K_0 be the expansion constant from (2).

G4. There exists $K_0 > K_1 > 1$ such that if we let $a_n = \sum_i \delta_{i,\max}$, the sum taken over those i that satisfy $\delta_{i,\min} \leq \frac{K_1^n}{K_0^n}$, then $\sum_n a_n < \infty$.

Condition G4 implies that

$$-\sum_i \delta_{i,\max} \log \delta_{i,\min} < \infty. \tag{4}$$

On the other hand, (4) implies G4 for any $K_0 > K_1 > 1$. So, if G4 holds for some K_1 as above , then it also holds for all $K_0 > K_1 > 1$.

REMARK 2. Condition G4 is satisfied if there are constants $0 < C_1, C_2$, $0 < a < b < 1$ such for every i and for every $z \in E_i$, one has $C_1 a^i \leq \delta_z(E_i) \leq C_2 b^i$.

THEOREM 1. *If F is a piecewise smooth mapping as above satisfying the geometric conditions* G1–G4, *the hyperbolicity conditions* H1–H3, *and the distortion conditions* D1, *then, F has an SRB measure whose basin has full Lebesgue measure in Q.*

§4. Distortions and ratios of derivatives

We proceed toward a sketch of the proof of Theorem 1.

We start with the following procedure of *width-reducing* .

Let ξ be a finite partition of Q into narrow full height rectangular posts $\bar{P}$, all of the same width χ, and let $\xi_1 = F^{-1}\xi$. Every element P_1 of ξ_1 is a full height curvilinear subpost of some orginal post E_i. Let $s_z \subset E_i$ be the horizontal crossection of E_i. The crossection s_z is mapped by F onto the full width curve γ in Q. Bounded distortions imply (see estimate (23) below) that there exists a constant C_{01} depending on C_0 such that for any vertical post $\bar{P}$ we have

$$\frac{|F^{-1}(\gamma \cap P)|}{|s_z|} \leq C_{01}|P \cap \gamma|. \tag{5}$$

It follows from (5) that for any $\varepsilon_1 > 0$ we can choose the original partition ξ so that

$$\frac{|s_z(P_1)|}{|s_z(E_i)|} \leq \varepsilon_1. \tag{6}$$

Thus for an arbitrarily small $d_0 > 0$ we can choose ε_1 such that the distortion estimates D1 restricted to the posts P_1 become

D1′. $\sup_{z \in P_1, i \geq 1} \dfrac{|f_{ijkl}(z)|}{|f_{i1x}(z)|}\delta_z(P_1) < d_0$.

We must estimate the distortion of iterates of our map F. For this purpose it is convenient to introduce certain affine coordinate systems centered at various points z, Fz in such a way that, with respect to these coordinates, the Jacobian matrix of F at z is diagonal. We will formulate our distortion estimates so that they are uniform for all such affine coordinate representations of F. Our assumption that f_i extends to the neighborhood $\mathcal{E}_i$ guarantees that f_i will be defined and C^2 smooth on each *admissible parallelogram* defined below.

Fix a post $P_1 \subset E_i$. Let $z \in P_1$ and let v, w be a pair of unit vectors with $v \in K^u_\alpha$, $w \in K^s_\beta$. If E is a parallelogram with edges parallel to v, w, then the edges parallel to v are called the *top* and *bottom* of E, and the edges parallel to w the *sides* of E. We define the width of E to be the common length of the top and bottom, and the height of E to be the common length of the sides. Let $s_{z,v}$ be the crossection of P_1 through z by the line spanned by v and let $l_{z,w}$ be the line through z spanned by w.

We fix some $T_0 > 1$ depending on hyperbolicity conditions, as described below. Let $E_{z,v,w}$ be a parallelogram of the smallest width with edges parallel to v, w such that

1. The left and right sides of $E_{z,v,w}$ do not meet the interior of P_1.
2. $\text{height}(E_{z,v,w}) = T\,\text{width}(s_{z,v})$, $0 < T \leq T_0$.
3. z divides the height of $E_{z,v,w}$ into equal parts.

Although z is not necessarily located in the center of $(s_{z,v})$, we say $E_{z,v,w}$ as indicated is an admissible parallelogram centered at z_0 inscribed in P_1 .

The hyperbolicity conditions imply that there exists n_0 such that for $v^{(u)} \in K^u_\alpha$, $v^{(s)} \in K^s_\beta$ we have

$$|DF^{n_0} v^{(u)}| \geq 2|v^{(u)}|, \tag{7}$$

$$|DF^{-n_0} v^{(s)}| \geq 2|v^{(s)}|. \tag{8}$$

Then there exists $T_0 \geq 1$ such that for $1 \leq n \leq n_0$ we have

$$|DF^n v^{(u)}| \geq 2T_0^{-1}|v^{(u)}|, \tag{9}$$

$$|DF^{-n} v^{(s)}| \geq 2T_0^{-1}|v^{(s)}|. \tag{10}$$

We use this T_0 in the above definition of the admissible parallelograms.

Since the tangent lines to the vertical boundaries of the posts belong to the stable cone and we choose v in the unstable cone, the admissible horizontal directions are transversal to the boundaries and $E_{z,v,w} \cap P_1 = E^{\text{dyn}}_{z,v,w}$ is a full width curvilinear subrectangle of P_1.

Given a point z and a pair of linearly independent vectors v, w, consider the associated unit vectors $\bar{v} = v/|v|$, $\bar{w} = w/|w|$. Let $e_1 = \binom{1}{0}$, $e_2 = \binom{0}{1}$ denote the standard unit vectors in $\mathbf{R}^2$. Consider the affine automorphism $A_{z,v,w}$ of $\mathbf{R}^2$ such that $A_{z,v,w}(z) = z$, $DA_{z,v,w}(e_1) = \bar{v}$, $DA_{z,v,w}(e_2) = \bar{w}$.

Some of these affine automorphisms will give us coordinate changes which will be useful to control distortions.

For $z \in P_1$, $v \in K^u_\alpha$, $w \in K^s_\beta$, write $f = f_i$ and consider the local coordinate representation

$$\tilde{f}_{z,v,w} = A^{-1}_{fz, Df_z(v), w} \circ f \circ A_{z,v,Df^{-1}_{fz}w}.$$

The map $\tilde{f}_{z,v,w}$ sends z to $f_i z$ and its Jacobian matrix at z is a diagonal matrix. We say that $\tilde{f}_{z,v,w}$ is an adapted representation of f_i or that we have a system of adapted coordinates for f_i. We also call $\tilde{f}_{z,v,w}$ the (z, v, w)-adapted representation of f_i.

We shall use admissible parallelograms $E_{z,v,Df^{-1}_{fz}w}$ with the sides parallel to adapted axes.

For a local diffeomorphism f and a parallelogram E on which f is defined and C^2 smooth, we define the following quantities:

$$\varepsilon_{12}(f, E) = \sup_{z \in E} \frac{|f_{1y}(z)|}{|f_{1x}(z)|}, \tag{11}$$

$$\varepsilon_{21}(f, E) = \sup_{z \in E} \frac{|f_{2x}(z)|}{|f_{1x}(z)|}, \tag{12}$$

$$\varepsilon_{22}(f, E) = \sup_{z \in E} \frac{|f_{2y}(z)|}{|f_{1x}(z)|}, \tag{13}$$

$$\Phi_{11}(f, E) = \max\left(\sup_{z \in E} \frac{|f_{1xx}(z)|\Delta(E)}{|f_{1x}(z)|}, \sup_{z \in E} \frac{|f_{2xx}(z)|\Delta(E)}{|f_{1x}(z)|} \right), \tag{14}$$

$$\Phi_{12}(f, E) = \max\left(\sup_{z \in E} \frac{|f_{1xy}(z)|\Delta(E)}{|f_{1x}(z)|}, \sup_{z \in E} \frac{|f_{2xy}(z)|\Delta(E)}{|f_{1x}(z)|} \right), \tag{15}$$

$$\Phi_{22}(f, E) = \max\left(\sup_{z \in E} \frac{|f_{1yy}(z)|\Delta(E)}{|f_{1x}(z)|}, \sup_{z \in E} \frac{|f_{2yy}(z)|\Delta(E)}{|f_{1x}(z)|} \right), \tag{16}$$

where $\Delta(E) = \max\{\text{height}(E), \text{width}(E)\}$ is *the size of* E.

Finally, we set

$$\Theta_{jk}(f_i) = \sup_{z \in P_1 \in E_i, v \in K_\alpha^u, w \in K_\beta^s} \Phi_{jk}(\tilde{f}_{z,v,w}, E_{z,v,Df_{fz}^{-1}w}) \tag{17}$$

for $(j,k) = (1,1), (1,2)$, or $(2,2)$.

An upper bound on $\Theta_{jk}(f_i)$ is a uniform upper bound on the distortions of the local coordinate representations of f_i, using those affine coordinate systems centered at $(z, f_i z)$ that diagonalize the Jacobian matrix of f_i at z.

We claim that $\Theta_{jk}(f_i)$ are uniformly bounded by a constant depending on distortion conditions and on hyperbolicity conditions . First we notice that second derivatives with respect to various adapted coordinates differ by a bounded factor. Next, hyperbolicity conditions imply that the vectors belonging to the unstable cone are expanded proportionally to $|f_{1x}|$. Then there exists $p_1 > 0$ such that for all i at the center z_0 of any adapted coordinate system we have

$$|f_{1x,\text{adapted}}(z_0)| > p_1 |f_{1x,\text{standard}}(z_0)|.$$

Finally, for an arbitrary point z of an admissible parallelogram we have a similar estimate because the difference of derivatives at z_0 and at z is estimated in terms of distortions (see below), which we made arbitrarily small. That gives

$$|f_{1x,\text{adapted}}(z)| > p_0 |f_{1x,\text{standard}}(z)|. \tag{18}$$

It remains to estimate the difference between the width of an admissible parallelogram and the width of the corresponding dynamically defined horizontal crossection s_z of P_1. That difference depends on the horizontal deviation of the boundary curves of P_1 within admissible height, which by definition is less than $(1/2)T_0|s_z|$. The horizontal deviation is a function of y given by

$$\delta x(y) = \int \left| \frac{f_{1y}}{f_{1x}} \right| dy. \tag{19}$$

In an adapted coordinate system, $f_{1y}(z_0) = 0$ and $\left| \frac{f_{1y}}{f_{1x}} \right|$ is estimated using distortion conditions D1 as in (21) below. Since we made the distortions arbitrarily small by reducing the width we get that the horizontal shift is arbitrarily small compared to the width of the horizontal section. Therefore

$$\text{width}(E_{z,v,Df_{fz}^{-1}w}) \le (1+\varepsilon)\delta_z(E^{\text{dyn}}_{z,v,Df_{fz}^{-1}w}) \tag{20}$$

for any $z \in P_1$.

This implies that D1 holds for any adapted coordinate system with another small constant depending on d_0. We shall keep the notation d_0 for that constant. So we get $\sup_{ijk} \Theta_{jk}(f_i) < d_0$.

§5. Fluctuation of derivatives

Let $z \in E_i$ and let $\hat{E}$ be an admissible parallelogram containing z (perhaps centered at a different point z_0 of E_i). Let $\hat{\Delta}$ be the size of $\hat{E}$. Write f for f_i. We can estimate $\varepsilon_{ij}(f)$ in terms of Φ_{ij} by the mean value theorem using the fact that in adapted coordinates $f_{1y}(z_0) = f_{2x}(z_0) = 0$. We get

(21) $$\frac{|f_{1y}(z)|}{|f_{1x}(z)|} \le \frac{|f_{1xy}(\tau)|}{|f_{1x}(z)|}\hat{\Delta} + \frac{|f_{1yy}(\tau)|}{|f_{1x}(z)|}\hat{\Delta},$$

(22) $$\log|f_{1x}(\tau) - \log f_{1x}(z)| \le \frac{|f_{1xx}(\tau_1)|}{|f_{1x}(\tau_1)|}\hat{\Delta} + \frac{|f_{1xy}(\tau_1)|}{|f_{1x}(\tau_1)|}\hat{\Delta}.$$

So we get

(23) $$\frac{|f_{1x}(\tau)|}{|f_{1x}(z)|} \le \exp(\Phi_{11}(f) + \Phi_{12}(f)).$$

From (21), (23) we get

(24) $$\varepsilon_{12} \le (\Phi_{12}(f) + \Phi_{22}(f))\exp(\Phi_{11}(f) + \Phi_{12}(f)).$$

Similarly from

(25) $$\frac{|f_{2x}(z)|}{|f_{1x}(z)|} \le \frac{|f_{2xx}(\tau)|}{|f_{1x}(z)|}\hat{\Delta} + \frac{|f_{2xy}(\tau)|}{|f_{1x}(z)|}\hat{\Delta}$$

we get

(26) $$\varepsilon_{21} \le (\Phi_{11}(f) + \Phi_{12}(f))\exp(\Phi_{11}(f) + \Phi_{12}(f)).$$

To estimate $\frac{|f_{2y}(z)|}{|f_{1x}(z)|}$, we use (9), (10) and get $\frac{|f_{2y}(z_0)|}{|f_{1x}(z_0)|} \le \tau_0$, where $\tau_0 = T_0^2/4$.

For other points $z \in \hat{E}$ we get as above

(27) $$\varepsilon_{22} \le (\tau_0 + \Phi_{12}(f) + \Phi_{22}(f))\exp(\Phi_{11}(f) + \Phi_{12}(f))$$

Finally, we see that after restricting to small parallelograms E^1 obtained after width-reducing, we have

(28) $$\frac{|f_{1x}(\tau)|}{|f_{1x}(z)|} \le \exp(2d_0),$$

(29) $$\varepsilon_{12}, \varepsilon_{21} \le (2d_0)\exp(2d_0),$$

(30) $$\varepsilon_{22} \le (\tau_0 + 2d_0)\exp(2d_0).$$

So by width-reducing we make $\varepsilon_{12}, \varepsilon_{21}$ arbitrarily small, ratios of derivatives arbutrarily close to one, and ε_{22} arbitrarily close to τ_0.

Estimates for $\Phi_{ij}(f)$ allow us to get bounds for the fluctuation of derivatives of f. Namely, for two unit tangent vectors both within a small cone about an adapted x axis, $v^1 = (v_x^1, v_y^1)$ at z_1, $v^2 = (v_x^2, v_y^2)$ at z_2, $z_1, z_2 \in \hat{E}$ as above, we have the following result.

LEMMA 1. *We have*

(31) $$\frac{|Df_{z_1}(v^1)|}{|Df_{z_2}(v^2)|} \le 1 + k_1\Phi_{11}\frac{|z_1 - z_2|}{\hat{\Delta}} + k_2(|v^1 - v^2|)$$

where $|v| = |v_x| + |v_y|$ *and the distances are measured with respect to the adapted coordinate system under consideration and* k_1, k_2 *are positive constants.*

To prove Lemma 1, we present the numerator of (31) as

$$Df_{z_2}(v^2) + Df_{z_2}(v^1 - v^2) + (Df_{z1} - Df_{z_2})(v^1)$$

and we take $|f_{1x}(z_2)(v_x^2)|$ out of the denominator. Then (28), (29), (30) give (31).

§6. Distortions of compositions

We estimate Φ_{kl} for compositions of hyperbolic maps. Let g be the map f_i and let f be the map f_j. The map $h = f_j \circ f_i = f \circ g$ is defined in $E_{ij} \equiv E_i \cap f_i^{-1} E_j$, which is a full height subpost of E_i. Let $z_0 \in E_i \cap f_i^{-1} E_j$. Earlier we estimated distortions on admissible parallelograms inscribed in the posts P_1 , but now we need to know the distortions on the preimages of such posts. As ξ is not a Markov partition, the posts P_1 can be in arbitrary positions relative to the posts P. To overcome that problem, we shall vary the initial partition ξ which we use for width reducing. Thus, instead of the fixed partition ξ we consider a family of partitions ξ_t obtained by translating each post by t along the x-axis. The distortion estimates obtained above do not depend on t.

We shall also use initial partitions with elements of two sizes. Let χ be the width of an element of ξ which is a standard rectangular post. Let ζ be a similar partition of Q into rectangular full height posts P_ζ that are so narrow that the width of any crossection spanned by $v \in K_\alpha^u$ of any preimage $F^{-k}(P_\zeta)$, $1 \leq k \leq n_0$, is less than $\chi/2$.

Any admissible parallelogram inscribed in $F^{-k}(P_\zeta)$ has width less than $(1 + \varepsilon)(\chi/2)$ and can be put in the middle of an element P_t of an appropriate partition ξ_t. There we can use distortion estimates (20)–(30).

For $v \in K_\alpha^u$, $w \in K_\beta^s$, let us consider the (z_0, v, w)-adapted representation $\tilde{h}$ for h. Thus, $\tilde{h} = A^{-1}_{h(z_0), Dh(v), w} \circ h \circ A_{z_0, v, Dh^{-1}(w)}$. We can express $\tilde{h}$ as $\tilde{h} = \tilde{f} \circ \tilde{g}$, where $\tilde{g}$ is the $(z_0, v, Df^{-1}w)$-adapted representation of g, and $\tilde{f}$ is the $(gz_0, Dg(v), w)$-adapted representation of f. Let $P_1 = f_j^{-1}(P_\zeta)$, $P_2 = f_i^{-1}(P_1) = f_i^{-1} \circ f_j^{-1}(P_\zeta)$.

Let E_h be the minimal parallelogram centered at z_0 with edges parallel to v, $Dh^{-1}(w)$ such that

1. the left and right sides of E_h do not meet the interior of P_2;
2. $\text{height}(E_h) = \text{width}(s_{z_0}(P_2))$.

Then E_h is contained in an admissible parallelogram E_g inscribed in $f_i^{-1} P_t$ with the height of E_g equal to the width of $s_{z_0}(f_i^{-1} P_t)$ centered at z_0 with edges parallel to $v, Dh^{-1}w$.

Although the vectors in K_β^s can expand under a single iterate, we know that they cannot expand by more than $T_0/2$. Respectively, the image $g(E_h \cap P_2)$ is contained in an admissible parallelogram E_f inscribed in P_1, centered at $g(z_0)$, and having edges parallel to $Dg(v), Df^{-1}(w)$. Let $\delta_h = \text{width}(E_h)$, $\delta_f = \text{width}(E_f)$, $\delta_g = \text{width}(E_g)$.

As in (19), we get that lengths of horizontal crossections δ_z of E_h and the width of E_h are related by

$$\delta_z \leq \delta_h \leq (1 + \varepsilon_h)\delta_z, \tag{32}$$

where ε_h is a small constant.

Since the map Dg restricted to E_h is close to a constant diagonal the horizontal crossections are mapped into curves close to horizontal crossections of E_f. Any horizontal crossection s_{E_f} of E_f lies in the middle of the corresponding crossection s_P of some rectangular post $P \in \xi_t$. Let us denote by $\delta_{f(\text{rltv})}$ the ratio $\frac{|s_{E_f}|}{|s_P|}$. By the choice of partitions ζ and ξ we have

$$\delta_{f(\text{rltv})} \leq (1/2)(1 + \varepsilon). \tag{33}$$

When we pullback, the ratios are multiplied by a factor due to distortion which is close to one because E_f is a small parallelogram. We obtain the following estimate:

$$\Phi_{jk}(g \mid E_h) \leq (1+\bar{\varepsilon})\delta_{f\,(\mathrm{rltv})}\Phi_{jk}(g \mid E_g). \tag{34}$$

Then we have

$$|\tilde{h}_{1x}| \geq |\tilde{f}_{1x}\tilde{g}_{1x}| \left(1 - \frac{|\tilde{g}_{2x}\tilde{f}_{1y}|}{|\tilde{g}_{1x}\tilde{f}_{1x}|}\right), \tag{35}$$

where the partial derivatives of g are computed on E_h. We define

$$\eta = \left(1 - \frac{|\tilde{g}_{2x}\tilde{f}_{1y}|}{|\tilde{g}_{1x}\tilde{f}_{1x}|}\right)^{-1} \tag{36}$$

and get, as in (25) and (26),

$$\eta < 1 + \varepsilon_{12}(\tilde{f})\delta_{f\,(\mathrm{rltv})}(1+\bar{\varepsilon})(\Phi_{11}(\tilde{g}) + \Phi_{12}(\tilde{g}))\exp(\Phi_{11}(\tilde{g}) + \Phi_{12}(\tilde{g})). \tag{37}$$

We also introduce

$$\gamma = \sup_{z_1, z_2 \in E_h} \frac{|\tilde{g}_{1x}(z_1)|}{|\tilde{g}_{1x}(z_2)|} \tag{38}$$

and get, as in (22) and (23),

$$\gamma \leq \exp\left[(\Phi_{11}(\tilde{g}) + \Phi_{12}(\tilde{g}))\delta_{f\,(\mathrm{rltv})}(1+\bar{\varepsilon})\right]. \tag{39}$$

Using the chain rule for partial derivatives, we get

$$\begin{aligned}
\Phi_{11}(\tilde{h}) \leq \eta\big[&\Phi_{11}(\tilde{f})\gamma(1+\varepsilon_h) + 2\Phi_{12}(\tilde{f})\varepsilon_{21}(\tilde{g})\gamma + \Phi_{22}(\tilde{f})\varepsilon_{21}(\tilde{g})^2\gamma \\
&+ \Phi_{11}(\tilde{g})\delta_{f\,(\mathrm{rltv})}(1+\bar{\varepsilon})\big], \\
\Phi_{12}(\tilde{h}) \leq \eta\big[&\Phi_{11}(\tilde{f})\varepsilon_{12}(\tilde{g})\delta_{f\,(\mathrm{rltv})}(1+\bar{\varepsilon})\gamma + \Phi_{12}(\tilde{f})[\varepsilon_{12}(\tilde{g})\varepsilon_{21}(\tilde{g})\gamma + \varepsilon_{22}(\tilde{g})\gamma] \\
&+ \Phi_{22}(\tilde{f})\varepsilon_{22}(\tilde{g})\varepsilon_{21}(\tilde{g})\gamma + \Phi_{12}(\tilde{g})\delta_{f\,(\mathrm{rltv})}(1+\bar{\varepsilon})\big].
\end{aligned}$$

The quantity $\Phi_{22}(\tilde{h})$ can be expressed similarly.

Note that in the formula for $\Phi_{11}(\tilde{h})$ the value $\Phi_{11}(\tilde{f})$ is multiplied by factors that are of the form $(1 + \mathrm{const}\,\delta_{f\,(\mathrm{rltv})})$ and by $(1+\varepsilon_h)$ from (32).

So when the domains of $\tilde{f}$ and of $\tilde{h}$ are small, these factors are close to one. In the formula for $\Phi_{12}(\tilde{h})$, distortions of f are multiplied by small factors in each term except for ε_{22} estimated in (30) and the same is true for $\Phi_{22}(\tilde{h})$.

Taking the suprema over adapted coordinates, we get analogous estimates for $\Theta_{jk}(h)$.

In a similar way we estimate distortions for compositions $F^{n_0} = f_{i_1} \circ \cdots \circ f_{i_{n_0}}$ and taking into account that the widths of the initial rectangles P can be taken arbitrary small, we get for distortions of compositions F^{n_0}

$$\sup_{ijk} \Theta_{jk}(F^{n_0}) < d_1, \tag{40}$$

where $d_1 = d_1(d_0)$ and $\lim_{d_0 \to 0} d_1 = 0$.

By construction, we have in adapted coordinates

$$\left|\frac{F^{n_0}_{2y}}{F^{n_0}_{1x}}\right| \le \frac{1}{4}. \tag{41}$$

Now if we consider the iterates $F^{n_0 m}$ of the map F^{n_0}, we get the uniform decrease of $\Theta_{12}(F^{n_0 m}), \Theta_{22}(F^{n_0 m})$.

This implies, by induction, the following estimate similar to the one-dimensional estimate of distortion for the compositions of hyperbolic maps [**5**, Lemma 1].

PROPOSITION 2. *For an arbitrary composition $F^k = f_{i_k} \circ \cdots \circ f_{i_1}$ restricted to the preimages $f_{i_k} \circ \cdots \circ f_{i_1}^{-1}(P)$ of the elements P of the partition ζ we have*

$$\Theta_{11}(F^k) \le c_1 d_1, \tag{42}$$

$$\max\left(\Theta_{12}(F^k), \Theta_{22}(F^k)\right) \le c_1 d_1 (q)^k, \tag{43}$$

where d_1 is from (40) and $c_1 > 0$, $0 < q < 1$ are determined by the initial parameters from the hyperbolicity and distortion conditions.

§7. Sinai local measures

According to Proposition 1, to any infinite sequence $(\ldots i_{-n} \ldots i_{-1} i_0)$ there corresponds an unstable manifold $W^u_{(\ldots i_{-n} \ldots i_{-1} i_0)}$ of full width in E_{i_0}. Moreover, $f_{i_0} W^u_{(\ldots i_{-n} \ldots i_{-1} i_0)}$ is a full width unstable manifold in Q. The curve $W^u_{(\ldots i_{-n} \ldots i_{-1} i_0)}$ as well as all its preimages have tangents inside $K^{(u)}_\alpha$.

Let $\tilde{E}_{-n} = E_{i_{-n}} \cap F^{-1} E_{i_{-n+1}} \cap \cdots \cap F^{-n} E_{i_0}$ and $\tilde{W}^u_{(-n)} = F^{-n} W^u_{(\ldots i_{-n} \ldots i_{-1} i_0)} \cap \tilde{E}_{-n}$. First we notice that the maps F^n from $\tilde{W}^u_{(-n)}$ onto the full width unstable manifolds in Q all have uniformly bounded distortions. For a given $k > 0$ we fix two points $z_1, z_2 \in \tilde{W}^u_{(-k)}$ and connect their images Z_1, Z_2 by a chain $Y_0 = Z_1, Y_1, \ldots, Y_m = Z_2$ of $m \le N + 1$ points, $N = \operatorname{card} \zeta$, such that Y_i, Y_{i+1} belong to the same elements of the partition ζ. Then z_1, z_2 are connected by the chain of preimages y_i, and for y_i, y_{i+1}, Proposition 2 and Lemma 1 give the uniform bound on the ratio of derivatives. Namely, applying Lemma 1 to the pairs $(y_0, v_1), (y_2, v_1), \ldots, (y_{m-1}, v_1), (y_m, v_2)$ we get from (31)

$$\frac{|DF^k_{z_1}(v^1)|}{|DF^k_{z_2}(v^2)|} \le (1 + k_1 \Phi_{11})^{N+1} (1 + k_2(|v^1 - v^2|) = C_1. \tag{44}$$

The estimate (44) implies that for any iterate of F the ratios of distances between points are uniformly preserved up to some constant which depends on C_1. Then we fix a large k, take $\tilde{E}_{-k} = E_{i_{-k}} \cap F^{-1} E_{i_{-k+1}} \cap \cdots \cap F^{-k} E_{i_0}$, and denote by $\tilde{\delta}_{-k}$, $\delta_{i_{-k}}$ the widths of $\tilde{E}_{-k}, E_{i_{-k}}$ respectively. Let $\tilde{W}^u_{(-k)} = F^{-k} W^u_{\ldots i_{-n} \ldots i_{-k} \ldots i_{-1} i_0} \cap \tilde{E}_{-k}$.

We want to estimate the ratios of derivatives at $z_1, z_2 \in \tilde{W}^u_{(-k)}$ and v_{z_1}, v_{z_2} tangent to $\tilde{W}^u_{(-k)}$ at z_1, z_2.

Using the hyperbolicity conditions and bounded distortion we get that the ratio of widths of $\tilde{E}_{-k}$ and $E_{i_{-k}}$ satisfies the estimate

$$\frac{\tilde{\delta}_{-k}}{\delta_{i_{-k}}} \le C_2 K_0^{-k}. \tag{45}$$

To compare unit tangent vectors v_{z_1}, v_{z_2}, we go backward to some high order preimage $\tilde{W}^{(u)}_{(-m)}$ of $\tilde{W}^{(u)}_{(-k)}$, take the preimages w_1, w_2 of z_1, z_2, and take unit horizontal vectors u_1, u_2 at w_1, w_2. Then v_{z_1}, v_{z_2} will be within the respective const $K_0^{-2(m-k)}$ cones around $DF^{m-k}u_1, DF^{m-k}u_2$. These last vectors have components $F^{m-k}_{1x}(w_j)$, $F^{m-k}_{2x}(w_j)$, $j = 1, 2$. We can assume that the points w_1, w_2 belong to an admissible parallelogram $\hat{E}_{(m)}$ that is mapped by F^m onto a full width curvilinear subrectangle of an element of the initial partition ζ. Since the elements of ζ have fixed widths, by using again the bounded distortion property we get

$$\frac{|w_1 - w_2|}{|\hat{E}_{(m)}|} \le C_3 K_0^{-k}. \tag{46}$$

We can take w_1 as the origin of an adapted coordinate system for $\hat{E}_{(m)}$ so that $F^{m-k}_{2x}(w_1) = 0$. Then

$$\left|\frac{F^{m-k}_{2x}(w_2)}{F^{m-k}_{1x}(w_2)}\right| \le \frac{C_3}{K_0^k},$$

and, respectively,

$$|v_{z_1} - v_{z_2}| \le \frac{C_3}{K_0^k}. \tag{47}$$

Using (45), (46), (47), and (31) we get

$$\frac{|DF_{z_1}(v^1)|}{|DF_{z_2}(v^2)|} \le \exp\left(\frac{C_4}{K_0^k}\right). \tag{48}$$

Let us denote by $D^u F(z)$ the derivative of F along the unstable manifold $W^u(z)$ at z. Then (48) implies that for any two points $z_1, z_2 \in \tilde{W}^u_{(-n)}$ the following limit exists:

$$\lim_{n\to\infty} \frac{\prod_{s=1}^n D^u F(F^{-s} z_2)}{\prod_{s=1}^n D^u F(F^{-s} z_1)}. \tag{49}$$

Considering $\xi(z_1, z_2)$ obtained in the preceding limits as functions of z_1, we get, up to constants, densities of special measures on the unstable manifolds. The family of these measures defined on local unstable manifolds is invariant in the following sense. If we have two Lebesgue measurable subsets of a local unstable manifold $W^u(z)$, and their images are subsets of the local unstable manifold $W^u(F(z))$, then the ratios of local measures of these subsets are preserved, see for example **[15**, Lecture 16]. We call these measures *Sinai local measures* or just *local measures*. Let $\rho_{W^u(z)}$ denote the normalized Sinai local measure on $W^u(z)$.

§8. Construction of an SRB measure

A global SRB measure is obtained by averaging the iterates of a local measure on an arbitrary unstable manifold. Let $W_0 = W^u_{(\dots i_{-n} \dots i_{-1} i_0)}$ and let μ_0 be the local measure on W_0. Then the measures $\mu_n = F^n_* \mu_0$ are defined on $F^n(W_0)$ by $\mu_n(A) = \mu_0(F^{-n}(A))$. For any E_i we fix full height curve transversal to unstable manifolds, for example, some stable manifold γ_i. Every iterate of W_0 is a union of unstable manifolds that are full width in Q. Each of these manifolds intersects γ_i at a unique point z. The piece of that manifold cut by E_i is the local unstable manifold denoted by $W_i(z)$. Then on each E_i the sequence of discrete factor measures m_{in} is defined by assigning to $z \in \gamma_i$ the measure $\mu_n(W_i(z))$.

Then we consider sequences of measures

$$\lambda_n = \frac{1}{n}\sum_{k=0}^{n-1}\mu_k$$

and

$$\lambda_n^i = \frac{1}{n}\sum_{k=0}^{n-1} m_{ik}$$

for each i and choose a subsequence n_k such that the corresponding measures weakly converge to

$$\lambda = \lim_{k\to\infty}\lambda_{n_k}, \quad \lambda^i = \lim_{k\to\infty}\lambda^i_{n_k}.$$

Then λ is an F-invariant measure. The following lemma is a modification of [**15**, Lecture 17, Theorem 5].

LEMMA 2. *λ is a Gibbsian measure , i.e., the conditional measures that λ generates on the local unstable manifolds coincide with the local measures.*

OUTLINE OF THE PROOF. Let

$$z_{(\dots i_{-n}\dots i_{-1} i_0)} = W^u_{(\dots i_{-n}\dots i_{-1} i_0)} \cap \gamma_{i_0}, \qquad X_{i_0} = \bigcup z_{(\dots i_{-n}\dots i_{-1} i_0)}.$$

Let $\varphi(x)$ be a continuous function with the support inside E_{i_0}. The uniformly bounded distortions imply that

$$\psi(z) = \int \varphi(y)\, d\rho_{W_{i_0}(z)}(y)$$

is a continuous function of $z \in X_{i_0}$.

Since we are dealing with an infinite number of E_i, the union of the strips S_i, as well as the intersection of that union with γ_{i_0}, are not closed sets, and the same is true for $\bigcup_i F^n(E_i)$ for any n. Respectively, X_{i_0} is not a closed subset of γ_{i_0} and $\psi(z)$ is not continuous on the closure $\bar{X}_{i_0}$. However, since the measure of the union of E_i with $i > n$ tends to zero as n tends to ∞, the bounded distortion conditions imply that $Y_{i_0} = \bar{X}_{i_0} \setminus X_{i_0}$ has uniform measure zero with respect to all μ_k in the following sense: for every $\varepsilon > 0$ there exists an open cover U_{ε, i_0} of Y_{i_0} such that $\mu_k(U_\varepsilon) \le \varepsilon$ for all k.

With this modification the proof of Lemma 2 is similar to the proof of Theorem 5 in [**15**, Lecture 17].

REMARK. When constructing λ we implicitly use the following result.

For any smooth curve $Y = y(x)$ transversal to the stable foliation with angles uniformly bounded away from zero, $Y \cap \left(\bigcup_i E_i\right)$ has full Lebesgue measure in Y.

In order to get this property we notice that G4 implies that the measure of $Y \cap \left(\bigcup_i E_i\right)$ depends continuously on the curve belonging to some smooth foliation. Then the property follows from G2 using the Fubini theorem.

§9. Absolute continuity and corollaries

The key property of SRB measures — absolute continuity of the projection along the stable foliation — is proved similarly to [3, 9], but again requires a modification because of the infinite number of E_i in the initial partition.

Let us take two smooth curves W_1, W_2 with tangent lines within K^u_α in the same E_i and iterate them forward long enough. After n iterates the distance between the pieces of images W_{n1}, W_{n2} that had the same itinerary $[i_1, i_2, \dots i_n]$ will be less than const K_0^{-n}. When we prove the absolute continuity property we compare the Lebesgue measures of preimages $W_{1,[i_1,i_2,\dots i_n]} = F^{-n}(W_{n1}) \in W_1$ and $W_{2,[i_1,i_2,\dots i_n]} = F^{-n}(W_{n2}) \in W_2$. By the uniformly bounded distortion, we have

$$\text{(50)} \qquad \operatorname{mes}(W_{k,[i_1,i_2,\dots i_n]}) \in [c_1, c_2] \operatorname{mes}(W_{nk}) \left(\prod_{s=1}^{n} D^u F^s(x_k) \right)^{-1}, \quad k = 1, 2,$$

where $x_1 \in W_{1,[i_1,i_2,\dots i_n]}$, $x_2 \in W_{2,[i_1,i_2,\dots i_n]}$ belong to the same stable manifold.

Formula (50) implies that the ratios of the measures of $W_{k,[i_1,i_2,\dots i_n]}$ are expressed up to a uniform constant by

$$\text{(51)} \qquad \frac{\prod_{s=1}^{n} D^u F(F^s x_1) \operatorname{mes} W_{n2}}{\prod_{s=1}^{n} D^u F(F^s x_2) \operatorname{mes} W_{n1}}.$$

However, contrary to the classical case, the ratios of unstable Jacobians at $F^s(x_k)$ are close to 1 only if the distances between $F^s(x_k)$ are small compared to the widths of those E_{i_s} that contain $F^s(x_k)$. Similarly, the ratios of the measures of W_{nk} are close to one under the same condition.

Here we use the geometric condition G4. According to this condition, if we avoid at step s those E_i that have widths less than $K_1^s K_0^{-s}$, we delete at this step the subset of W_k of the relative measure less than the corresponding term of a converging series. So we get at the limit a Cantor set C^1_k of positive measure in W_k, $k = 1, 2$. On this set the ratios (51) are uniformly bounded between two positive constants r_{11}, r_{12}. If π is the projection along stable manifolds, then it follows from [3] that for any subset of positive Lebesgue measure $M_1 \subset C^1_1$ and for $M_2 = \pi M_1 \subset C^1_2$ the ratios of Lebesgue measures of M_1 and M_2 are uniformly bounded.

The remaining points in W_k belong to the preimages of E_i with large i. For these sets we take several extra forward iterates and repeat the previous arguments. We obtain Cantor sets C^2_k of positive measure, disjoint from C^1_k, with uniformly bounded ratios (51), but with different constants r_{21}, r_{22}. Repeating this construction, we get

$$W_k = \bigcap_{j=1}^{\infty} C^j_k \mod 0,$$

where C^j_k are disjoint Cantor subsets of W_k of positive measure with uniform estimates r_{j1}, r_{j2} of ratios (51) on C^j_k, $k = 1, 2$. That proves absolute continuity of π.

The ergodicity of λ follows from the absolute continuity of π and from the "Bernoulli" topological structure of the map F (see, for example, [3]). The same arguments show that any two measures constructed by the above averaging procedure starting from different unstable manifolds, coincide. Finally, the absolute continuity of π implies that for Lebesgue a.e. point in Q and any continuous function φ, (1) holds with $\gamma = \mu$.

§10. Further ergodic properties of (F, λ)

10.1. Let us denote by α the original partition of the attractor Λ

$$\alpha = (\Lambda \cap E_i)_{i=1}^{\infty}.$$

Let

$$\eta = \bigvee_{-\infty}^{0} F^i \alpha.$$

Up to a set of measure zero, the elements of η coincide with the intersections of full height stable manifolds with the attractor. The partition η satisfies the following properties:

$$F\eta \succ \eta, \qquad \vee_n F^n \eta = \varepsilon, \qquad \bigwedge_n F^n \eta = \nu.$$

So we get from **[16]** that (F, λ) is a K-system.

PROPOSITION 3. *The map (F, λ) is Bernoulli.*

The following *weak Markov property* was introduced in **[11]**. It was used to prove the Bernoulli property of Anosov flows (see **[4, 11]**).

Let β be any partition,

$$\beta_k^l = \bigvee_{k \le i \le l} F^i \beta.$$

We say that β is weak Markov (WM) if for any $\varepsilon \ge 0$ there exists $N = N(\varepsilon)$, a set $P = P(\varepsilon)$ of atoms of β_0^∞, $\lambda(P) \ge 1 - \varepsilon$, and a set $M = M(\varepsilon)$ of atoms of $\beta_{-\infty}^N$, $\lambda(M) \ge 1 - \varepsilon$, such that if $\bar{x}, \bar{y} \in P \cap x_0^N$, $x_0^N \in \beta_0^N$, then for any set $A \subset M$ of atoms of $\beta_{-\infty}^N$ one has

$$\left| \frac{\lambda(A|\bar{x})}{\lambda(A|\bar{y})} - 1 \right| \le \varepsilon. \tag{52}$$

We take for β our original partition α of the attractor. Then the corresponding α_0^∞ is the partition into unstable manifolds $W^u_{\dots i_{-n} \dots i_0}$, elements of α_0^N are $S_{i_{-n} \dots i_0} = F^n E_{i_{-n}} \cap F^{n-1} E_{i_{-(n-1)}} \cap \dots \cap F E_{i_{-1}} \cap E_{i_0}$, and $\alpha_{-\infty}^N$ is the partition into pieces of stable manifolds within α_0^N.

In order to get P, we first delete an open cover U as in Lemma 2. Then $\bigcup_{i \le i_0} E_i \setminus U$ is a closed set and we can cover it by a finite number of strips $S_{i_{-n} \dots i_0}$, $n \le N$, such that for any two points belonging to the same stable manifold $z_1, z_2 \in W_0^s \cap S_{i_{-n} \dots i_0}$ the ratio of densities of local measures (49) differs from 1 by less that ε. The union of the unstable manifolds $W^u_{\dots i_{-n} \dots i_0}$ that belong to the above strips $S_{i_{-n} \dots i_0}$, $n \le N$, constitutes P. For M we take the union of stable leaves outside of $\bigcup_{i \ge i_0} E_i$ and within $S_{i_{-n} \dots i_0}$. Then, the WM property is satisfied.

Using the WM property for α we get from **[1**, Proposition 2.2**]** that every partition $\zeta_k = \{E_1, E_2, \dots, E_k, \bigcup_{i \ge k} E_i\}$ is weak Bernoulli. As $k \to \infty$, the partitions ζ_k converge to the generating partition α and we get Bernoulli property for (F, λ), see **[8]**.

10.2. Entropy formula. By construction, the measures of E_i satisfy

$$c_1 \delta_{i,\min} < \lambda(E_i) < c_2 \delta_{i,\max}. \tag{53}$$

Using (4), (56) we get that the entropy of the generating partition α is finite and

$$h_\lambda(F) \leq H(\alpha) < \infty. \tag{54}$$

Similarly,

$$(c_1\delta_{i,\min})^{-1} > |D^u F \mid E_i| > (c_2\delta_{i,\max})^{-1}$$

implies

$$\int \log |D^u F| d\lambda < \infty. \tag{55}$$

Let $\xi = \bigvee_{k=0}^{\infty} F^k\alpha$. The elements of ξ coincide with the local unstable manifolds $W_i(z)$. Respectively, ξ is an increasing partition with respect to F^{-1} and it has the same K-properties as the above η.

Since α is generating and $\alpha \prec \xi$, we get, using the properties of the entropy, that

$$h_\lambda(F) = H(F^{-1}\alpha|\xi) = H(F^{-1}\xi|\xi). \tag{56}$$

Then the arguments of Theorem 5.1 from **[16]** (proved for systems with smooth invariant measure) or similar arguments from Section 4 of **[7]** (where the smoothness of invariant measure is not assumed) give

$$H(F^{-1}\xi|\xi) = \int \log |D^u F| d\lambda, \tag{57}$$

which proves the entropy formula.

Acknowledgements. We want to thank D. Rudolph and F. Ledrappier for useful discussions during the preparation of this paper.

References

1. R. Adler, *Afterword to R. Bowen, Invariant measures for Markov maps of the interval*, Comm. Math. Phys. **69** (1979), no. 1, 1–17.
2. V. M. Alekseev, *Quasi-random dynamical systems*, Math. USSR-Sb. **5** (1968), 73–128.
3. D. V. Anosov and Ya. G. Sinai, *Some smooth ergodic systems*, Russian Math. Surveys **22** (1967), 103–167.
4. L. Bunimovich, *On a class of special flows*, Math. USSR-Izv. **8** (1974), 219–232.
5. M. V. Jakobson, *Absolutely continuous invariant measures for one-parameter families of one-dimensional maps*, Comm. Math. Phys. **81** (1981), 39–88.
6. M. Hirsch and C. Pugh, *Stable manifolds and hyperbolic sets*, Proceedings of Symposia in Pure Mathematics **14** (1970), 133–164.
7. F. Ledrappier and J.-M. Strelcyn, *A proof of the estimation from below in Pesin's entropy formula*, Ergodic Theory Dynamical Systems **2** (1982), 203–219.
8. D. Ornstein, *Ergodic theory, randomness, and dynamical systems*, Yale University Press, New Haven, CT, 1975.
9. C. Pugh and M. Shub, *Ergodic attractors*, Trans. Amer. Math. Soc. **312** (1989), no. 1, 1–54.
10. Ya. B. Pesin, *Dynamical systems with generalized hyperbolic attractors: hyperbolic, ergodic and topological properties*, Ergodic Theory Dynamical Systems **12** (1992), 123–151.
11. M. Ratner, *Anosov flows with gibbs measures are also bernoullian*, Israel J. Math. **17** (1974), 380–391.
12. M. Rychlik, *Mesures invariantes et principe variationel pour les applications de Lozi*, C. R. Acad. Sci. Paris **296** (1983), 19–22.
13. E. A. Sataev, *Invariant measures for hyperbolic maps with singularities*, Russian Math. Surveys **47** (1992), 191–251.
14. S. Smale, *Diffeomorphisms with many periodic points*, Differential and combinatorial topology, Princeton University Press, Princeton, NJ, 1965, pp. 63–80.
15. Ya. G. Sinai, *Topics in ergodic theory*, Princeton University Press, Princeton, NJ, 1994.

16. ______, *Classical systems with countable Lebesgue spectrum.* II, Izv. Akad. Nauk. SSSR, Ser. Math. **30** (1966), 15–68; English transl., Amer. Math. Soc. Transl. Ser. 2, vol. 68, Amer. Math. Soc., Providence, RI, 1968, pp. 34–88.
17. P. Walters, *Invariant measures and equilibrium states for some mappings which expand distances*, Trans. Amer. Math. Soc. **236** (1978), 121–153.

MICHAEL JAKOBSON, DEPARTMENT OF MATHEMATICS, UNIVERSITY OF MARYLAND, COLLEGE PARK, MD 20742

E-mail address: `mvy@math.umd.edu`

SHELDON E. NEWHOUSE, MATHEMATICS DEPARTMENT, MICHIGAN STATE UNIVERSITY, E. LANSING, MI 48824-1027

E-mail address: `newhouse@math.msu.edu`

Amer. Math. Soc. Transl.
(2) Vol. **171**, 1996

Thermodynamic Formalism for Random Transformations and Statistical Mechanics

K. Khanin and Y. Kifer

On the 60th birthday of our teacher Ya. G. Sinai, a pioneer in applications of statistical mechanics ideas to dynamical systems

Abstract. We prove the existence and uniqueness of equilibrium states in the relativized variational principle for random transformations that expand only "in average". We construct equilibrium states under substantially weaker conditions than in **[K2]**. We give proofs both using the random Ruelle-Perron-Frobenius operator and also direct statistical mechanics methods to corresponding random symbolic systems.

1. Introduction

The main motivation for the present paper is the following setup (see **[K2]**). Let M be a compact connected d-dimensional C^1 Riemannian manifold and $\mathcal{F}$ is the space of C^1 endomorphisms f of M such that $fM = M$ and the differential Df of f is nondegenerate for all $f \in \mathcal{F}$. Such a setup includes also the space $\Omega = \mathcal{F}^{\mathbb{F}}$ of bi-infinite sequences $\omega = (\dots, \omega_{-1}, \omega_0, \omega_1, \dots)$, $\omega_i \in \mathcal{F}$, and a probability measure P on Ω invariant with respect to the shift transformation $\theta : \Omega \to \Omega$, $(\theta\omega)_i = \omega_{i+1}$. Set $F(\omega) = \omega_0 \in \mathcal{F}$, $\omega \in \Omega$. Then $F : \Omega \to \mathcal{F}$ is a Borel map with respect to the product topology on Ω. We will study the compositions

$$(1.1) \qquad F^n(\omega) = F(\theta^{n-1}\omega)F(\theta^{n-2}\omega)\cdots F(\omega), \qquad F^{-n}(\omega) = (F^n(\omega))^{-1}.$$

Let $\tau : M \times \Omega \to M \times \Omega$ be the skew-product transformation defined by $\tau(x, \omega) = (F(\omega)x, \theta\omega)$. We denote the space of probability measures μ on $M \times \Omega$ by $\mathcal{P}(M \times \Omega)$ and we write $\mu \in \mathcal{P}_P(M \times \Omega)$ if the marginal of μ on Ω is P. Such measures are determined by an essentially unique measurable map μ^ω from Ω to the space $\mathcal{P}(M)$ of probability measures on M such that

$$\mu(dx, d\omega) = \mu^\omega(dx)P(d\omega)$$

and μ is τ-invariant if and only if $F(\omega)\mu^w = \mu^{\theta\omega}$ for P-almost all (a.a.) $\omega \in \Omega$. The relativized variational principle proved in **[LW]** and extended in **[B1]** states that for

1991 *Mathematics Subject Classification*. Primary 58F15; Secondary 82B44.
Key words and phrases. Thermodynamic formalism, random transformations, statistical mechanics.

any family $\varphi = \{\varphi_\omega, \omega \in \Omega\}$ of continuous functions φ_ω on M such that $\|\varphi_\omega\| = \sup_{x\in M} |\varphi_\omega(x)| \in L^1(\Omega, P)$ one has

$$(1.2) \qquad Q_P^{(r)}(\varphi) = \sup\left\{\int \varphi d\mu + h_\mu^{(r)}(\tau) : \mu \in \mathcal{P}_P(M \times \Omega) \text{ is } \tau\text{-invariant}\right\},$$

where $Q_P^{(r)}(\varphi)$ is the relativized topological pressure of φ and $h_\mu^{(r)}(\tau)$ is the relativized entropy (see **[B1]**). A τ-invariant measure $\mu_\varphi \in \mathcal{P}_P(m \times \Omega)$ is called an equilibrium state for φ if

$$(1.3) \qquad Q_P^{(r)}(\varphi) = \int \varphi d\mu_\varphi + h_{\mu_\varphi}^{(r)}(\tau).$$

In this paper we establish the existence and uniqueness of equilibrium states for the relativized variational principle under substantially weaker and more natural assumptions than in **[K2]** and **[BG2]**, which were dealing essentially with expanding $F(\omega)$ for all ω. We do it both by modifying the proof from **[K2]**, which employs the random Ruelle-Perron- Frobenius (RPF) operator, and by passing to a symbolic representation that enables us to use some statistical mechanics technique. Another natural extension lies in the basis of the relativized approach and can be expressed by the following general principle: the results should not depend substantially on corrections by functions depending only on ω. Namely, let $\|\varphi_\omega\| \notin L^1(\Omega, P)$, but there exists a function $\psi(\omega)$ on Ω such that $\tilde{\varphi}_\omega^\psi = \varphi_\omega + \psi(\omega)$ satisfies $\|\tilde{\varphi}_\omega^\psi\| \in L^1(\Omega, P)$ which, is possible if and only if

$$(1.4) \qquad \delta_\varphi(\omega) = \sup_x \varphi_\omega(x) - \inf_x \varphi_\omega(x) \in L^1(\Omega, P).$$

Then we can write (1.2) and (1.3) for $\tilde{\varphi}^\psi$ in place of φ and call $\mu_{\tilde{\varphi}^\psi}$ the equilibrium state for φ, as well as for $\tilde{\varphi}^\psi$. This is natural since if $\|\tilde{\varphi}_\omega^{\psi_1}\|, \|\tilde{\varphi}_\omega^{\psi_2}\| \in L^1(\Omega, P)$, then

$$(1.5) \qquad Q_P^{(r)}(\tilde{\varphi}^{\psi_1}) - Q_P^{(r)}(\tilde{\varphi}^{\psi_2}) = \int (\psi_1 - \psi_2) dP,$$

and so $\mu_{\tilde{\varphi}^\psi}$ is the same for all such ψ. If $Q_P^{(r)}(\tilde{\varphi}^\psi) = \infty$ for any such ψ, then it is also convenient to modify the definitions of both the pressure and the entropy so that we still could speak about equilibrium states that would coincide with the old ones if $Q_P^{(r)}(\tilde{\varphi}^\psi) < \infty$.

Let $\|\zeta\|$ denote the Riemannian norm of the element ζ in the tangent bundle TM of M and for each $f \in \mathcal{F}$ let

$$\|Df^{-1}\| = \sup\{\|Df^{-1}\zeta\| : \zeta \in TM, \|\zeta\| = 1\}.$$

The following is the main assumption of this paper:

$$(1.6) \qquad \log\|DF^{-1}\| \in L^1(\Omega, P) \text{ and } \alpha = E_P(\log\|DF^{-1}\| \, \mathcal{J}) < 0,$$

where $E_P(\cdot|\cdot)$ denotes the conditional expectation corresponding to P and $\mathcal{J}$ is the completion with respect to P of the σ-field of θ- invariant subsets of Ω. Note that by the Birkhoff ergodic theorem, with probability one we have

$$(1.7) \qquad \alpha(\omega) = \lim_{|n|\to\infty} \frac{1}{n} \sum_{k=0}^{n-1} \log\|DF^{-1}(\theta^k\omega)\|.$$

The second condition in (1.6) says that $F(\omega)$ expands "in average".

Let Φ_β be the space of real functions $\varphi = \{\varphi_\omega, \omega \in \Omega\}$ on $M \times \Omega$ satisfying

$$|\varphi_\omega(x) - \varphi_\omega(y)| \leq C_\varphi(\omega)|x-y|^\beta \tag{1.8}$$

with $\beta > 0$ independent of ω and

$$\log C_\varphi \in L^1(\Omega, P). \tag{1.9}$$

Fix a point $x_0 \in M$ and set

$$\tilde{\varphi}_\omega(x) = \varphi_\omega(x) - \varphi_\omega(x_0). \tag{1.10}$$

If (1.4) is true, then

$$\|\tilde{\varphi}_\omega\| \in L^1(\Omega, P).$$

Denote by $n_F(\omega)$ the number of elements in $F^{-1}(\omega)x$, $x \in M$ (it does not depend on x since M is connected), and put

$$\tilde{Q}_P^{(r)}(\varphi) = Q_P^{(r)}(\tilde{\varphi} - \log n_F). \tag{1.11}$$

We shall see that this number is finite if (1.4) is satisfied. It is easy to derive from the condition (1.6) (cf. (3.1)–(3.3) in Section 3) that for any finite measurable partition ζ of M with elements of sufficiently small diameter, the intersection $\vee_{n=0}^{\infty} F^{-n}(\omega)\zeta$ generates the entire Borel σ-algebra $\mathcal{B}_M$ of M and for any τ-invariant $\mu = \{\mu^\omega, \omega \in \Omega\} \in \mathcal{P}_P(M \times \Omega)$ (see **[B1]**, **[B2]**) we have

$$h_\mu^{(r)}(\tau) = \int H_{\mu^\omega}(\zeta | F^{-1}(\omega)\mathcal{B}_M) dP(\omega), \tag{1.12}$$

where $H_{\mu^\omega}(\zeta|\mathcal{A}) = -\int \sum_i \mu^\omega(A_i|\mathcal{A}) \log \mu^\omega(A_i|\mathcal{A}) d\mu^\omega$ and $\mu^\omega(\cdot|\mathcal{A})$ is the conditional μ^ω-probability with respect to a σ-algebra $\mathcal{A}$. Set

$$\tilde{h}_\mu^{(r)}(\tau) = \int \Big(H_{\mu^\omega}(\zeta | F^{-1}(\omega)\mathcal{B}_M) - \log n_F(\omega) \Big) dP(\omega). \tag{1.13}$$

THEOREM A. *Let $\varphi \in \Phi_\beta$ and (1.4), (1.6) be satisfied. Then there exists a unique τ-invariant measure $\mu_\varphi \in \mathcal{P}_P(M \times \Omega)$ such that*

$$\begin{aligned} \tilde{Q}_P^{(r)}(\varphi) &= \int \tilde{\varphi} d\mu_\varphi - \tilde{h}_{\mu_\varphi}^{(r)}(\tau) \\ &= \sup \left\{ \int \tilde{\varphi} d\mu - \tilde{h}_\mu^{(r)}(\tau) : \mu \in \mathcal{P}_P(M \times \Omega) \text{ is } \tau\text{-invariant} \right\}. \end{aligned} \tag{1.14}$$

If, in addition,

$$\log n_F \in L^1(\Omega, P), \tag{1.15}$$

then

$$\tilde{h}_{\mu_\varphi}^{(r)}(\tau) = h_{\mu_\varphi}^{(r)}(\tau) - \int \log n_F dP \tag{1.16}$$

and

$$\tilde{Q}_P^{(r)}(\varphi) = Q_P^{(r)}(\tilde{\varphi}) - \int \log n_F dP, \tag{1.17}$$

and so μ_φ is the unique τ-invariant probability measure satisfying

$$Q_P^{(r)}(\tilde{\varphi}) = \int \tilde{\varphi} d\mu_\varphi - h_{\mu_\varphi}^{(r)}(\tau). \tag{1.18}$$

If, in addition to (1.15),

$$\|\varphi_\omega\| \in L^1(\Omega, P) \tag{1.19}$$

then μ_φ is the unique equilibrium state for the variational principle (1.2). *If the shift θ on (Ω, P) is ergodic (respectively, mixing) then the skew product transformation τ on $(M \times \Omega, \mu_\varphi)$ is ergodic (respectively, mixing).*

We shall construct the measures μ_φ similarly to **[K2]**, using the random RPF operator $\mathcal{L}_\varphi = \{\mathcal{L}_\varphi^\omega, \omega \in \Omega\}$ acting on the space $C(M)^\Omega$ of families $\{q_\omega, \omega \in \Omega\}$ of continuous functions q_ω on M by the formula

$$\mathcal{L}_\varphi^\omega q_\omega(x) = \sum_{y \in F^{-1}(\omega)x} e^{\varphi_\omega(y)} q_\omega(y). \tag{1.20}$$

The construction of μ_φ yields as a byproduct the following result (cf. **[K2]**).

THEOREM B. *Let* Jac $D_x F(\omega)$ *denote the Jacobian of the differential $D_x F(\omega)$ and suppose that*

$$\varphi_\omega(x) = -\log|\operatorname{Jac} D_x F(\omega)| \in \Phi_\beta, \tag{1.21}$$

φ satisfies (1.4), *and* (1.6) *holds. Then the measure μ_φ given by Theorem A has disintegrations μ_φ^ω, i.e., $d\mu_\varphi(x,\omega) = d\mu_\varphi^\omega(x) dP(\omega)$, which are equivalent to the Riemannian volume m on M. If, in addition,* (1.15) *is satisfied, then Theorems* C *and* D *of* **[K2]** *remain true yielding, in particular, relativized upper and lower large deviations bounds for the random measures*

$$\zeta_{x,\omega}^n = \frac{1}{n} \sum_{k=0}^{n-1} \delta_{\tau^k(x,\omega)}, \tag{1.22}$$

where δ_y is the unit mass sitting at y.

By modifying the proof in **[K2]**, we shall extend the results to our more general situation.

Another approach we deal with in this paper is based on the statistical mechanics considerations. The first construction of Gibbs measures in the dynamical systems setup was given by Sinai in [S]. Normally, one should first pass to the symbolic coding so that the action of $F(\omega)$ on M becomes conjugate to the full shift or subshift of finite type in the sequence space. Such coding is known to exist only in a one-dimensional situation or if all $F(\omega)$ belong to a compact set of expanding maps (see **[BG1]**). That is why we are using statistical mechanics ideology without passing to symbolic setup.

For fixed $\omega \in \Omega$ we first define the notion of ϕ-Gibbs measure on M corresponding to "potential" ϕ_ω. Our definition is motivated by the Dobrushin-Lanford-Ruelle condition in statistical mechanics. Take arbitrary $y \in M$, $n \in \mathbb{N}$ and denote by $F^{-n}(\omega)y$ the set of preimages of $y \in M$:

$$F^{-n}(\omega)y = \{x \in M : \; F^n(\omega)x = y\}. \tag{1.23}$$

Obviously, $F^{-n}(\omega)y$ is a discrete subset of M. In a nondegenerate situation its cardinality does not depend on y and is equal to

$$\prod_{i=0}^{n-1} n_F(\theta^i\omega).$$

Next we define a probability measure on $F^{-n}(\omega)y$ as follows:

$$P_n^{\omega}(x|y) = \frac{1}{Z_n^{\omega}(y)} \exp\left\{\sum_{i=0}^{n-1} \phi_{\theta^i\omega}(F^i(\omega)x)\right\}, \tag{1.24}$$

where

$$Z_n^{\omega}(y) = \sum_{x\in F^{-n}(\omega)y} \exp\left\{\sum_{i=0}^{n-1} \phi_{\theta^i\omega}(F^i(\omega)x)\right\}. \tag{1.25}$$

It is easy to see that the measures $P_n^{\omega}(\cdot|y)$ are defined in a consistent way. Namely, for $m > n$ and $z = F^{m-n}(\theta^n\omega)y$, the conditional distribution of $x \in F^{-m}(\omega)z$ under the condition $F^n(\omega)x = y$ coincides with $P_n^{\omega}(\cdot|y)$.

Consider a measure ν^{ω} on M. For almost all y with respect to the measure $F^n(\omega)\nu^{\omega}$ one can define conditional probabilities $\nu_n^{\omega}(\cdot|y)$ on M under condition $F^n(\omega)x = y$. Clearly, these conditional probabilities are supported on $F^{-n}(\omega)y$.

1.1 Definition. A measure ν^{ω} is called a ϕ-Gibbs measure if for all $n \in \mathcal{N}$:

$$\nu_n^{\omega}(\cdot|y) = P_n^{\omega}(\cdot|y) \quad \text{for } F^n(\omega)\nu^{\omega}\text{-a.a. } y. \tag{1.26}$$

Clearly, (1.26) is equivalent to the following condition

$$\int f(x)\nu^{\omega}(dx) = \int \left(\sum_{x\in F^{-n}(\omega)y} f(x)P_n^{\omega}(x|y)\right) F^n(\omega)\nu^{\omega}(dy), \tag{1.27}$$

where $f(x)$ is an arbitrary function from $C(M)$.

We shall prove in Section 4 that for P-a.a. $\omega \in \Omega$, a ϕ-Gibbs measure ν^{ω} exists and is unique. Next we define the ϕ-Gibbs measure on $M \times \Omega$ by

$$\nu(dx, d\omega) = \nu^{\omega}(dx)P(d\omega).$$

Strictly speaking, the symbolic counterpart of the measures $\nu^{\omega}(dx)$ are ϕ-Gibbs measures on the right half-line with the empty boundary condition on the left. That is why the measure ν on $M \times \Omega$ feels the boundary condition and it is not invariant under τ. Thus it is natural to consider the measures

$$\mu(dx, d\omega) = \mu^{\omega}(dx)P(d\omega) \in \mathcal{P}_P(M \times \Omega)$$

such that μ is τ-invariant and for P-a.a. $\omega \in \Omega$ the measure $\mu^{\omega}(dx)$ is absolutely continuous with respect to $\nu^{\omega}(dx)$. This brings us to the following definition.

1.2. Definition. A measure $\mu \in \mathcal{P}_p(M \times \Omega)$ is called a ϕ-Gibbs state if $\tau\mu = \mu$ and for P-a.a. ω,

$$\mu^{\omega}(dx) \prec \nu^{\omega}(dx).$$

THEOREM C. *Let $\phi \in \Phi_\beta$ and* (1.6) *be satisfied. Then there exists a unique ϕ-Gibbs state μ. If, in addition,* (1.4) *holds then μ satisfies the generalized variational principle* (1.14) *and thus coincides with the equilibrium state μ_ϕ constructed in Theorem A.*

We see that a ϕ-Gibbs state exists and is unique under milder conditions and thus it is a more general object than the equilibrium states. Next statement is analogous to Theorem B, but again without the condition (1.4).

THEOREM B$'$. a) *Let* Jac $D_xF(\omega)$ *denote the Jacobian of the differential $D_xF(\omega)$ and suppose that*

$$\phi_\omega(x) = -\log|\operatorname{Jac} D_xF(\omega)| \in \Phi_\beta,$$

and (1.6) *holds. Then there exists a unique τ-invariant measure $\mu \in \mathcal{P}_P(M \times \Omega)$, $\mu(dx, d\omega) = \mu^\omega(dx)P(d\omega)$, such that for P-a.a. ω disintegrations $\mu^\omega(dx)$ are absolutely continuous with respect to the Riemannian volume m on M.*

b) *Suppose $\chi(dx) = h(x)m(dx)$ is a measure on M absolutely continuous with respect to m. Then for P-a.a. ω the measures $F^n(\omega)\chi$ approach $\mu^{\theta^n\omega}(dx)$ in the following sense: for any $f(x) \in C(M)$,*

$$\left|\int f(x)(F^n(\omega)\chi)(dx) - \int f(x)\mu^{\theta^n\omega}(dx)\right| \to 0 \quad \text{as } n \to \infty. \tag{1.28}$$

c) *Suppose $\kappa \in \mathcal{P}_P(M \times \Omega)$, $\kappa(dx, d\omega) = \kappa^\omega(dx)P(d\omega)$, and for P-a.a. ω the measures $\kappa^\omega(dx)$ are absolutely continuous with respect to m. Then $\tau^n\kappa \to \mu$ weakly as $n \to \infty$.*

We want to mention that the notion of a ϕ-Gibbs measure is close to the notion of a g-measure introduced by M. Keane **[Ke]**. On the other hand, we think that our Definitions 1.1 and 1.2 are more convenient from the point of view of statistical mechanics.

In Section 5 we consider random subshifts of finite type. Here we start with a random alphabet $\{1, 2, \dots, l(\omega)\}$, $\omega \in \Omega$. Instead of M now we have a random compact set $X_\omega(A) = \{x = (\dots, x_{-1}, x_0, x_1, \dots) : x_i \in \{1, \dots, l(\theta^i\omega)\}$ and $a_{x_ix_{i+1}}(\theta^i\omega) = 1$ for all $i \in \mathbb{Z}\}$, where $A(\omega) = (a_{ij}(\omega)$, $\omega \in \Omega$, is a measurable family of $l(\omega) \times l(\theta\omega)$-matrices with the entries 0 and 1. Endomorphisms $F(\omega)$ are replaced by the random shift

$$\sigma(\omega) : X_\omega(A) \to X_{\theta\omega}(A), \ (\sigma(\omega)x)_i = x_{i+1}. \tag{1.29}$$

The transformation τ is defined on the skew-product space $\mathcal{X}(A) = \prod_{\omega\in\Omega} X_\omega(A)$: $\tau(x, \omega) = (\sigma(\omega)x, \theta\omega)$. A measurable family of random potentials $U_\omega(y)$ playing the same role as the functions φ_ω above are defined on the space of one-sided sequences

$$Y_\omega(A) = \{y = (y_0, y_1, \dots) : y_i \in \{1, \dots, l(\theta^i\omega)\}, \ a_{y_iy_{i+1}}(\theta^i\omega) = 1\}.$$

Denote by $\sigma_+(\omega)$ the random shift transformation acting on $Y_\omega(A)$ and by τ_+ the skew-shift transformation acting on $\mathcal{Y}(A) = \prod_{\omega\in\Omega} Y_\omega(A)$ by the formula $\tau_+(y, \omega) = (\sigma_+(\omega)y, \theta\omega)$.

Conditions (1.8), (1.9) are replaced by the analogous conditions

$$|U_\omega(y') - U_\omega(y'')| \le C_U(\omega)e^{-\tilde{\alpha}n}, \ \tilde{\alpha} > 0, \tag{1.30}$$

provided y' and y'' coincide on the first n positions $y'_i = y''_i$, $0 \le i \le n-1$, and

$$\log(1 + C_U(\omega)) \in L^1(\Omega, P). \tag{1.31}$$

Suppose that the matrices $A(\omega)$ satisfy the following natural conditions.

For P-a.a. ω :

(1.32)

a. For any t_1, $1 \leq t_1 \leq l(\omega)$, there exists t_2, $1 \leq t_2 \leq l(\theta\omega)$ such that $a_{t_1 t_2} = 1$.

b. For any t_2, $1 \leq t_2 \leq l(\theta\omega)$, there exists t_1, $1 \leq t_1 \leq l(\omega)$ such that $a_{t_1 t_2} = 1$.

c. There exists a random number $N(\omega) \in \mathbb{Z}_+$ such that all entries of the product matrix $A(\omega)A(\theta\omega)\dots A(\theta^{N-1}\omega)$ are positive.

Notice that the condition (1.32c) is weaker and more natural than the uniform aperiodicity condition from **[BG]**, which says that $A(\omega)A(\theta\omega)\dots A(\theta^{N-1}\omega)$ is positive for some N independent of ω.

Using the potentials U_ω one can naturally define for P-a.a. $\omega \in \Omega$ the notion of the U-Gibbs measure on $X_\omega(A)$, $Y_\omega(A)$ and of the U-Gibbs state on $\mathcal{X}(A)$, $\mathcal{Y}(A)$ (see Definitions 5.3, 5.4, 5.8). Denote by $\mathcal{P}_P(\mathcal{X}(A))$, $\mathcal{P}_P(\mathcal{Y}(A))$ the set of probability measures on $\mathcal{X}(A)$, $\mathcal{Y}(A)$ respectively with the marginal distribution P on Ω .

THEOREM D. *Suppose the conditions* (1.30)–(1.32) *hold. Then there exists a unique U-Gibbs state $\mu \in \mathcal{P}_P(\mathcal{X}(A))$. The measure μ is τ-invariant and ergodic with respect to τ. Moreover, μ is mixing if P is mixing with respect to θ. The projection μ_+ of the measure μ to $\mathcal{Y}(A)$ is the unique U-Gibbs state in $\mathcal{P}_P(\mathcal{Y}(A))$.*

One can define generalized variational principle for random subshifts of finite type almost in the same way as above (see (5.32)). In order to prove that a U-Gibbs state satisfies generalized variational principle, we impose two more conditions. One of them,

$$\delta_U(\omega) = \left[\sup_y U_\omega(y) - \inf_y U_\omega(y)\right] \in L^1(\Omega, P) \tag{1.33}$$

is a complete analog of (1.4). The second condition deals with the properties of matrices $A(\omega)$. Denote

$$l_m(\omega) = \min_{1\leq j\leq l(\theta\omega)} \sum_{i=1}^{l(\omega)} a_{ij}(\omega), \quad l^m(\omega) = \max_{1\leq j\leq l(\theta\omega)} \sum_{i=1}^{l(\omega)} a_{ij}(\omega).$$

Notice that $\sum_{i=1}^{l(\omega)} a_{ij}(\omega)$ is the number of letters $i \in \{1, \dots, l(\omega)\}$ that are A–consistent with a given $j \in \{1, \dots, l(\theta\omega)\}$. Suppose that

$$\bar{l}(\omega) \equiv [\log l^m(\omega) - \log l_m(\omega)] \in L^1(\Omega, P). \tag{1.34}$$

THEOREM E. *Suppose that the conditions* (1.30)–(1.34) *hold. Then the U-Gibbs state μ_+ is the unique τ_+-invariant measure in $\mathcal{P}_P(\mathcal{Y}(A))$ that satisfies the generalized variational principle* (5.32).

Acknowledgements. This paper was started during the visit of the first author to the Hebrew University. It was completed during the visit of both authors to the Department of Mathematics at Princeton University, whose support and excellent working conditions we gratefully acknowledge.

2. Preliminaries

First, notice that it suffices to establish Theorem A only when P is an ergodic invariant measure of θ: $\Omega \to \Omega$. Indeed, by Choquet's theorem (see **[P]**) any θ-invariant P can be written as the integral

$$P = \int_\Gamma P_\gamma d\nu_P(\gamma), \tag{2.1}$$

where P_γ, $\gamma \in \Gamma$, are ergodic θ-invariant probability measures parametrized by a set Γ and ν_P is a probability measure on Γ. By (1.6), for ν_P-a.a. $\gamma \in \Gamma$,

$$\log \|DF^{-1}\| \in L^1(\Omega, P_\gamma) \text{ and } \alpha_\gamma = \int \log \|DF^{-1}\| dP_\gamma < 0. \tag{2.2}$$

Next, if $\varphi = \{\varphi_\omega, \omega \in \Omega\}$ satisfies (1.11) and (1.12), then for ν_P-a.a. $\gamma \in \Gamma$,

$$\delta_\varphi, \log C_\varphi \in L^1(\Omega, P_\gamma). \tag{2.3}$$

Let $d\mu(x,\omega) = d\mu^\omega(x)dP(\omega)$ and $d\mu^{(\gamma)}(x,\omega) = d\mu^\omega(x)dP(\omega)$. Then $\mu = \int_\Gamma \mu^{(\gamma)} d\nu_P(\gamma)$. It is easy to see that both $h_\mu^{(r)}(\tau)$ and $\tilde{h}_\mu^{(r)}(\tau)$ are affine in P, i.e.,

$$h_\mu^{(r)}(\tau) = \int_\Gamma h_{\mu^{(\gamma)}}^{(r)}(\tau) d\nu_P(\gamma) \text{ and } \tilde{h}_\mu^{(r)}(\tau) = \int_\Gamma \tilde{h}_{\mu^{(\gamma)}}^{(r)}(\tau) d\nu_P(\gamma). \tag{2.4}$$

Thus the existence and uniqueness of equilibrium states $\mu_\varphi^{(\gamma)}$ and generalized equilibrium state $\tilde{\mu}_\varphi^{(\gamma)}$ for P corresponding to φ for P_γ and ν_P-a.a. $\gamma \in \Gamma$ yields the existence and uniqueness of the equilibrium state μ_φ and the generalized equilibrium state $\tilde{\mu}_\varphi$ for P corresponding to φ.

The following general result will play an important part in our considerations.

2.1. Proposition. *Let θ be an invertible measure preserving ergodic transformation of a probability space (Ω, P). Suppose that measurable functions a and b on Ω and a measurable subset $\Psi \subset \Omega$ satisfy the conditions*

$$Ea = \alpha < 0, \quad E|b| < \infty, \quad P(\Psi) = \beta > 0, \tag{2.5}$$

where $Ea = \int a\, dP$. Then for any $\varepsilon > 0$ there exists $R_\varepsilon > 0$ and a measurable set $\Psi_\varepsilon \subset \Omega$ such that

$$\Psi_\varepsilon \subset \Psi, \quad P(\Psi \backslash \Psi_\varepsilon) \leq \varepsilon, \tag{2.6}$$

and

$$A_k(\omega) \stackrel{\text{def}}{=} \sum_{\ell=-\infty}^{k} \exp\Big(b(\theta^\ell \omega) + \sum_{j=\ell}^{k} a(\theta^j \omega)\Big) \leq R_\varepsilon \text{ whenever } \theta^k \omega \in \Psi_\varepsilon. \tag{2.7}$$

In particular, for P-a.a. ω there exists a sequence $k_i = k_i(\omega), k_i \to \infty$ as $i \to \infty$, such that

$$A_{k_i} \leq R_\varepsilon, \quad \theta^{k_i}\omega \in \Psi, \quad \lim_{i\to\infty} \frac{k_{i+1}}{k_i} = 1. \tag{2.8}$$

PROOF. First, we show that for P-a.a. ω,

$$A_0(\omega) < \infty. \tag{2.9}$$

Indeed, by (2.5) and the ergodic theorem there exists $N^a = N^a(\omega)$ such that

$$\sum_{i=0}^{n} a(\theta^{-i}\omega) \leq \frac{1}{2} n\alpha \text{ for } n \geq N^a. \tag{2.10}$$

Again by (2.5),

$$\infty > E|b| \geq \frac{|\alpha|}{4} \sum_{i=1}^{\infty} P\left\{|b(\omega)| \geq \frac{|\alpha|}{4} i\right\} = \frac{|\alpha|}{4} \sum_{i=1}^{\infty} P\left\{|b(\theta^{-i}\omega)| \geq \frac{|\alpha|}{4} i\right\}, \tag{2.11}$$

and so by the Borel-Cantelli lemma for P-a.a. ω there exists $N^b(\omega) > 0$ such that

$$|b(\theta^{-i}\omega)| < \frac{|\alpha|}{4} i \text{ for } i \geq N^b(\omega). \tag{2.12}$$

Now (2.9) follows from (2.10) and (2.12). Observe that

$$A_k(\omega) = A_0(\theta^k\omega), \tag{2.13}$$

and so we obtain (2.7) by taking

$$\Psi_\varepsilon = \Psi \bigcap \{\omega : A_0(\omega) \leq R_\varepsilon\} \tag{2.14}$$

with sufficiently large R_ε. Next, take $\varepsilon = \frac{\beta}{2}$. Then $P(\Psi_\varepsilon) \geq \frac{\beta}{2}$. By the ergodic theorem for P-a.a. ω,

$$\lim_{n\to\infty} \frac{1}{n} \sum_{k=0}^{n-1} \mathbf{1}_{\Psi_{\frac{\beta}{2}}}(\theta^k\omega) = P(\Psi_{\frac{\beta}{2}}) \geq \frac{\beta}{2}, \tag{2.15}$$

where $\mathbf{1}_{\Psi_{\beta/2}}(\omega) = 1$ if $\omega \in \Psi_{\beta/2}$, and 0 otherwise. We denote $k_1 = k_1(\omega) = \min\{k \geq 0 \colon \theta^k\omega \in \Psi_{\beta/2}\}$ and $k_{i+1} = k_{i+1}(\omega) = \min\{k > k_i(\omega) \colon \theta^k\omega \in \Psi_{\beta/2}\}$. Then by (2.15) for P-a.a. ω,

$$\lim_{i\to\infty} \frac{i}{k_i(\omega)} = \frac{\beta}{2}, \tag{2.16}$$

which yields (2.8). □

2.2. Remark. We thank B. Weiss for useful remarks concerning the proof of Proposition 2.1.

3. Random transformations

Without loss of generality (as we explained in the previous section) below we shall assume that θ is an ergodic transformation of (Ω, P). Clearly,

$$\|DF^{-n}(\omega)\| \leq \prod_{j=0}^{n-1} \|DF^{-1}(\theta^j\omega)\|. \tag{3.1}$$

By (1.6) and the ergodic theorem, there exists a measurable function $C(\omega)$ such that for P-a.a. ω,

$$\prod_{j=0}^{n-1} \|DF^{-1}(\theta^j\omega)\| \le C(\omega)e^{\alpha n/2} \tag{3.2}$$

and, in addition, by Proposition 2.1 considered with $a(\omega) = \log\|DF^{-1}(\omega)\|$ $b(\omega) = \log C_\varphi(\omega)$, and $\Psi = \Psi_c = \{\omega : C(\omega) \le c\}$, for any $\varepsilon > 0$ there exist $R_\varepsilon > 0$ and a measurable set $\Psi_c^\varepsilon \subset \Psi_c$ such that $P(\Psi_c \setminus \Psi_c^\varepsilon) \le \varepsilon$ and

$$A_k(\omega) = \sum_{\ell=0}^{k-1} C_\varphi(\theta^\ell\omega)\|DF^{-(k-\ell)}(\theta^\ell\omega)\|^\beta \le R_\varepsilon \text{ whenever } \theta^k\omega \in \Psi_c^\varepsilon, \tag{3.3}$$

where C_φ and β are from (1.11). This gives a sequence $k_i = k_i(\omega) \to \infty$ satisfying (2.8).

Next we collect the assertions concerning the random RPF operator $\mathcal{L}_\varphi$ defined by (1.20).

3.1. THEOREM. *Suppose that* (1.6), (1.8), *and* (1.9) *hold true. Then for P-a.a.* ω *there exist unique* $\nu^\omega \in \mathcal{P}(M)$, $h_\omega \in C(M)$, *and* $\lambda^\omega > 0$ *depending measurably on* ω *such that*

(a) $(\mathcal{L}_\varphi^\omega)^*\nu^{\theta\omega} = \lambda^\omega\nu^\omega$;

(b) $\nu^\omega(h_\omega) = 1, \mathcal{L}_\varphi^\omega h_\omega = \lambda^w h_{\theta\omega}, h_\omega > 0$, *and*

$$|\log h_\omega(x) - \log h_\omega(x')| \le K_\varphi(\omega)\Big(d(x,x')\Big)^\beta, \tag{3.4}$$

where

$$K_\varphi(\omega) = \sum_{i=1}^\infty C_\varphi(\theta^{-i}\omega)\Big(\|DF^{-1}(\theta^{-i}\omega)\|\cdots\|DF^{-1}(\theta^{-1}\omega)\|\Big)^\beta \tag{3.5}$$

and C_φ, β *are the same as in* (1.8);

(c) *uniformly on* M *as* $n \to \infty$,

$$(\lambda^{\theta^{n-1}\omega}\cdots\lambda^{\theta\omega}\lambda^\omega)^{-1}(h_{\theta^n\omega})^{-1}(\mathcal{L}_\varphi^{\omega,n}q) \to \nu^\omega(q) \tag{3.6}$$

for any $q \in C(M)$, *where* $\nu^\omega(q) = \int q d\nu^\omega$ *and*

$$\mathcal{L}_\varphi^{\omega,n} = \mathcal{L}_\varphi^{\theta^{n-1}\omega} \circ \cdots \circ \mathcal{L}_\varphi^{\theta\omega} \circ \mathcal{L}_\varphi^\omega;$$

(d) *set* $\mu^\omega = h_\omega\nu^\omega$ *and* $g_\omega = (e^{\varphi_\omega}h_\omega)/\lambda^\omega(h_{\theta\omega} \circ F(\omega))$. *Then* $\mu^\omega \in \mathcal{P}(M)$, $F(\omega)\mu^{\theta\omega} = \mu^\omega$, μ^ω *is uniquely determined by*

$$(\mathcal{L}_{\log g_\omega}^\omega)^*\mu^{\theta\omega} = \mu^\omega, \tag{3.7}$$

and for any $q \in C(M)$ *uniformly on* M,

$$\mathcal{L}_{\log g_\omega}^{\omega,n} q \to \mu^\omega(q) \quad \text{as } n \to \infty; \tag{3.8}$$

(e) *for any bounded Borel function* q *on* M *and* $r \in C(M)$ *satisfying* $\nu^\omega(r) = 1$ *one has*

$$\nu^\omega\Big(r(q \circ F^n(\omega))\Big) - \mu^\omega\Big(q \circ F^n(\omega)\Big) \to 0 \text{ as } n \to \infty; \tag{3.9}$$

(f) *set* $B_x^\omega(\delta,n)=\{y:\operatorname{dist}\big(F^i(\omega)x,\ F^i(\omega)y\big)\le\delta$ *for all* $i=0,1,\dots,n\}$. *Then for all sufficiently small* $\delta>0$ *and all sufficiently large* D *there exists* $C=C^\omega_{\delta,\varepsilon,D}>0$ *such that for any* $x\in M$ *and any* k *satisfying the conditions* (3.3) *and* $\|DF(\theta^k\omega)\|\le D$, *we have*

$$C^{-1}\le\mu^\omega\Big(B_x^\omega(\delta,k)\Big)\exp\Big(\sum_{i=0}^{k}\big(\log\lambda^{\theta^i\omega}-\varphi_\omega(F^i(\omega)x)\big)\Big)\le C. \tag{3.10}$$

PROOF. The arguments of the proof are essentially similar to **[K2]** except for a few points discussed below. First, for all $\omega\in\Omega$ satisfying (a) one constructs $\lambda^\omega>0$ and $\nu^\omega\in\mathcal{P}(M)$ in the same way as in Lemma 2.2 of **[K2]**. By (1.9), in the same way as in (2.10) we have

$$\infty>E|\log C_\varphi|\ge\varepsilon\sum_{i=1}^{\infty}P\{|\log C_\varphi(\theta^{-i}\omega)|\ge\varepsilon i\} \tag{3.11}$$

for any $\varepsilon>0$. By the Borel-Cantelli lemma for P-a.a. ω there exists $n_\varepsilon(\omega)>0$ such that

$$C_\varphi(\theta^{-i}\omega)\le e^{\varepsilon i}\ \text{for all}\ i\ge n_\varepsilon(\omega). \tag{3.12}$$

Taking $\varepsilon<\frac{\alpha}{2}$ we derive from (3.2) and (3.12) that

$$K_\varphi(\omega)<\infty,\qquad P\text{-almost surely (a.s.)}. \tag{3.13}$$

Next, in the same way as in Proposition 2.5 of **[K2]**, we construct $h=\{h_\omega,\omega\in\Omega\}$ satisfying the assertion (b) of Theorem 3.1.

Let $g=\{g_\omega,\omega\in\Omega\}$ be defined as in (d) of Theorem 3.1. Then

$$\mathcal{L}^\omega_{\log g_\omega}\mathbf{1}(x)=\sum_{y\in F^{-1}(\omega)x}g(y)=\Big(\lambda^\omega h_{\theta\omega}(x)\Big)^{-1}\mathcal{L}^\omega_\varphi h_\omega(x)=1. \tag{3.14}$$

Set $\mathcal{L}^\omega=\mathcal{L}^\omega_{\log g_\omega}$ and $\mathcal{L}^{\omega,n}=\mathcal{L}^{\omega,n}_{\log g_\omega}$. The main point in the construction of the measures μ^ω in **[K2]** is the proof that for any $q\in C(M)$ the sequence $\{\mathcal{L}^{\omega,n}q:n\ge0\}$ is equicontinuous. Together with the uniform boundedness of the sequences, which is proved in our case in the same way as in **[K2]**, this yields the existence of a uniformly converging subsequence

$$\mathcal{L}^{\omega,n_i}q\to q^*_\omega\ \text{as}\ n_i\to\infty. \tag{3.15}$$

After obtaining this it is easy to see that in fact,

$$\lim_{n\to\infty}\mathcal{L}^{\omega,n}q=q^*_\omega, \tag{3.16}$$

and one puts $\mu^\omega(q)=q^*_\omega$ obtaining, finally, μ^ω by the Riesz theorem. We cannot prove the equicontinuity of the whole sequence $\mathcal{L}^{\omega,n}q$, $n\ge0$, which requires the uniform expanding of all $F(\omega),\omega\in\Omega$, and the Hölder equicontinuity of $\varphi_\omega,\omega\in\Omega$, but since we need only a converging subsequence it suffices to show that there exists an equicontinuous subsequence $\{\mathcal{L}^{\omega,k_i}q,i=1,2,\dots\}$, where we shall take $k_i=k_i(\omega)\to\infty$ as $i\to\infty$ satisfying (2.8) with $\Psi=\Psi_c$ and $A_k(\omega)$ the same as in (3.3).

For any $x, x' \in M$ with sufficiently small $\mathrm{dist}(x, x')$, we consider pairs $y \in F^{-k_i(\omega)}x, y' \in F^{-k_i(\omega)}x'$ belonging to the same branch of $F^{-k_i}(\omega)M$. By (3.1), (3.2), and (3.14),

(3.17)
$$
\begin{aligned}
&|(\mathcal{L}^{\omega,k_i}q)(x) - (\mathcal{L}^{\omega,k_i}q)(x')| \\
&\quad \leq \sum_{y \in F^{-k_i}(\omega)x} g_\omega(x) g_{\theta\omega}(F(\omega)y) \cdots g_{\theta^{k_i-1}\omega}(F^{k_i-1}(\omega)y)|q(y) - q(y')| \\
&\quad + \Big| \sum_{y \in F^{-k_i}(\omega)x} q(y')(g_\omega(y) g_{\theta\omega}(F(\omega)y) \cdots g_{\theta^{n-1}\omega}(F^{k_i-1}(\omega)y) \\
&\qquad - g_\omega(y') g_{\theta\omega}(F(\omega)y') \cdots g_{\theta^{n-1}\omega}(F^{k_i-1}(\omega)y'))\Big| \\
&\quad \leq \sup\{|q(v) - q(w)| : \mathrm{dist}(v, w) \leq C(\omega)e^{\alpha k_i/2}\,\mathrm{dist}(x, x')\} \\
&\qquad + \|q\| \sum_{y \in F^{-k_i}(\omega)x} g_\omega(y') \cdots g_{\theta^{n-1}\omega}(F^{k_i-1}(\omega)y') G^\omega_{k_i}(y, y') \\
&\quad \leq \gamma^\omega_{k_i}(x, x') + \|q\| G^\omega_{k_i}(y, y'),
\end{aligned}
$$

where

$$
G^\omega_k(y, y') = \left\| \frac{g_\omega(y) \cdots g_{\theta^{k-1}\omega}(F^{k-1}(\omega)y)}{g_\omega(y') \cdots g_{\theta^{k-1}\omega}(F^{k-1}(\omega)y')} - 1 \right\| \tag{3.18}
$$

and

$$
\gamma^\omega_{k_i}(x, x') \to 0 \text{ as } d(x, x') \to 0. \tag{3.19}
$$

By (1.8), (3.1)–(3.5), (3.13), and the definition of g_ω, for any k satisfying (3.3) we have

(3.20)
$$
\begin{aligned}
G^\omega_k(y, y') &= \Big| \exp\Big((\log h_\omega(y) - \log h_\omega(y')) + (\log h_{\theta^k\omega}(x') \\
&\quad - \log h_{\theta^k\omega}(x)) + \sum_{i=0}^{k-1} (\varphi_{\theta^i\omega}(F^i(\omega)y) - \varphi_{\theta^i\omega}(F^i(\omega)y'))\Big) - 1 \Big| \\
&\leq \Big| \exp\Big(((d(x, x'))^\beta (K_\varphi(\omega)(C(\omega))^\beta e^{\alpha\beta k/2} \\
&\quad + K_\varphi(\theta^k\omega) + \sum_{i=0}^{k-1} C_\varphi(\theta^i\omega)\|DF^{-(k-i)}(\theta^i\omega)\|^\beta)\Big) - 1 \Big| \\
&\leq \Big| \exp\Big((d(s, s'))^\beta (K_\varphi(\omega)(C(\omega))^\beta e^{\alpha\beta k/2} + c + R_\varepsilon)\Big) - 1 \Big| \to 0
\end{aligned}
$$

as $d(x, x') \to 0$, uniformly in k satisfying $\theta^k\omega \in \Psi^\varepsilon_c$. This provides an equicontinuous sequence $\mathcal{L}^{\omega,k_i}q$, $i = 1, 2, \dots$ with $k_i = k_i(\omega)$ satisfying (2.8). Together with other arguments from **[K2]** and a simple remark that by (3.1) and (3.2) for any $\varepsilon > 0$ there exists $n_\varepsilon = n_\varepsilon(\omega)$ such that for each $x \in M$ and all $n \geq n_\varepsilon$ the set $DF^{-n}(\omega)x$ is ε-dense, this yields the construction of the measures μ^ω and completes the proof of (a) – (d) of Theorem 3.1.

The assertion (e) follows, since by (a) we have

$$
\begin{aligned}
\int r(q \circ F^n(\omega))d\nu^\omega &= (\lambda^\omega \cdots \lambda^{\theta^{n-1}\omega})^{-1} \int \mathcal{L}_\varphi^{\omega,n}(r(q \circ F^n(\omega)))d\nu^{\theta^n\omega} \\
&= \int q(\lambda^\omega \cdots \lambda^{\theta^{n-1}\omega})^{-1} \mathcal{L}_\varphi^{\omega,n} r d\nu^{\theta^n\omega} \qquad (3.21)\\
&= \int q(\lambda^\omega \cdots \lambda^{\theta^{n-1}\omega})^{-1}(h_{\theta^n\omega})^{-1} \mathcal{L}_\varphi^{\omega,n} r d(F^n(\omega)\mu^\omega),
\end{aligned}
$$

which, by (3.6), yields (3.9). It remains to obtain (f). Since $\|DF(\theta^k\omega)\| \le D < \infty$, there exists $\delta_0 > 0$ (depending on D) such that for any $\delta < \delta_0$ the sets $B = B_x^\omega(\delta, k)$ contain at most one point from $F^{-k}(\omega)z$ for any $x,\ z \in M$, and so for any k satisfying (3.3) with $R = R_\varepsilon$ we have

$$
\mathcal{L}_\varphi^{\omega,k}(h_\omega \mathbf{1}_B)(z) = \sum_{y \in F^{-k}(\omega)z} e^{S_k\varphi_\omega(y)} h_\omega(y)\mathbf{1}_B(y) \le e^{(S_k\varphi_\omega(x)+R)}\|h_\omega\|, \qquad (3.22)
$$

where $S_n\varphi_\omega = \sum_{j=0}^{n-1} \varphi_\omega \circ F^j(\omega)$ and $z \in B_{F^k(\omega)x}(\delta)$ is such that $F^{-k}(\omega)z \cap B \neq 0$.

Thus,

$$
\begin{aligned}
\mu^\omega(B) = \nu(h_\omega \mathbf{1}_B) &= (\lambda^\omega \cdots \lambda^{\theta^{k-1}\omega})^{-1}\nu^\omega(\mathcal{L}_\varphi^{\omega,k}(h_\omega \mathbf{1}_B)) \qquad (3.23)\\
&\le (\lambda^\omega \cdots \lambda^{\theta^{k-1}\omega})^{-1} e^{(S_k\varphi_\omega(x)+R\delta^\beta)}\|h_\omega\|
\end{aligned}
$$

implying the right-hand side of (3.7). Next,

$$
F^k(\omega)B_x^\omega(\delta, k) \supset B_{F^k(\omega)x}(\delta/R)
$$

and there exists $n_\delta = n_\delta(\omega)$ such that for any $z \in M$,

$$
F^{-n_\delta}(\omega)z \cap B_{F^k(\omega)x}(\delta/R) \neq \emptyset.
$$

Thus

$$
F^{-(n_\delta+k)}z \cap B \neq \emptyset;
$$

let y' be a point from this intersection. Then

$$
\begin{aligned}
\mathcal{L}_\varphi^{\omega,k+n_\delta}(h_\omega \mathbf{1}_B)(z) &\ge \exp((S_{k+n_\delta}\varphi_\omega(y'))h_\omega(y')) \qquad (3.24)\\
&\le (\inf h)e^{S_k\varphi_\omega(x)}\exp(-n_\delta\|\varphi_\omega\| - R\delta^\beta).
\end{aligned}
$$

Hence

(3.25)

$$
\begin{aligned}
\mu^\omega(B) &= (\lambda^\omega \cdots \lambda^{\theta^{k+n_\delta-1}\omega})^{-1}\nu^\omega(\mathcal{L}_\varphi^{\omega,k+n_\delta}(h_\omega \mathbf{1}_B)) \\
&\le \exp(-n_\delta\|\varphi_\omega\| - R\delta^\beta)(\inf h)\exp\left(-\sum_{i=0}^{k}(\varphi_\omega(F^i(\omega)x) - \log\lambda(\theta^i\omega))\right)
\end{aligned}
$$

completing the proof of (3.10), as well as of Theorem 3.1. □

3.2 THEOREM. *The measure $\mu_\varphi = \{\mu^\omega, \omega \in \Omega\}$ constructed in Theorem 3.1 is ergodic with respect to the skew product transformation τ (recall that P is assumed to be ergodic with respect to the shift θ). If P is mixing with respect to θ, then μ_φ is mixing with respect to τ.*

PROOF. Note that

$$\tau^{-1}A = \{(\omega, x) : (\theta\omega, F(\omega)x) \in A\} = \{(\omega, x) : F(\omega)x \in A_{\theta\omega}\},$$

where $A_\omega = \{x : (\omega, x) \in A\}$. Thus

$$A = \tau^{-1}A \text{ if and only if } F(\omega)A_\omega = A_{\theta\omega}, \forall\omega \in \Omega. \tag{3.26}$$

Then also $F^n(\omega)A_\omega = A_{\theta^n\omega}$, and so

$$A_\omega \in \bigcap_{n=0}^{\infty} F^{-n}(\omega)\mathcal{B} \stackrel{\text{def}}{=} \mathcal{B}_\infty^\omega, \tag{3.27}$$

where $\mathcal{B}$ is the Borel σ-field on M. In order to establish the ergodicity of $\mu_\varphi = \{\mu^\omega, \omega \in \Omega\}$ with respect to τ it suffices to show that with probability one $\mathcal{B}_\infty^\omega$ consists of sets μ^ω whose measure equal 0 or 1. Indeed, denoting $\mathcal{L}^\omega = \mathcal{L}_{\log g_\omega}^\omega$ we have

(3.28)

$$\mu^\omega(A_\omega) = \int \mathbf{1}_{A_\omega} d\mu^\omega = \int \mathbf{1}_{A_\omega} d(\mathcal{L}^\omega)^*\mu^{\theta\omega} = \int \mathcal{L}^\omega \mathbf{1}_{A_\omega} d\mu^{\theta\omega} = \mu^{\theta\omega}(A_{\theta\omega}),$$

since

$$\mathcal{L}^\omega \mathbf{1}_{A_\omega}(x) = \sum_{y\in F^{-1}(\omega)x} g_\omega(y)\mathbf{1}_{A_\omega}(y) = \mathbf{1}_{A_{\theta\omega}}(x) \sum_{y\in F^{-1}(\omega)x} g_\omega(y) = \mathbf{1}_{A_{\theta\omega}}(x).$$

Thus, $f(\omega) = \mu^\omega(A_\omega)$ is θ-invariant and from the ergodicity of θ it follows that $\mu^\omega(A_\omega) = \text{const}$ P-a.s. If we know that $\mathcal{B}_\infty^\omega$ consists only of sets of μ^ω-measure 0 or 1, we obtain by (3.27) that either $\mu^\omega(A_\omega) = 1$ P-a.s. or $\mu^\omega(A_\omega) = 0$ P-a.s., and so either $\mu_\varphi(A) = 1$ or $\mu_\varphi(A) = 0$, which would imply the ergodicity of μ_φ.

In order to establish the μ^ω-triviality of $\mathcal{B}_\infty^\omega$ it suffices to show that for any $q \in C(M)$,

$$\mu^\omega(q|\mathcal{B}_\infty^\omega) = \mu^\omega(q)\ \mu^\omega - a.s, \tag{3.29}$$

where $\mu^\omega(q|\cdot)$ denotes the conditional expectation for μ^ω. Since

$$\mu^\omega(q|F^{-N}(\omega)\mathcal{B}) = (\mathcal{L}^{\omega,N}q) \circ F^N(\omega), \tag{3.30}$$

where $\mathcal{L}^{\omega,N} = \mathcal{L}_{\log g_\omega}^{\omega,N}$ (see **[W]** or **[K2]**) we have

(3.31)

$$I_N^\omega = \int |\mu^\omega(q|F^{-N}(\omega)\mathcal{B}) - \mu^\omega(q)|d\mu^\omega = \int |(\mathcal{L}^{\omega,N}q) \circ F^N(\omega) - \mu^\omega(q)|d\mu^\omega$$
$$= \int |\mathcal{L}^{\omega,N}q - \mu^\omega(q)|dF^N(\omega)\mu^\omega \le \|\mathcal{L}^{\omega,N}q - \mu^\omega(q)\|,$$

and so by (3.8),

$$\lim_{N\to\infty} I_N^\omega = 0. \tag{3.32}$$

On the other hand,

$$I_N^\omega = \int \mu^\omega\left(\left|\mu^\omega(q|F^{-N}(\omega)\mathcal{B}) - \mu^\omega(q)\right| \mathcal{B}_\infty^\omega\right)d\mu^\omega \le \int |\mu^\omega(q|\mathcal{B}_\infty^\omega) - \mu^\omega(q)|d\mu^\omega$$

which, together with (3.32), gives (3.29), completing the proof of ergodicity of μ_φ. Observe that the ergodicity of μ_φ can be also derived directly from the uniqueness of equilibrium states and the affine character of the relativized entropy (see **[B2]**).

Next suppose that P is mixing. It suffices to show that

$$\lim_{n\to\infty}\int r(q\circ\tau^n)d\mu_\varphi=\int rd\mu_\varphi\int qd\mu_\varphi \tag{3.33}$$

for product functions $r(x,\omega)=\tilde r(x)\mathbf{1}_U(\omega)$ and $q(x,\omega)=\tilde q(x)\mathbf{1}_V(\omega)$, where $\tilde r$, $\tilde q$ are continuous functions on M and U, V are Borel subsets of Ω. Indeed, linear combinations of such functions approximate all Borel functions on $M\times\Omega$. Next,

$$\mu_\varphi(r(q\circ\tau^n))=\int\limits_{U\cap\theta^{-n}\omega V}\mu^\omega(\tilde r(\tilde q\circ F^n(\omega)))dP(\omega) \tag{3.34}$$

and

$$\begin{aligned}\mu^\omega(\tilde r(\tilde q\circ F^n(\omega)))&=\int\tilde r(\tilde q\circ F^n(\omega))d(\mathcal{L}^{\omega,n})^*\mu^{\theta^n\omega}\\ &=\int\mathcal{L}^{\omega,n}(\tilde r(\tilde q\circ F^n(\omega)))d\mu^{\theta^n\omega}=\int\tilde q(\mathcal{L}^{\omega,n}\tilde r)d\mu^{\theta^n\omega}.\end{aligned} \tag{3.35}$$

By (3.8) and (3.35),

$$\begin{aligned}|\mu^\omega(\tilde r(\tilde q\circ F^n(\omega)))-\mu^\omega(\tilde r)\mu^{\theta^n\omega}(\tilde q)|&=\left|\int\tilde q(\mathcal{L}^{\omega,n}\tilde r-\mu^\omega(\tilde r))d\mu^{\theta^n\omega}\right|\\ &\le\|\tilde q\|\|\mathcal{L}^{\omega,n}\tilde r-\mu^\omega(\tilde r)\|\to 0\ \text{ as }\ n\to\infty.\end{aligned} \tag{3.36}$$

Set $\psi(\omega)=\mathbf{1}_U(\omega)\mu^\omega(\tilde r)$ and $\eta(\omega)=\mathbf{1}_V(\omega)\mu^\omega(\tilde q)$. From the mixing property of P we have

$$\begin{aligned}\int\limits_{U\cap\theta^{-n}V}\mu^\omega(\tilde r)\mu^{\theta^n\omega}(\tilde q)dP(\omega)&=\int\limits_\Omega\psi(\eta\circ\theta^n)dP\\ &\to\int\limits_\Omega\psi dP\int\limits_\Omega\eta dP=\mu_\varphi(r)\mu_\varphi(q)\text{ as }n\to\infty,\end{aligned} \tag{3.37}$$

which, together with (3.34) and (3.36), yields (3.33), completing the proof of Theorem 3.2. □

Next we shall show that under the conditions of Theorem A the measure μ_φ constructed above is the unique equilibrium state in the variational principle (1.14). In view of (3.2)–(3.3) any finite measurable partition ζ of M with elements of sufficiently small diameter is a generator, i.e., for P-a.a. ω the intersection $\bigvee_{n=0}^{\infty}F^{-n}(\omega)\zeta$ generates $\mathcal{B}_M$. In the same way as in Proposition 3.1 of **[K2]** we derive from (d) of Theorem 3.1 that $\mu_\varphi=\{\mu^\omega,\omega\in\Omega\}$ is the unique τ-invariant measure satisfying the following condition. For any other τ-invariant measure $m=\{m^\omega,\omega\in\Omega\}$,

$$\begin{aligned}0&=H_{\mu^\omega}(\zeta^\omega|F^{-1}(\omega)\mathcal{B}_M)+\int\log g_\omega(x)d\mu^\omega(x)\\ &\ge H_{m^\omega}(\zeta^\omega|F^{-1}(\omega)\mathcal{B}_M)+\int\log g_\omega(x)dm^\omega(x)\ P\text{-a.s.},\end{aligned} \tag{3.38}$$

where $\zeta=\{\zeta^\omega,\omega\in\Omega\}$ is any family of finite measurable partitions of M such that $\bigvee_{n=0}^{\infty}F^{-n}(\omega)\zeta^\omega$ generates $\mathcal{B}_M$ and for any $z\in M$ and $\omega\in\Omega$ different points in $F^{-1}(\omega)z$ belong to different elements of ζ^ω.

Since by (a) of Theorem 3.1

$$\lambda^\omega = \lambda^\omega \nu^\omega(1) = \left((\mathcal{L}_\varphi^\omega)^* \nu^{\theta\omega}\right)(1) = \nu^{\theta\omega}(\mathcal{L}_\varphi^\omega 1), \tag{3.39}$$

we have

$$n_F(\omega)\exp(\inf_x \varphi_\omega(x)) \leq \lambda^\omega \leq n_F(\omega)\exp(\sup_x \varphi_\omega(x)), \tag{3.40}$$

implying that

$$-\delta_\varphi(\omega) \leq \log\lambda^\omega - \log n_F(\omega) - \log\varphi_\omega(x_0) \leq \delta_\varphi(\omega), \tag{3.41}$$

where δ_φ is given in (1.4). In the same way as in Theorem 3.2 of **[K2]** we conclude that

$$\int (\log\lambda^\omega - \log n_F(\omega) - \log g_\omega(x_0))dP(\omega) = \tilde{Q}_P^{(r)}(\varphi), \tag{3.42}$$

which is finite in view of (1.4) and (3.41). Next, set $\Omega_n = \{\omega \in \Omega : n-1 < \|DF(\omega)\| \leq n\}$, $n = 1, 2, \ldots$. It is easy to see that one can choose a sequence of finite measurable partitions $\zeta_n, n = 1, 2, \ldots$ of M with elements of small diameter such that $\bigvee_{\ell=0}^{\infty} F^{-\ell}(\omega)\zeta_n$ generates $\mathcal{B}_M$ for each n and P-a.a. ω and that for every $\omega \in \Omega_n$ and $z \in M$ different points in $F^{-1}(\omega)z$ belong to different elements of ζ_n. Moreover, one can choose the partitions ζ_n in such a way that the number of elements in ζ_n does not exceed Cn with some $C > 0$ independent of n. The partition $\zeta = \{\zeta_n \times \Omega_n, n = 1, 2, \ldots\}$ of $M \times \Omega$ is a $\mathcal{B}_\Omega$-generator (see **[K1]**), and so

$$h_m^{(r)}(\tau) = \int H_{m^\omega}(\zeta^\omega | F^{-1}(\omega)\mathcal{B}_M)dP(\omega), \tag{3.43}$$

where $\zeta^\omega = \zeta_n$ for $\omega \in \Omega$ and $m \in \mathcal{P}_P(M \times \Omega)$ is τ-invariant.

Let $\varphi \equiv 0$. Then $\mathcal{L}_\varphi^\omega 1 = n_F(\omega)1$, and so by the uniqueness, $h_\omega \equiv 1$, $\lambda^\omega = n_F(\omega)$, and $g_\omega = 1/n_F(\omega)$. Using (3.38) in this situation we derive that

$$\eta_m(\omega) = H_{m^\omega}(\zeta^\omega | F^{-1}(\omega)\mathcal{B}_M) - \log n_F(\omega) \leq 0 \tag{3.44}$$

for any τ-invariant $m \in \mathcal{P}_P(M \times \Omega)$ and ζ^ω as above. Next, for μ and m as above set

$$\gamma_\mu(\omega) = \int \log h_\omega(x)d\mu^\omega(x), \gamma_m(\omega) = \int \log h_\omega(x)dm^\omega(x)$$

and $\Psi_N = \{\omega : |\gamma_\mu(\omega)|, |\gamma_m(\omega)| \leq N\}$. Since we do not know the integrability of γ_μ, γ_m, we proceed as follows. Note that

$$\int \log h_{\theta^k\omega}(F^k(\omega)x)d\mu^\omega(x) = \int \log h_{\theta^k\omega}(x)d\mu^{\theta^k\omega}(x) = \gamma_\mu(\theta^k\omega), \tag{3.45}$$

since $F(\omega)\mu^\omega = \mu^{\theta\omega}$, and the same holds with μ replaced by m. Take a sequence $k_i = k_i(\omega) \to \infty$ such that $\theta^{k_i}\omega \in \Psi_N$. Considering (3.38) for $\theta^\ell\omega, \ell = 0, 1, \ldots$ instead of ω, summing over $\ell = 0, \ldots, k_i - 1$, dividing by k_i, and letting $i \to \infty$, we derive from (1.4), (3.41)–(3.45), the definition of g_ω, and the ergodic theorem that (1.14) holds and μ_φ is the only τ-invariant measure in $\mathcal{P}_P(M \times \Omega)$ satisfying (1.4). This completes the proof of Theorem A. □

3.3. Remark. Essentially the same proof leads to the following generalization of Theorem A. Let M be a compact Hausdorff metric space and $\mathcal{F}$ the space of continuous endomorphisms f of M such that $fM = M$ and each f^{-1} has $n(f) < \infty$ branches $f_1^{-1}, \dots, f_{n(f)}^{-1}$ so with f restricted to each f_i^{-1}, M is a homeomorphism onto M. Assume also that for any $f \in \mathcal{F}$,

$$K_{f^{-1}} = \sup\left\{\frac{\operatorname{dist}(x,y)}{\operatorname{dist}(fx,fy)} : x,y \in f_i^{-1}M, x \neq y \right.$$

$$\left. \text{for some } i = 1,\dots,k(f)\right\} < \infty; \tag{3.46}$$

K_f is the Lipshitz constant of branches of f^{-1}. Set $K(\omega) = K_{F^{-1}(\omega)}$, where $F(\omega)$: $\Omega \to \mathcal{F}, F(\omega) = \omega_0 \in \mathcal{F}$ in the same way as in the Introduction. Instead of (1.6) we assume now that

$$\log K \in L^1(\Omega, P) \text{ and } \alpha = E_P(\log K|\mathcal{J}) < 0, \tag{3.47}$$

where $E_P(\cdot|\mathcal{J})$ is the same as in (1.6). In this setup we can obtain the result similar to Theorem A by replacing in all arguments $\|DF^{-n}(\omega)\|$ with $K_n(\omega) = \prod_{j=0}^{n-1} K(\theta^j\omega)$. To extend our method to random subshifts of finite type considered in **[BG2]** we must now generalize the above arguments to random maps of compact bundles. Namely, we should have not just one space M, but a measurable family of compact metric spaces $X_\omega, \omega \in \Omega$, imbedded into one compact space X such that $F(\omega)X_\omega = X_{\theta\omega}$. In the case of random subshifts of finite type $X_\omega = \{x = (x_0, x_1, \cdots) : x_i \in (1, \dots, \ell(\theta^i\omega))$ and $a_{x_i x_{i+1}}(\theta^i\omega) = 1$ for all $i = 0, 1, \dots\}$, where $A(\omega) = (a_{ij}(\omega)), \omega \in \Omega$, is a measurable family of $\ell(\omega) \times \ell(\theta\omega)$-matrices with 0 and 1 entries. The fibers X_ω are imbedded into $\mathcal{Z} = \bar{\mathcal{F}}_+^{\mathcal{F}_+} = \{x = (x_0, x_1, \dots) : x_i \in \bar{\mathcal{F}}_+\}$ where $\bar{\mathcal{F}}_+$ is the one point compactification of $\mathcal{F}_+ = \{1, 2, \dots\}$. For $x, y \in Z$ set $d(x,y) = 2^{-N_{xy}}$, where $N_{xy} = \min\{i \geq 0 : x_i \neq y_i\}$. In this metric $F^{-1}(\omega)$ is uniformly contracting with the coefficient $\frac{1}{2}$. However, $\mathcal{Z}$ is not compact with respect to this metric which complicates the situation. It turns out that a more suitable metric here is the metric

$$\rho(x,y) = \sum_{n=0}^{\infty} 2^{-n}\left|\frac{1}{x_i} - \frac{1}{y_i}\right|$$

considered in **[W]** and **[BG]**, which makes $\mathcal{Z}$ compact. Using the properties of metric ρ (see **[W]** and **[BG]**) one can extend the thermodynamic formalism arguments to this situation as well. Note that the uniform aperiodicity condition on $A(\omega)$ from **[BG]** saying that $A(\omega) \cdots A(\theta^{N-1}\omega)$ is a positive matrix for some N independent of ω, can be relaxed assuming that $N = N(\omega)$ is a random variable and considering the sequence $k_i = k_i(\omega) \to \infty$ such that $\theta^{k_i}\omega \in \{\omega : N(\omega) = \ell\}$ for a fixed ℓ satisfying $P\{N = \ell\} > 0$. All this can be done using the random RPF-operator in the same way as above. We shall treat this setup in the next section by statistical mechanics arguments.

3.4. Remark. Under more restrictive conditions one can derive the thermodynamic formalism results for random expanding transformations via Markov partitions $\zeta^\omega = \{C_1^\omega, \dots, C_{\kappa(\omega)}^\omega\}$ considered in **[BG1]**. These partitions consist of closed sets, and by definition, satisfy the conditions

$$\bigcup_i C_i^\omega = M, \qquad \operatorname{int} C_i^\omega \neq \emptyset,$$
$$\text{if } F(\omega) \operatorname{int} C_i^\omega \cap \operatorname{int} C_j^{\theta\omega} \neq \emptyset, \text{ then } F(\omega)\bar{C}_i^\omega \supset \bar{C}_j^{\theta\omega}. \tag{3.48}$$

Such partitions are known to exist if all $F(\omega)$ belong to a compact set of expanding maps (see **[BG1]**), and the construction can be extended to the noncompact case as well. In the one-dimensional situation one can take C_i^ω to be different connected branches of $F^{-1}(\omega)$. This enables one to transfer the Gibbs measures constructed on the symbolic spaces X_ω from Remark 3.3 to the space M on which random transformations act. In our case the spaces X_ω will be characterised by matrices $A(\omega) = (a_{ij}(\omega))$ such that

$$a_{ij}(\omega) = \begin{cases} 1, & \text{if } F(\omega)C_i^\omega \cap C_j^{\theta\omega} \neq \emptyset, \\ 0, & \text{otherwise.} \end{cases} \tag{3.49}$$

A key ingredient here is the proof that the boundaries of Markov partitions have zero measure with respect to any Gibbs measure constructed by a family of Hölder continuous functions. The main nesessary property of such measures $\mu = \{\mu^\omega\}$ on the symbolic spaces X_ω is the fact that they are invariant ergodic measures of the skew product transformation τ and that they are positive on open sets. Let ϕ^ω be the natural map from X_ω to M given by $\phi^\omega(x) = z$, $x = (x_0, x_1, \dots)$, $x_i \in \{1, \dots, l(\theta^i\xi)\}$ if $F^i(\omega)z \in C_{x_i}^{\theta^i\omega}$, i.e., $z \in \bigcap_{i=0}^\infty (F^i(\omega))^{-1} C_{x_i}^{\theta^i\omega}$. By the definition of a Markov partition, $F(\omega)\partial\zeta^\omega \subset \partial\zeta^{\theta\omega}$, where ∂ denotes the boundary. Denote

$$\partial(\omega) = \bigcup_{n\geq 0} (F^n(\omega))^{-1}\partial\zeta^{\theta^n\omega} = \lim_{n\to\infty} (F^n(\omega))^{-1}\partial\zeta^{\theta^n\omega}.$$

Since $F(\omega)\partial(\omega) = \partial(\theta\omega)$, we have $\sigma(\phi^\omega)^{-1}\partial(\omega) = (\phi^{\theta\omega})^{-1}\partial(\theta\omega)$, where $\sigma : X_\omega \to X_{\theta\omega}$ is the shift on the symbolic space. By the ergodicity of μ it follows that either $\mu^\omega((\phi^\omega)^{-1}\partial(\omega)) = 0$ with probability one or $\mu^\omega((\phi^\omega)^{-1}\partial(\omega)) = 1$ with probability one. But

$$\begin{aligned} \mu^\omega\left((\phi^\omega)^{-1}\partial(\omega)\right) &= \lim_{n\to\infty} \mu^\omega\left((\phi^\omega)^{-1}F^n(\omega)^{-1}\partial\zeta^{\theta^n\omega}\right) \\ &= \lim_{n\to\infty} \mu^\omega\left((F(\omega)\phi^\omega)^{-1}F^{n-1}(\theta\omega)^{-1}\partial\zeta^{\theta^n\omega}\right) \\ &= \lim_{n\to\infty} \mu^\omega\left(\sigma^{-1}(\phi^{\theta\omega})^{-1}(F^{n-1}(\theta\omega))^{-1}\partial\zeta^{\theta^n\omega}\right) \\ &= \lim_{n\to\infty} \mu^{\theta\omega}\left((\phi^{\theta\omega})^{-1}(F^{n-1}(\theta\omega))^{-1}\partial\zeta^{\theta^n\omega}\right) \\ &= \lim_{n\to\infty} \mu^{\theta^n\omega}\left((\phi^{\theta^n\omega})^{-1}\partial\zeta(\theta^n\omega)\right). \end{aligned} \tag{3.50}$$

Observe that

$$\int \mu^{\theta^n\omega}\left((\phi^{\theta^n\omega})^{-1}\partial\zeta^{\theta^n\omega}\right)dP(\omega) = \int \mu^\omega\left((\phi^\omega)^{-1}\partial\zeta^\omega\right)dP(\omega),$$

and so

$$\int \mu^\omega\left((\phi^\omega)^{-1}\partial(\omega)\right)dP(\omega) = \int \mu^\omega\left((\phi^\omega)^{-1}\partial\zeta^\omega\right)dP(\omega).$$

If the left-hand side of this formula equals one, then the right-hand side equals one too, and so $\mu^\omega\Big((\phi^\omega)^{-1}\partial\zeta^\omega\Big) = 1$. But $(\phi^\omega)^{-1}\partial\zeta^\omega$ is a closed proper subset of X_ω and μ^ω is positive on open sets, which is a contradiction.

3.5. Remark. One can further generalize the class of functions for which the construction of Gibbs measures holds. Namely, similarly to **[W]** we can assume Hölder continuity only outside a nowhere dense set, so that, for instance, the functions should be Hölder continuous only in the interior of elements of Markov partitions.

3.6. Remark. Fast decay of correlations type property (3.36) (see also (4.31) below) leads also to a "relativized" version of the Central Limit Theorem saying that for P-a.a. ω,

$$\lim_{n\to\infty}\mu^\omega\left\{x : \frac{1}{n^{1/2}\sigma}\sum_{i=0}^{n-1}\left(\varphi\circ\tau^i(x,\omega) - \int\varphi_{\theta^i\omega}d\mu^{\theta^i\omega}\right) \le a\right\} = (2\pi)^{-\frac{1}{2}}\int_{-\infty}^{a} e^{-\xi^2/2}d\xi,$$

where

$$\sigma^2 = \lim_{n\to\infty}\frac{1}{n}\int\left(\sum_{i=0}^{n-1}(\varphi\circ\tau^i(x,\omega) - \int\varphi_{\theta^i\omega}d\mu^{\theta^i\omega})\right)^2 d\mu^\omega(x)dP(\omega).$$

For Markov measures of finite type subshifts this result was derived essentially in **[C]**. It can be extended to general Gibbs measures in a similar way, since (3.36) enables one to make approximations in the distributional sense by sums of independent not identically distributed random variables and a Lindenberg type theorem yields the result.

4. φ-Gibbs states

We start with the φ-Gibbs measures defined by (1.26). Existence of φ-Gibbs measures μ^ω follows from the standard compactness arguments. Taking any limit point of a sequence of discrete measures $P_n^\omega(\cdot|y_n)$, one obviously get a φ-Gibbs measure. Uniqueness easily follows from the result below.

4.1. THEOREM. *There exists a set $\Omega' \subset \Omega$, $P(\Omega') = 1$ such that for all $\omega \in \Omega'$ and all $f \in C(M)$,*

$$\sup_{y_1,y_2}\left|\sum_{x\in F^{-n}(\omega)y_1} f(x)P_n^\omega(x|y_1) - \sum_{x\in F^{-n}(\omega)y_2} f(x)P_n^\omega(x|y_2)\right| \to 0 \tag{4.1}$$

as $n \to \infty$.

PROOF. In some sense this lemma has been proved above (see Theorem 3.1). However, we want to emphasize the probabilistic point of view. For finite n there is no problem to pass to the symbolic coding for $x \in F^{-n}(\omega)y$. One can naturally label points x by a sequence $(\sigma_1, \sigma_2, \ldots, \sigma_n)$, where $1 \le \sigma_i \le n_F(\theta^i\omega)$. It is just a natural labeling for different branches of $F^{-n}(\omega)$ and one can construct it in a nonunique way. The simple way to construct labeling is to fix some $z \in M$ and to connect any point $y \in M$ with z by a, say, geodesic curve $\gamma(y, z)$.

Next we consider the sequence of sets

$$F^{-1}(\theta^{n-1}\omega)z, F^{-2}(\theta^{n-2}\omega)z, \ldots, F^{-n}(\omega)z$$

and label inductively all points of these sets by sequences of words

$$\{(\sigma_{n-1})\}, \{(\sigma_{n-2}, \sigma_{n-1})\}, \dots, \{(\sigma_0, \dots, \sigma_{n-1})\}$$

in the following way. First, we label points from each $F^{-1}(\theta^{n-k}\omega)z$, $k = 1, \dots, n$, arbitrarily by different integers from 1 to $n_F(\theta^{n-k}\omega)$. Suppose that all points in each $F^{-j}(\theta^{n-j}\omega)z$, $j = 1, \dots, k$, are already labeled. Let $v \in F^{-k}(\theta^{n-k}\omega)z$ be labeled by a word $(\sigma_{n-k}, \dots, \sigma_{n-1})$. If $w \in F^{-1}(\theta^{n-k-1}\omega)v$, then we label w by the word $(\sigma_{n-k-1}, \sigma_{n-k}, \dots, \sigma_{n-1})$ provided $F^{-1}(\theta^{n-k-1}\omega)\gamma(v, z)$ connects w with a point from $F^{-1}(\theta^{n-k-1}\omega)z$ whose label is σ_{n-k-1}. Now for an arbitrary point y we connect it with z by $\gamma(y, z)$. Then every preimage of y connected by preimages of $\gamma(y, z)$ with the corresponding preimage of z inherits its coding.

Denote by $x(\sigma_0, \dots, \sigma_{n-1}|y)$ points in $F^{-n}(\omega)y$ corresponding to $(\sigma_0, \dots, \sigma_{n-1})$. It is clear that points with the same sequence $(\sigma_0, \dots, \sigma_{n-1})$ and different y_1, y_2 are close to each other. Indeed, such points are connected by the preimage of the curve $(\gamma(y_1, z) \cup \gamma(y_2, z))$. Thus it follows from (3.2) that for all $y_1, y_2 \in M$,

$$\text{dist}(x(\sigma_0, \dots, \sigma_{n-1}|y_1), x(\sigma_0, \dots, \sigma_{n-1}|y_2)) \le 2LC(\omega)e^{\alpha n/2}, \tag{4.2}$$

where $L = \max l(\gamma(y, z))$ and $l(\gamma)$ is the Riemannian length. Since any $x \in M$ is a preimage of $y = F^{-n}(\omega)$, it follows from (4.2) that for any $\varepsilon > 0$ there exists $n(\varepsilon, \omega)$ such that for all $n > n(\varepsilon, \omega)$ and all $y \in M$ the sets $\{x(\sigma_0, \dots, \sigma_{n-1}|y)\}$ form an ε-net. Since we need to check (4.1) for continuous functions, it is enough to show that for fixed k and for P-a.a. ω,

$$\sup_{y_1, y_2}\Bigg| \sum_{\sigma_{k+1}, \dots, \sigma_{n-1}} P_n^\omega(x(\sigma_0, \dots, \sigma_k, \sigma_{k+1}, \dots, \sigma_{n-1}|y_1)) - \sum_{\sigma_{k+1}, \dots, \sigma_{n-1}} P_n^\omega(x(\sigma_0, \dots, \sigma_k, \sigma_{k+1}, \dots \sigma_{n-1}|y_2))\Bigg| \to 0 \text{ as } n \to \infty. \tag{4.3}$$

This can be shown using the standard methods of one-dimensional statistical mechanics. Denote by $P_1^\omega(\sigma_0, \dots, \sigma_{n-1})$, $P_2^\omega(\sigma_0, \dots, \sigma_{n-1})$ the probability distributions on the words $(\sigma_0, \dots, \sigma_{n-1})$ generated by $P_n^\omega(x|y_1), P_n^\omega(x|y_2)$. For two probability distributions on the discrete set W denote by $\rho(P_1, P_2)$ the variational distance between P_1 and P_2:

$$\rho(P_1, P_2) = \frac{1}{2} \sum_{w \in W} |P_1(w) - P_2(w)|. \tag{4.4}$$

4.2. Lemma. *Suppose that n is chosen in such a way that*

$$A_n(\omega) = \sum_{l=0}^{n} C_\varphi(\theta^l \omega) \|DF^{-(n-l)}(\theta^l \omega)\|^\beta \le R. \tag{4.5}$$

Then

$$\rho(P_1^\omega, P_2^\omega) \le \gamma_R = 1 - e^{-2R(2L)^\beta}. \tag{4.6}$$

PROOF. We have

$$(4.7)\qquad \begin{aligned} P_1^{\omega}(\sigma_0,\dots,\sigma_{n-1}) &= \frac{1}{Z_n^{\omega}(y_1)}\exp\left\{\sum_{i=0}^{n-1}\varphi_{\theta^i\omega}(F^i(\omega)x(\sigma_o,\dots,\sigma_{n-1}|y_1))\right\},\\ P_2^{\omega}(\sigma_0,\dots,\sigma_{n-1}) &= \frac{1}{Z_n^{\omega}(y_2)}\exp\left\{\sum_{i=0}^{n-1}\varphi_{\theta^i\omega}(F^i(\omega)x(\sigma_o,\dots,\sigma_{n-1}|y_2))\right\}. \end{aligned}$$

Since

$$\operatorname{dist}\left(F^i(\omega)x(\sigma_0,\dots,\sigma_{n-1}|y_1), F^i(\omega)x(\sigma_0,\dots,\sigma_{n-1}|y_2)\right) \le 2L\|DF^{-(n-i)}(\theta^i\omega)\|,$$

we get

$$(4.8)\qquad e^{-R(2L)^{\beta}} \le \frac{\exp(\sum_{i=0}^{n-1}\varphi_{\theta^i\omega}(F^i(\omega)x((\sigma_0,\dots,\sigma_{n-1}|y_1))))}{\exp(\sum_{i=0}^{n-1}\varphi_{\theta^i\omega}(F^i(\omega)x((\sigma_0,\dots,\sigma_{n-1}|y_2))))} \le e^{R(2L)^{\beta}},$$

$$e^{-R(2L)^{\beta}} \le \frac{Z_n^{\omega}(y_1)}{Z_n^{\omega}(y_2)} \le e^{R(2L)^{\beta}}.$$

Hence,

$$(4.9)\qquad e^{-2R(2L)^{\beta}} \le \frac{P_1^{\omega}(\sigma_0,\dots,\sigma_{n-1})}{P_2^{\omega}(\sigma_0,\dots,\sigma_{n-1})} \le e^{2R(2L)^{\beta}}.$$

Denote

$$\begin{aligned} P_{\max}^{\omega}(\sigma_0,\dots,\sigma_{n-1}) &= \max(P_1^{\omega}(\sigma_0,\dots,\sigma_{n-1}), P_2^{\omega}(\sigma_0,\dots,\sigma_{n-1})),\\ P_{\min}^{\omega}(\sigma_0,\dots,\sigma_{n-1}) &= \min(P_1^{\omega}(\sigma_0,\dots,\sigma_{n-1}), P_2^{\omega}(\sigma_0,\dots,\sigma_{n-1})). \end{aligned}$$

Then,

$$\begin{aligned} \rho(P_1^{\omega}, P_2^{\omega}) &= \frac{1}{2}\sum_{\sigma_0,\dots,\sigma_{n-1}}(P_{\max}^{\omega}(\sigma_0,\dots,\sigma_{n-1}) - P_{\min}^{\omega}(\sigma_0,\dots,\sigma_{n-1}))\\ &= \frac{1}{2}\sum_{\sigma_0,\dots,\sigma_{n-1}} P_{\max}^{\omega}(\sigma_0,\dots,\sigma_{n-1})\left(1 - \frac{P_{\min}^{\omega}(\sigma_0,\dots,\sigma_{n-1})}{P_{\max}^{\omega}(\sigma_0,\dots,\sigma_{n-1})}\right)\\ &\le \frac{1}{2}(1-e^{-2R(2L)^{\beta}})\sum_{\sigma_0,\dots,\sigma_{n-1}} P_{\max}^{\omega}(\sigma_0,\dots,\sigma_{n-1}) \le 1 - e^{-2R(2L)^{\beta}}. \quad \square \end{aligned}$$

Denote by $\sigma(V)$ the "spin" configuration in $V \subset \mathbb{Z}_+$. Consider conditional probabilities $P_1^{\omega}(\sigma[0,n_1-1]|\sigma_1[n_1,n-1])$, $P_2^{\omega}(\sigma[0,n_1-1]|\sigma_2[n_1,n-1])$ and suppose that the conditions $\sigma_1([n_1,n-1]), \sigma_2([n_1,n-1])$ coincide in the first s positions: $\sigma_1(i) = \sigma_2(i), n_1+1 \le i \le n_1+s-1$.

4.3. LEMMA. *Suppose that*

$$(4.10)\qquad A_{n_1}(\omega) = \sum_{l=0}^{n_1-1} C_{\varphi}(\theta^l\omega)\|DF^{-(n-l)}(\theta^l\omega)\|^{\beta} \le R$$

and

$$(4.11)\qquad \frac{2LC(\theta^{n_1}\omega)e^{s\alpha/2}}{1-e^{\alpha/2}} < \delta^{1/\beta}.$$

Then

(4.12)
$$\rho(P_1^\omega(\sigma[0,n_1-1]|\sigma_1[n_1,n-1]),P_2^\omega(\sigma[0,n_1-1]|\sigma_2[n_1,n-1]))\le\gamma_\delta=1-e^{-2R\delta}.$$

PROOF. The proof of Lemma 4.3 is similar to the proof of Lemma 4.2. Indeed, both $P_1^\omega(\sigma[0,n_1-1]|\sigma_1[n_1,n-1])$ and $P_2^\omega(\sigma[0,n_1-1]|\sigma_2[n_1,n-1])$ are conditional probabilities on the set of preimages $F^{-n}(\omega)$ of the points $z_1(\sigma_1[n_1,n-1]|y_1)$, $z_2(\sigma_2[n_1,n-1]|y_2)$. But now the curve that connects these points and whose preimages connect their preimages has length less than

$$2L\sum_{i=s}^{n-n_1}\|DF^{-i}(\theta^{n_1}\omega)\|\le 2L\sum_{i=s}^{n-n_1}C(\theta^{n_1}\omega)e^{\alpha i/2}\le 2LC(\theta^{n_1}\omega)\frac{e^{\alpha s/2}}{1-e^{\alpha/2}}\le\delta^{1/\beta}.$$

Here we used the estimate (3.2). Repeating (4.7)–(4.9) we get (4.12). □

4.4. COROLLARY. *Suppose* $s=0$ *and*

$$\frac{2LC(\theta^{n_1}\omega)}{1-e^{\alpha/2}}<L_1^{1/\beta}. \tag{4.13}$$

Then

$$\rho(P_1^\omega(\cdot|\sigma_1),P_2^\omega(\cdot|\sigma_2))\le\gamma=1-e^{2RL_1}.$$

Fix an arbitrarily small δ and choose s in such a way that the inequality (4.11) holds with positive probability. It follows from Proposition 2.1 that with P–probability 1, one can find an infinite sequence $n_1(j)\to\infty$, $j\ge 1$, such that $n_1(1)>s$, $n_1(j+1)-n_1(j)>s$, and (4.10), (4.11), (4.13) hold for all $n_1(j)$. Denote by $v(j)$ a spin configuration $\sigma([n_1(j),n_1(j+1)-1])$ in the volume $[n_1(j),\ n_1(j+1)-1]$, $j\ge 1$, and let $v(0)\equiv\sigma([0,n_1(1)-1]$. Consider two probability distributions $P_1(v(0),\dots,v(J-1))$, $P_2(v(0),\dots,v(J-1))$ corresponding to the preimages of the points y_1,y_2. Below, ω is fixed and we omit it in our notations. Denote by $P_1^{(j)},P_2^{(j)}$ the projections of P_1,P_2 to the space of spin variables $(v(0),\dots,v(j))$. It follows easily from Lemma 4.3 and Corollary 4.4 that for all j,

$$\rho(P_1^{(j)}(v(0),\dots,v(j-1)|v_1(j)),P_2^{(j)}(v(0),\dots,v(j-1)|v_2(j)))\le\gamma$$

and

$$\rho(P_1^{(j)}(v(0),\dots,v(j-1)|v(j)),P_2^{(j)}(v(0),\dots,v(j-1)|v(j)))\le\gamma_\delta.$$

Denote $\rho_j=\rho(P_1^{(j)},P_2^{(j)})$. The following statement was proven by Dobrushin **[D]**.

4.5. LEMMA (Dobrushin). *For all j, we have*

$$\rho_j\le\rho_{j+1}\gamma+(1-\rho_{j+1})\gamma_\delta. \tag{4.14}$$

Iterating the inequality (4.14) we get

$$\rho_j\le\gamma_\delta\sum_{i=0}^{J-j-2}(\gamma-\gamma_\delta)^i+(\gamma-\gamma_\delta)^{J-j-1}\rho_{J-1} \tag{4.15}$$

and

$$\rho_0\le\frac{\gamma_\delta}{1-\gamma+\gamma_\delta}+(\gamma-\gamma_\delta)^{J-1}. \tag{4.16}$$

This obviously implies (4.3) since $\gamma_\delta \to 0$ as $\delta \to 0$. Theorem 4.1 is proven. □

It follows from the uniqueness of φ-Gibbs measures that with P–probability 1, $P_n^\omega(\cdot|y_n) \to \nu^\omega$ weakly as $n \to \infty$ for any sequence $y_n \in M$.

Consider now the φ-Gibbs state $\mu(dx, d\omega) = \mu^\omega(dx)P(d\omega)$. We construct the measure $\mu^\omega(dx)$ as a limit of $F^m(\theta^{-m}\omega)\nu^{\theta^{-m}\omega}$ as $m \to \infty$. Consider first discrete measures $P_{m+n}^{\theta^{-m}\omega}(\cdot|y)$ and their shifts $F^m(\theta^{-m}\omega)P_{m+n}^{\theta^{-m}\omega}(\cdot|y)$.

We consider two measures corresponding to $m = M_1, M_2$. Suppose that $M_1 > M_2$. Fix $0 < m_1 < M_2 - k$ and consider the measures $F^{M_1-m_1}(\theta^{-M_1}\omega)$ $P_{M_1+n}^{\theta^{-M_1}\omega}(\cdot|y)$ and $F^{M_2-m_1}(\theta^{-M_2}\omega)P_{M_2+n}^{\theta^{-M_2}\omega}(\cdot|y)$. Construct the symbolic coding starting from $(-m_1)$. It means that each point of $F^{-(M_1+n)}(\theta^{-M_1}\omega)y$ can be represented as $x(\sigma_{-M_1}, \sigma_{-M_1+1}, \ldots, \sigma_{-m_1-1}|u)$, where $u \in F^{-(m_1+n)}(\theta^{-m_1}\omega)y$.

Denote by $P_1(u|\sigma_{-m_1-1}, \ldots, \sigma_{-M_1}), P_2(u|\sigma_{-m_1-1}, \ldots, \sigma_{-M_2})$ the conditional distributions on u constructed from $P_{M_1+n}^{\theta^{-M_1}\omega}(\cdot|y)$ and $P_{M_2+n}^{\theta^{-M_2}\omega}(\cdot|y)$.

Set

$$\sigma' = (\sigma'_{-m_1-1}, \ldots, \sigma'_{-M_1}), \qquad \sigma'' = (\sigma''_{-m_1-1}, \ldots, \sigma''_{-M_2}),$$
$$\sigma'(k) = (\sigma_{-m_1-1}, \ldots, \sigma_{-m_1-k}, \sigma'_{-m_1-k-1}, \ldots, \sigma'_{-M_1}),$$
$$\sigma''(k) = (\sigma_{-m_1-1}, \ldots, \sigma_{-m_1-k}, \sigma''_{-m_1-k-1}, \ldots, \sigma''_{-M_2}).$$

4.6. LEMMA. *Suppose that*

$$\sum_{l=m_1+1}^{\infty} C_\varphi(\theta^{-l}\omega)\|DF^{-(l-m_1)}(\theta^{-l}\omega)\|^\beta \le R, \tag{4.17}$$

$$\sum_{l=m_1+k+1}^{\infty} C_\varphi(\theta^{-l}\omega)\|DF^{-(l-m_1-k)}(\theta^{-l}\omega)\|^\beta \le R, \tag{4.18}$$

and

$$\|DF^{-k}\theta^{(-m_1-k+1)}\omega\|^\beta \le \delta. \tag{4.19}$$

Then,

$$\rho(P_1(u|\sigma'), P_2(u|\sigma'')) \le \overline{\gamma}_R = 1 - e^{4RL^\beta}, \tag{4.20}$$

$$\rho(P_1(u|\sigma'_k), P_2(u|\sigma''_k)) \le \overline{\gamma}_\delta = 1 - e^{4RL^\beta\delta}. \tag{4.21}$$

PROOF. Take $z \in M$ as in the construction of the symbolic coding. Then

$$\begin{aligned} P_i(u|\sigma^{\prime,\prime\prime}) = \frac{1}{Z_i(\sigma^{\prime,\prime\prime})} \exp\Bigg(&\sum_{l=m_1+1}^{M_i} [\varphi_{\theta^{-l}\omega}(u^{\prime,\prime\prime}_{-l}) - \varphi_{\theta^{-l}\omega}(z^{\prime,\prime\prime}_{-l})] \\ &+ \sum_{s=-m_1}^{n} \varphi_{\theta^s\omega}(F^{s+m_1}(\theta^{-m_1}\omega)u)\Bigg), \quad i = 1, 2, \end{aligned} \tag{4.22}$$

where $u'_{-l}, u''_{-l}, z'_{-l}, z''_{-l}$ are preimages of u, z corresponding to the sequences σ', σ''. Since

$$\sum_{l=m_1+1}^{M_1} |\varphi_{\theta^{-l}\omega}(u'_{-l}) - \varphi_{\theta^{-l}\omega}(z'_{-l})| \le RL^\beta,$$

$$\sum_{l=m_1+1}^{M_2} |\varphi_{\theta^{-l}\omega}(u''_{-l}) - \varphi_{\theta^{-l}\omega}(z''_{-l})| \le RL^\beta,$$

we get

$$e^{-4RL^\beta} \le \frac{P_1(u|\sigma')}{P_2(u|\sigma'')} \le e^{4RL^\beta}, \tag{4.23}$$

which proves (4.20). In the second case σ'_k and σ''_k coincide at the first k positions. Thus the first k terms in (4.22) are the same for P_1 and P_2 and we get from (4.18), (4.19) that

$$e^{-4RL^\beta\delta} \le \frac{P_1(u|\sigma'_k)}{P_2(u|\sigma''_k)} \le e^{4RL^\beta\delta} \tag{4.24}$$

which proves the second inequality. □

Now we again apply Proposition 2.1 and get with probability 1 an infinite sequence $-\overline{m}_1(j)$ such that (4.17), (4.18) hold for $m_1 = \overline{m}_1(j), k = \overline{m}_1(j+1) - \overline{m}_1(j)$. Denote by $m_1(j)$ a subsequence of $-\overline{m}_1(j)$ such that for $m_1 = m_1(j), k = m_1(j+1) - m_1(j)$ (4.19) holds. Suppose that $M_1, M_2 > m_1(J)$. Set

$$P_i(j) = F^{M_i - m_1(j)}(\theta^{-M_i}\omega)P^{\theta^{-M_i}\omega}_{M_i+n}(\cdot|y), \quad i = 1, 2,$$
$$\rho_j = \rho(P_1(j), P_2(j)).$$

Using Lemmas 4.5, 4.6 we obtain the following estimates:

$$\rho_j \le \overline{\gamma}_R \rho_{j+1} + \overline{\gamma}_\delta(1 - \rho_{j+1}), \quad j \ge 1. \tag{4.25}$$

Since $\lim_{\delta\to 0}\overline{\gamma}_\delta = 0$ and δ are arbitrarily small, iterating (4.25) we get the following statement.

4.7. PROPOSITION. a) *For P-a.a.* $\omega \in \Omega$ *uniformly on* n *and* y *there exists the limit*

$$\lim_{m\to\infty} F^m(\theta^{-m}\omega)P^{\theta^{-m}\omega}_{m+n}(\cdot|y) = \overline{P}_n{}^{\omega}(\cdot|y). \tag{4.26}$$

Moreover, uniformly with respect to n *and* y,

$$\rho\left(F^m(\theta^{-m}\omega)P^{\theta^{-m}\omega}_{m+n}(\cdot|y), \overline{P}^{\omega}_n(\cdot|y)\right) \to 0 \text{ as } m \to \infty. \tag{4.27}$$

b) *The measure* $\overline{P}_n(\cdot|y)$ *is absolutely continuous with respect to* $P_n(\cdot|y)$ *and*

$$e^{-D(\omega)} \le \frac{\overline{P}^{\omega}_n(x|y)}{P^{\omega}_n(x|y)} \le e^{D(\omega)}, \tag{4.28}$$

where

$$D(\omega) = 2L^\beta \sum_{l=1}^{\infty} C(\theta^{-l}\omega)\|DF^l(\theta^{-l}\omega)\|^\beta.$$

In a similar way it is possible to prove the following lemma.

4.8. LEMMA. *For P-a.a. $\omega \in \Omega$ uniformly in n and y, we have*

$$\lim_{m\to\infty} \rho\left(F^m(\omega)P^{\omega}_{m+n}(\cdot|y), \overline{P}^{\theta^m\omega}_n(\cdot|y)\right) = 0.$$

Now we can easily prove the existence and uniqueness of φ-Gibbs state μ. Consider $F^m(\theta^{-m}\omega)P^{\theta^{-m}\omega}_{m+n}(\cdot|y_n)$. Since the limit (4.26) is uniform and the sequence $F^m(\theta^{-m}\omega)P^{\theta^{-m}\omega}(\cdot|y_n)$ tends weakly to $F^m(\theta^{-m}\omega)\nu^{\theta^{-m}\omega}$ as $n \to \infty$, we have weak convergence of $\overline{P}^{\omega}_n(\cdot|y_n)$ as $n \to \infty$. Denote $\mu^\omega = \lim_{n\to\infty} \overline{P}^{\omega}_n(\cdot|y_n)$ and $\mu = \mu^\omega P(d\omega)$. Obviously with P–probability 1 the measure μ^ω is the unique Gibbs measure corresponding to the set of conditional probabilities $\overline{P}^{\omega}_n(\cdot|y)$. It is easy to check that for P-a.a. ω,

$$F(\omega)\mu^\omega = \mu^{\theta\omega}, \tag{4.29}$$

which implies τ-invariance of μ. It follows from (4.28) that for P-a.a. ω the measures μ^ω and ν^ω are equivalent and

$$e^{-D(\omega)} \le \frac{d\mu^\omega}{d\nu^\omega} \le e^{D(\omega)}. \tag{4.30}$$

The ergodicity follows from uniqueness of μ. Indeed, since μ^ω is the unique Gibbs measure for $\overline{P}_n(\cdot|y)$, the σ–algebra at the infinity B^{ω}_{∞} is trivial for P-a.a. ω. This implies ergodicity of μ (see Theorem 3.2.). Thus the first part of Theorem C is proven.

Next, we shall prove Theorem B′ and then the variational principle. Notice that for all $\omega \in \Omega$ the Riemannian volume m is the φ-Gibbsian measure for

$$\varphi_\omega(x) = -\log|\operatorname{Jac} D_x F(\omega)|.$$

Thus part a) of Theorem B′ follows from the first part of Theorem C.

Let us prove the estimate (1.28). Set $\rho^\omega(x) = dm/d\mu^\omega$. Then

$$\begin{aligned}
&\left|\int f(x)(F^n(\omega)\chi(dx)) - \int f(y)\mu^{\theta^n\omega}(dy)\right| \\
&\quad= \left|\int f(F^n(\omega)x)h(x)\rho^\omega(x)\mu^\omega(dx) - \int f(y)\mu^{\theta^n\omega}(dy)\right| \\
&\quad= \left|\int\Big(\sum_{x\in F^{-n}(\omega)y} f(F^n(\omega)x)h(x)\rho^\omega(x)\overline{P}^{\omega}_n(x|y)\Big)\mu^{\theta^n\omega}(dy)\right. \\
&\qquad\qquad \left. - \int f(y)\mu^{\theta^n\omega}(dy)\right| \\
&\quad= \left|\int f(y)\Big(\sum_{x\in F^{-n}(\omega)y} h(x)\rho^\omega(x)\overline{P}^{\omega}_n(x|y) - 1\Big)\mu^{\theta^n\omega}(dy)\right|.
\end{aligned} \tag{4.31}$$

If $h(x)\rho^\omega(x) \in C(M)$, then $\sum_{x\in F^{-n}(\omega)y} h(x)\rho^\omega(x)\overline{P}_n(x|y) \to 1$ uniformly in y, which implies convergence to zero in (4.31). In the general case one should first approximate $h(x)\rho^\omega(x)$ by a bounded measurable function $h_N(x)$, $|h_N(x)| \le N$, and then take a sequence of bounded continuous functions that tends to $h_N(x)$ uniformly on a set of large μ–measure. The last statement of the Theorem B′ can be proven analogously.

Finally we show that the φ-Gibbs state μ satisfies the generalized variational principle (1.14). For any τ invariant measure χ define

$$\xi_n^\chi(\omega|y) = \sum_{x\in F^{-n}(\omega)y}\Big(\sum_{i=0}^{n-1}\tilde\varphi_{\theta^i\omega}(F^i(\omega)x)\Big)\chi_n^\omega(x|y),$$

$$\tilde h_n^\chi(\omega|y) = -\sum_{x\in F^{-n}(\omega)y}\chi_n^\omega(x|y)\log\chi_n^\omega(x|y) - \sum_{i=0}^{n-1}\log n_F(\theta^i\omega),$$

and

(4.32) $$\xi_n^\chi(\omega) = \int\xi_n^\chi(\omega|y)\chi^{\theta^n\omega}(dy), \quad \tilde h^\chi(\omega) = \int h_n^\chi(\omega|y)\chi^{\theta^n\omega}(dy).$$

Since

(4.33) $$\xi_{n+m}^\chi(\omega) = \xi_n^\chi(\omega) + \xi_m^\chi(\theta^n\omega), \quad \tilde h_{n+m}^\chi(\omega) = \tilde h_n^\chi(\omega) + \tilde h_m^\chi(\theta^n\omega),$$

we have for P-a.a. ω

(4.34) $$\frac{1}{n}\xi_n^\chi(\omega) \to \int\tilde\varphi d\chi, \quad \frac{1}{n}\tilde h_n^\chi(\omega) \to \tilde h_\chi^{(r)}(\tau) \text{ as } n\to\infty.$$

Define

(4.35)
$$H_n^\omega(x) = \sum_{i=0}^{n-1}\tilde\varphi_{\theta^i\omega}(F^i(\omega)x), \quad \tilde Z_n^\omega(y) = \Big(\prod_{i=0}^{n-1}n_F(\theta^i\omega)\Big)^{-1}\sum_{x\in F^{-n}(\omega)y}\exp H_n^\omega(x),$$

and

(4.36) $$\tilde Z_n(\omega) = \int\tilde Z_n^\omega(y)\nu^{\theta^n\omega}(dy), \quad \tilde q_n(\omega) = \log\tilde Z_n(\omega).$$

It is easy to see that

(4.37) $$\tilde q_{n+m}(\omega) = \tilde q_n(\omega) + \tilde q_m(\theta^n\omega).$$

If (1.4) holds, then $\tilde q_1(\omega) \in L^1(\omega, P)$ and with P–probability 1 we have

(4.38) $$\frac{1}{n}\tilde q_n(\omega) \to \tilde Q_P^{(r)}(\varphi) \text{ as } n\to\infty.$$

Choose once again a sequence $n_i \to \infty$ such that

(4.39) $$\sum_{l=0}^{n_i-1}C_\varphi(\theta^l\omega)\|DF^{-(n_i-l)}(\theta^l\omega)\|^\beta \le R.$$

Then

(4.40) $$e^{-2R(2L)^\beta} \le \frac{\tilde Z_{n_i}^\omega(y_1)}{\tilde Z_{n_i}^\omega(y_2)} \le e^{2R(2L)^\beta}$$

and for all $y \in M$,

(4.41) $$|\log\tilde Z_{n_i}^\omega(y) - \tilde q_{n_i}(\omega)| \le e^{2R(2L)^\beta}.$$

It is well known that for any measure χ^ω on M

(4.42) $$\xi_n^\chi(\omega|y) + \tilde h_n^\chi(\omega|y) \le \log\tilde Z_n^\omega(y)$$

and equality in (4.42) corresponds to $\chi^\omega = \nu^\omega$.

Integrating (4.42) over y, dividing by n, and taking limit as $n = n_i \to \infty$ we get

$$\int \tilde{\varphi} d\chi + \tilde{h}_\chi^{(r)}(\tau) \leq \tilde{Q}_P^{(r)}(\varphi). \tag{4.43}$$

Suppose now that $\chi = \mu$. It follows from (4.28) that

$$|\tilde{h}_n^\nu(\omega|y) - \tilde{h}_n^\mu(\omega|y)| \leq D(\omega). \tag{4.44}$$

Using Lemma 4.8 we have

$$\frac{1}{n}(\xi_n^\nu(\omega|y) - \xi_n^\mu(\omega|y)) \to 0 \text{ as } n \to \infty \tag{4.45}$$

uniformly in y. Since

$$\xi_{n_i}^\nu(\omega|y) + \tilde{h}_{n_i}^\nu(\omega|y) = \log \tilde{Z}_{n_i}^\omega(y),$$

we get

$$\int \tilde{\varphi} d\mu + \tilde{h}_\mu^{(r)}(\tau) = \tilde{Q}_P^{(r)}(\varphi),$$

which completes the proof of Theorem C. □

5. Random subshifts of finite type (r.s.f.t.)

In this section we shall construct Gibbs states for r.s.f.t and prove that the Gibbs states are unique maximizing measures in the relativized variational principle. Set

$$\begin{aligned} X_\omega &= \{x = (x_{-1}, x_0, x_1, \dots) : \ x_i \in \{1, \dots, l(\theta^i\omega)\}, \ -\infty < i < +\infty\}, \\ Y_\omega &= \{y = (y_0, y_1, \dots) : \ y_i \in \{1, \dots, l(\theta^i\omega)\}, \ i \geq 0\}. \end{aligned} \tag{5.1}$$

Clearly $X_\omega(A) \subset X_\omega$, $Y_\omega(A) \subset Y_\omega$. Sequences from $X_\omega(A)$, $Y_\omega(A)$ are called A-consistent. Obviously, X_ω, Y_ω are compact sets in the product topology of the discrete topologies on different components. Since $X_\omega(A)$, $Y_\omega(A)$ are closed subsets they are also compact. We consider X_ω, $X_\omega(A)$, Y_ω, $Y_\omega(A)$ as the measurable spaces with the Borel σ-algebras generated by cylinder sets. Thus we also have the natural skew-product σ-algebras defined on the skew-product spaces:

$$\mathcal{X} = \prod_{\omega\in\Omega} X_\omega, \ \mathcal{Y} = \prod_{\omega\in\Omega} Y_\omega, \ \mathcal{X}(A) = \prod_{\omega\in\Omega} X_\omega(A), \ \mathcal{Y}(A) = \prod_{\omega\in\Omega} Y_\omega(A). \tag{5.2}$$

Denote by $\sigma(\omega)$, $\sigma_+(\omega)$ the shift transformations on X_ω, Y_ω acting by the formulas

$$(\sigma(\omega)x)_i = x_{i+1}, \quad i \in \mathbb{Z}, \qquad (\sigma_+(\omega)y)_i = y_{i+1}, \quad i \in \mathbb{Z}_+, \tag{5.3}$$

and consider the skew-product transformations τ, τ_+ on $\mathcal{X}$, $\mathcal{Y}$ defined as follows:

$$\tau(x,\omega) = (\sigma(\omega)x, \theta\omega), \tau_+(y,\omega) = (\sigma_+(\omega)y, \theta\omega). \tag{5.4}$$

Obviously $\sigma(\omega)(X_\omega(A)) = X_{\theta\omega}(A)$, $\sigma_+(\omega)(Y_\omega(A)) = Y_{\theta\omega}(A)$, $\tau\mathcal{X}(A) = \mathcal{X}(A)$,
$\tau_+\mathcal{Y}(A) = \mathcal{Y}(A)$. This simply means that the property of the sequence to be A-consistent is shift-invariant.

5.1. DEFINITION. Transformations τ acting on $\mathcal{X}(A)$ and τ_+ acting on $\mathcal{Y}(A)$ are called random subshifts of finite type corresponding to $(l(\omega), A(\omega))$.

Denote by $\mathcal{P}_P(\mathcal{X}(A))$, $\mathcal{P}_P(\mathcal{Y}(A))$ the spaces of probability measures on $\mathcal{X}(A)$, $\mathcal{Y}(A)$ with the marginal distribution P on Ω. For $\mu \in (\mathcal{P}(\mathcal{X}(A))$, $\mu_+ \in (\mathcal{P}_P(\mathcal{Y}(A))$ we have

$$\mu(dx, d\omega) = \mu^\omega(dx)P(d\omega), \quad \mu_+(dx, d\omega) = \mu_+^\omega(dx)P(d\omega), \tag{5.5}$$

where disintegrations μ^ω, μ_+^ω are defined for P-a.a. ω. Obviously, μ^ω, μ_+^ω are τ, τ_+–invariant if and only if for P-a.a. ω,

$$\sigma(\omega)\mu^\omega = \mu^{\theta\omega}, \quad \sigma_+(\omega)\mu_+^\omega = \mu_+^{\theta\omega}. \tag{5.6}$$

Fix $\omega \in \Omega$ such that potentials $U_i^\omega \equiv U^{\theta^i\omega}$ are defined for all $i \in \mathbb{Z}^1$ and consider the corresponding one-dimensional system of statistical mechanics. For each point $i \in \mathbb{Z}^1$, the spin variable x_i takes values from the set $\{1, \dots, l_i(\omega) \equiv l(\theta^i\omega)\}$. We shall consider only A–consistent configurations which means that $a_{x_i x_{i+1}}(\theta^i\omega) = 1$. The set of potentials is given by $U_i^\omega(\pi(\sigma^i(\omega)x))$, where $\pi : X_\omega(A) \to Y_\omega(A)$ is the natural projection, $(\pi x)_i = x_i$, $i \in \mathbb{Z}_+$, and $\pi(\sigma^i(\omega)x) = (x_i, x_{i+1}, \dots)$. We shall interpret $U_i^\omega(x_i, x_{i+1}, \dots)$ as an energy of interaction between spin x_i and all the spins to the right of x_i.

For any subset $W \subset \mathbb{Z}^1$ denote by $X_\omega(W, A)$ the set of spin configurations in W that are A-consistent and have A-consistent continuation to the whole $X_\omega(A)$. Fix $\overline{x} \in X_\omega(\mathbb{Z}^1 \setminus V, A)$ and denote by $X_\omega(V, A, \overline{x})$ the set of configurations from $X_\omega(V, A)$ that are A-consistent with $\overline{x}$. If $\min(i : i \in V) > -\infty$, one has the lexicographical order on $X_\omega(V, A)$. We denote by $x_m(V)$ and $x_m(V, \overline{x})$ the minimal elements in $X_\omega(V, A)$ and $X_\omega(V, A, \overline{x})$. Since ω is fixed, we shall often omit it in the notations.

Consider a finite volume $V = [u, v] \subset \mathbb{Z}^1$ and fix configuration of spins outside V: $\overline{x} = \overline{x}(\mathbb{Z}^1 \setminus V) \subset X_\omega(\mathbb{Z}^1 \setminus V, A)$. Take any $x = x(V) \in X_\omega(V, A, \overline{x})$ and set $z(x, \overline{x}) = (x \cup \overline{x}) \in X_\omega(A)$. Define $H_i(x, \overline{x})$, $-\infty < i \le v$ by the formula

$$H_i(x, \overline{x}) = \begin{cases} U_i(\pi(\sigma^i z)), & u \le i \le v, \\ U_i(\pi(\sigma^i z)) - U_i(\pi(\sigma^i z_m)), & i < u, \end{cases} \tag{5.7}$$

where $z_m(\overline{x}) = x_m(V, \overline{x}) \cup \overline{x} \in X_\omega(A)$. Put

$$H_V(x, \overline{x}) = \sum_{i=-\infty}^{v} H_i(x, \overline{x}). \tag{5.8}$$

Here H_V is the energy of interactions of spins in V, including the interaction with the boundary configuration $\overline{x}$. The term $U_i(\pi(\sigma^i z_m))$ subtracted in (5.7) corresponds to self-interaction of the spins of $\overline{x}$. Define also the energy of interactions with the "empty" boundary condition by

$$H_V(x) = \sum_{i=u}^{v} U_i(x(i) \cup x'_m), \tag{5.9}$$

where $x(i) = x([i, v])$ and x'_m is the minimal configuration on $[v + 1, \infty]$ consistent with x.

5.2. LEMMA. *For P-a.a.* $\omega \in \Omega$ *the following holds: for arbitrary* $U = [u, v] \subset \mathbb{Z}^1$ *the series* (5.8) *converges uniformly in* $\overline{x}$.

PROOF. It is enough to consider $i < u$. It follows from (1.30) that

$$H_i(x, \overline{x}) \le C_U(\theta^i \omega) e^{-(u-i)\tilde{\alpha}}. \tag{5.10}$$

The condition (1.31) implies (3.12) for $C_U(\theta^i \omega)$. Thus for $i \le -n_\varepsilon(\omega)$ we have

$$|H_i(x, \overline{x})| \le e^{-\varepsilon i} e^{-(u-i)\tilde{\alpha}} = e^{-\varepsilon u - (u-i)(\tilde{\alpha} - \varepsilon)}, \tag{5.11}$$

which yields the uniform convergence of the series (5.8). □

Now we can define the family of Gibbs conditional distributions. For an arbitrary finite $V = [u, v] \subset \mathbb{Z}^1$ and the boundary condition $\overline{x} \in X_\omega(\mathbb{Z}^1 \setminus V, A)$ define

$$\begin{aligned} P_V^\omega(x(V)|\overline{x}) &= \frac{1}{Z_V^\omega(\overline{x})} \exp(H_V^\omega(x(V), \overline{x})), \quad x(V) \in X_\omega(V, A, \overline{x}), \\ Z_V^\omega(\overline{x}) &= \sum_{x(V) \in X_\omega(V, A, \overline{x})} \exp(H_V^\omega(x(V), \overline{x})). \end{aligned} \tag{5.12}$$

Fix ω such that the family (5.12) is defined.

5.3. DEFINITION. A probability distribution μ^ω on $X_\omega(A)$ is called an U-Gibbs measure if for all finite $V = [u, v]$ and for μ^ω-a.a. $\overline{x} \in X_\omega(\mathbb{Z}^1 \setminus V, A)$ we have

$$\mu^\omega(x(V)|\overline{x}) = P_V^\omega(x(V)|\overline{x}). \tag{5.13}$$

We also define the notion of Gibbs measure on $Y_\omega(A)$ with the empty boundary condition on the left. Let $V = [0, v]$, $\overline{y} = \overline{y}([v+1, \infty)) \in X_\omega([v+1, \infty), A)$. Define

$$\begin{aligned} P_{V,l}^\omega(x(V)|\overline{y}) &= \frac{1}{Z_{V,l}^\omega(\overline{y})} \exp\left(\sum_{i=0}^{v} H_i(x(V), \overline{y})\right), \quad x(V) \in X_\omega(V, A, \overline{y}), \\ Z_{v,l}^\omega &= \sum_{x(V) \in X^\omega(V, A, \overline{y})} \exp\left(\sum_{i=0}^{v} H_i^\omega(x(V), \overline{y})\right). \end{aligned} \tag{5.14}$$

5.4. DEFINITION. A probability distribution ν^ω on $Y_\omega(A)$ is called an U-Gibbs measure with the left empty boundary condition (l.e.b.c.) if for all $V = [0, v]$ and ν^ω-a.a. $\overline{y} = \overline{y}([v+1, \infty)) \in X_\omega([v+1, \infty), A)$,

$$\nu^\omega(x(V)|\overline{y}) = P_{V,l}^\omega(x(V)|\overline{y}). \tag{5.15}$$

Consider the Gibbs measure μ^ω and its marginal distributions $\mu_V^\omega(x(V))$. Take $V_n = [-n, n] \supset V$, $\overline{x}_n \in X_\omega(\mathbb{Z}^1 \setminus V_n, A)$ and consider $P_{V_n}^\omega(x(V_n)|\overline{x}_n)$ and its marginal distribution $P_{V_n,V}^\omega(x(V)|\overline{x}_n) = \sum_{x(V_n \setminus V)} P_{V_n}^\omega(x(V) \cup x(V_n \setminus V)|\overline{x}_n)$. Denote by $\rho_n(V, \overline{x})$ the variational distance between $\mu_V^\omega(\cdot)$ and $P_{V_n,V}^\omega(\cdot|\overline{x}_n)$ given by

$$\rho_n(V, \overline{x}_n) = \frac{1}{2} \sum_{x(V)} |\mu_V^\omega(x(V)) - P_V^\omega(x(V)|\overline{x}_n)| \tag{5.16}$$

and put

$$\rho_n(V) = \sup_{\overline{x}_n \in X_\omega(\mathbb{Z}^1 \setminus V_n, A)} \rho_n(V, \overline{x}_n). \tag{5.17}$$

5.5. DEFINITION. a) We say that μ^ω is ergodic if the σ-algebra at infinity is trivial. b) We say that μ^ω is mixing if $\rho_n(V) \to 0$ as $n \to \infty$ for all $V = [u, v] \subset \mathbb{Z}^1$.

With obvious changes, Definition 5.5 can be applied to Gibbs distribution with l.e.b.c.

5.6. THEOREM. *For P-a.a. $\omega \in \Omega$ there exists a unique Gibbs measure μ^ω and Gibbs measure with l.e.b.c. ν_l^ω. Measures μ^ω and ν_l^ω are ergodic and mixing.*

5.7. Remark. It is possible to prove Theorem 5.1 in the same way we proved Theorem C. Namely, using the estimates for variational distance. However, the situation now is simpler since we have symbolic coding from the very beginning. Thus, we were able to define the notion of the Gibbs measure on $\mathcal{X}(A)$ directly. Using general facts of the theory of Gibbs measures we can present more conceptual and simple proof of Theorem 5.1.

PROOF OF THE THEOREM 5.6. The existence of the Gibbs distribution μ^ω follows from the standard compactness arguments. It is well known (see **[R]**) that the set of Gibbs measures forms a Choquet simplex and its extremal points are mutually singular ergodic Gibbs measures (pure phases). We show that any two Gibbs measures are absolutely continuous with respect to each other. The uniqueness and ergodicity follow immediately since this implies that there is only one extremal point of our Choquet simplex. Mixing also follows from uniqueness. Indeed, if $\limsup \rho_n(V) > 0$ for some V, then one can find a subsequence of intervals V_{n_k} and of boundary conditions $\overline{x}_{n_k} \in X_\omega(\mathbb{Z}^1 \setminus V_{n_k}, A)$ such that $P^\omega_{V_{n_k}}(\cdot|\overline{x}_{n_k})$ tends to different Gibbs distribution, which is impossible.

Consider the random variable

$$H_-(\omega) = \sum_{i<0} C_U(\theta^i\omega)e^{i\tilde{\alpha}}, \tag{5.18}$$

which exists according to Lemma 5.2. Set

$$M(\omega) = \sup_{y\in Y_\omega(A)} |U_\omega(y)|,\ l(N,\omega) = \prod_{i=0}^{N(\omega)-1} l_i(\omega), \quad M(N,\omega) = \sum_{i=0}^{N(\omega)-1} M(\theta^i\omega), \tag{5.19}$$

where $N(\omega)$ is the random variable defined in (1.32).

Consider the set $\Psi(H_1, L_1, M_1, N_1) = \{\omega \in \Omega : H_-(\omega) \le H_1,\ l(N,\omega) \le L_1,\ M(N,\omega) \le M_1,\ N(\omega) \le N_1\}$ and constants $H_1,\ L_1,\ M_1,\ N_1$ such that $P(\Psi(H_1, L_1, M_1, N_1)) > 0$. Then for P-a.a. ω there exists an infinite sequence of negative integers $u_n \to \infty$ and an infinite sequance of positive integers $v_n \to \infty$ such that $\theta^{u_n}\omega, \theta^{v_n}\omega \in \Psi(H_1, L_1, M_1, N_1)$.

Take $V_n = [u_n, v_n + N(\theta^{v_n}\omega)]$ and an arbitrary boundary condition $\overline{x}_n \in X_\omega(\mathbb{Z}^1 \setminus V_n, A)$. Suppose that $v_n - u_n > N_1$ and thus $u_n + N(\theta^{u_n}\omega) < v_n$. Set $W_n = [u_n + N(\theta^{u_n}\omega), v_n] \subset V_n$. Clearly, for any $x(W_n) \in X_\omega(W_n, A)$ and $\overline{x}_n$ there exists at least one extension $x(V_n \setminus W_n)$ such that $x(W_n) \cup x(V_n \setminus W_n) \cup \overline{x}_n \in X_\omega(A)$, and the maximal number of such configurations $x(V_n \setminus W_n)$ is bounded by L_1^2. We also have

$$\left|H_{V_n}\left(x(W_n) \cup x(V_n \setminus W_n), \overline{x}_n\right) - H_{W_n}(x(W_n))\right| \le 2M_1 + 2H_1. \tag{5.20}$$

Suppose that μ is an U-Gibbs measure. Then

$$
(5.21)\qquad \begin{aligned} \mu_{V_n}(x(V_n)) &= \int \frac{1}{Z_{V_n}(\overline{x}_n)} \exp(H_{V_n}(x(V_n), \overline{x}_n) d\mu(\overline{x}_n) \\ &= \frac{\exp(H_{W_n}(x(W_n))}{Z_{W_n}} \int \frac{\xi}{\eta} d\mu(\overline{x}_n), \end{aligned}
$$

where

$$
(5.22)\qquad \begin{aligned} Z_{W_n} &= \sum_{x(W_n)\in X_\omega(W_n,A)} \exp(H_{W_n}(x(W_n))), \\ \xi &= \exp(H_{V_n}(x(V_n), \overline{x}_n) - H_{W_n}(x(W_n))), \; \eta = Z_{V_n}(\overline{x}_n)/Z_{W_n}. \end{aligned}
$$

It follows form the estimates above that

$$
(5.23)\qquad \exp(-2M_1 - 2H_1) \le \xi \le \exp(2M_1 + 2H_1)
$$

and

$$
(5.24)\qquad \exp(-2M_1 - 2H_1) \le \eta \le L_1^2 \exp(2M_1 + 2H_1).
$$

Hence,

$$
(5.25)\qquad \mu_{V_n}(x(V_n)) = \frac{\zeta}{Z_{W_n}} \exp(H_{W_n}(x(W_n))),
$$

where

$$
(5.26)\qquad \frac{1}{L_1^2} \exp(-4M_1 - 4H_1) \le \zeta \le \exp(4M_1 + 4H_1).
$$

Thus for any two Gibbs measures μ', μ'' their marginal distributions are equivalent and

$$
(5.27)\qquad \frac{1}{L_1^2} \exp(-8M_1 - 8H_1) \le \frac{\mu'_{V_n}(x(V_n))}{\mu''_{V_n}(x(V_n))} \le L_1^2 \exp(8M_1 + 8H_1).
$$

Since the densities in (5.27) are uniformly bounded and $V_n \uparrow \mathbb{Z}^1$, we get equivalence of μ' and μ''. The proof for Gibbs measures with l.e.b.c. is the same. □

Set $\mu_+^\omega = \pi\mu^\omega$, where μ^ω is a Gibbs measure. We call μ_+^ω a U-Gibbs measure on $Y_\omega(A)$.

5.8. Remark. Consider discrete conditional probabilities $\mu_{+,n}^\omega(\cdot|\overline{y}_n)$, $\nu_{l,n}^\omega(\cdot|\overline{y}_n)$ on $Y_\omega(A)$ under the condition $\sigma_+^n(\omega)y = \overline{y}_n$, where $\sigma_+^n(\omega) = \sigma_+(\theta^{n-1}\omega)\ldots\sigma_+(\omega)$. It follows from the proof of Theorem 5.6 that $\mu_{+,n}^\omega(\cdot|\overline{y}_n)$ and $\nu_{l,n}^\omega(\cdot|\overline{y}_n)$ are absolutely continuous with respect to each other and there exists a random variable $D_U(\omega)$ such that

$$
(5.28)\qquad e^{-D_U(\omega)} \le \frac{\mu_{+,n}^\omega(\cdot|\overline{y}_n)}{\nu_{l,n}^\omega(\cdot|\overline{y}_n)} \le e^{D_U(\omega)}
$$

uniformly with respect to n and $\overline{y}_n$.

5.9. DEFINITION. a) A measure μ on $\mathcal{X}(A)$ is called an U-Gibbs state if $\mu = \mu^{\omega} dP(\omega)$ and μ^{ω} is a Gibbs measure on $X_{\omega}(A)$ for P-a.a. ω.

b) A probability measure $\mu_+(\nu_l)$ on $\mathcal{Y}(A)$ is called a U-Gibbs state (with l.e.b.c.) if $\mu_+ = \mu_+^{\omega} dP(\omega)$, $(\nu_l = \nu_l^{\omega} dP(\omega))$ and $\mu_+(\nu_{l.}^{\omega})$ is an U-Gibbs measure (with l.e.b.c.) on $Y_{\omega}(A)$ for P-a.a. ω.

The following statement follows immediately from Theorem 5.6.

5.10. PROPOSITION. *The U-Gibbs states μ on $\mathcal{X}(A)$ and μ_+, ν_l on $\mathcal{Y}(A)$ exist and are unique.*

Obviously, the states μ and μ_+ are invariant under τ and τ_+ respectively. The ergodicity and mixing follows from Theorem 5.6 (see Theorem 3.2). Theorem D is completely proven.

Next we introduce the generalized variational principle and prove Theorem E. Denote by $y_m(\omega)$ the minimal element in $Y_{\omega}(A)$ and set

$$\tilde{U}^{\omega}(y) = U^{\omega}(y) - U(y_m(\omega)). \tag{5.29}$$

Consider a τ_+-invariant measure $\chi \in \mathcal{P}_P(\mathcal{Y}(A))$. For a fixed ω let $\chi_n^{\omega}(y|\overline{y}_n)$ be the conditional distribution on $Y_{\omega}(A)$ for $\chi^{\theta^n \omega}$-a.a. $\overline{y}_n$. Take the entropy of the discrete measure $\chi^{\omega}(\cdot|\overline{y}_n)$, subtract $(\sum_{i=0}^{n-1} \log l^m(\theta^i \omega))$, and integrate it over $\overline{y}_n$, i.e., define

$$\begin{aligned} \tilde{h}_n^{\chi}(\omega|\overline{y}_n) &= - \sum_{y:\sigma_+^n(\omega)y=\overline{y}_n} \log \chi_n^{\omega}(y|\overline{y}_n)\chi_n^{\omega}(y|\overline{y}_n) - \sum_{i=0}^{n-1} \log l^m(\theta^i \omega), \\ \tilde{h}_n^{\chi}(\omega) &= \int \tilde{h}_n^{\chi}(\omega|\overline{y}_n) d\chi^{\theta^n \omega}(\overline{y}_n). \end{aligned} \tag{5.30}$$

One can easily check that $\tilde{h}_{m+n}^{\chi}(\omega) = \tilde{h}_n^{\chi}(\omega) + \tilde{h}_m^{\chi}(\theta^n \omega)$. Since $\tilde{h}_n^{\chi}(\omega) \leq 0$, there exists a nonrandom limit

$$\tilde{h}_{\chi}^{(r)}(\tau_+) = \lim_{n\to\infty} \frac{1}{n} \tilde{h}_n^{\chi}(\omega), \tag{5.31}$$

where $\tilde{h}_{\chi}^{(r)}(\tau_+)$ can take value $(-\infty)$. Definition (5.30) is analogous to (1.13), (4.32). Consider the following variational problem: find τ_+-invariant probability measures $\overline{\chi}$ on $\mathcal{P}_P(\mathcal{Y}(A))$ such that

$$\begin{aligned} &\int \tilde{U} d\overline{\chi} + \tilde{h}_{\overline{\chi}}^{(r)}(\tau_+) \\ &= \sup \left(\int \tilde{U}^{\omega}(y)\chi(dy, d\omega) + \tilde{h}_{\chi}^{(r)}(\tau_+),\ \chi \in \mathcal{P}_P(\mathcal{Y}(A)) \text{ is } \tau_+ \text{ -invariant} \right). \end{aligned} \tag{5.32}$$

Below we will show that the U-Gibbs state μ_+ is the unique τ_+-invariant measure in $\mathcal{P}_p(\mathcal{Y}(A))$ that satisfies variational principle (5.32). The proof is quite standard and similar to the proof of variational principle in Section 4. First we define the generalized pressure. Similarly to (4.35) and (4.36), set

$$\begin{aligned} \tilde{Z}_n^{\omega}(\overline{y}_n) &= \frac{1}{\prod_{i=0}^{n-1} l^m(\theta^i \omega)} \sum_{y:\sigma_+^n(\omega)y=\overline{y}_n} \exp\left(\sum_{i=0}^{n-1} \tilde{U}^{\theta^i \omega}(\sigma_+^i(\omega)y) \right), \\ \tilde{Z}_n^{\omega} &= \int \tilde{Z}_n^{\omega}(\overline{y}_n)\nu_l^{(\theta^i \omega)}(d\overline{y}_n),\ \ \tilde{q}_n(\omega) = \log \tilde{Z}_n^{\omega}, \end{aligned} \tag{5.33}$$

where ν_l is the U-Gibbs state with l.e.b.c. Since $\tilde{q}_{n+m}(\omega) = \tilde{q}_n(\omega) + \tilde{q}_m(\theta^n \omega)$ and $|\tilde{q}_1(\omega)| \le \delta_U(\omega) + \bar{l}(\omega)$, there exists a finite limit

$$(5.34) \qquad \lim_{n\to\infty} \frac{1}{n}\overline{q}_n(\omega) \equiv \tilde{Q}_P^{(r)}(U).$$

Basically the same arguments as in Section 4 show that for any τ_+-invariant $\chi \in \mathcal{P}_P(\mathcal{Y}(A))$ we have

$$(5.35) \qquad \int \tilde{U}^\omega(y)\chi(dy, d\omega) + \tilde{h}_\chi^{(r)}(\tau_+) \le \tilde{Q}_P^{(r)}(U).$$

The only difference is the choice of sequence $n_i(\omega) \to \infty$ (see (4.39)). One should take $n_i = v_i + N(\theta^{v_i}\omega)$, where $\theta^{v_i}(\omega) \in \Psi(H_1, L_1, M_1, N_1)$ as in the proof of Theorem 5.6.

It follows from (5.28) that

$$(5.36) \qquad |\tilde{h}_n^{\nu_l}(\omega|\overline{y}_n) - \tilde{h}_n^{\mu_+}(\omega|\overline{y}_n)| \le D_U(\omega).$$

Moreover, there exists a constant $\overline{D}_U(\omega)$ such that for $n = n_i$,

$$(5.37) \qquad |\tilde{h}_{n_i}^{\nu_l}(\omega|\overline{y}'_{n_i}) - \tilde{h}_{n_i}^{\nu_l}(\omega|\overline{y}''_{n_i})| \le \overline{D}_U(\omega).$$

Similarly to Section 4, denote

$$(5.38) \qquad \xi_n^{\nu_l}(\omega|\overline{y}_n) = \sum_{y:\sigma_+^n(\omega)y=\overline{y}_n} \left(\sum_{i=0}^{n-1} \tilde{U}^{\theta^i\omega}(\sigma_+^i(\omega)y)\right)\nu_{l,n}^\omega(y|\overline{y}_n).$$

Then

$$(5.39) \qquad \frac{1}{n}\xi_n^{\nu_l}(\omega|\overline{y}_n) = \frac{1}{n}\log \tilde{Z}_n^\omega(\overline{y}_n) - \frac{1}{n}\tilde{h}_n^{\nu_l}(\omega|\overline{y}_n).$$

Put $n = n_i$ and integrate (5.39) over $\overline{y}_n$ with respect to the measure $\sigma_+^n \nu_l^\omega$. We get

$$(5.40) \qquad \int \frac{1}{n_i}\xi_{n_i}^{\nu_l}(\omega|\overline{y}_{n_i})\sigma_+^{n_i}\nu_l^\omega(d\overline{y}_n) = \int \frac{1}{n_i}\left(\log \tilde{Z}_{n_i}^\omega(\overline{y}_{n_i}) - \tilde{h}_{n_i}^{\nu_l}(\omega|\overline{y}_{n_i})\right)\sigma_+^{n_i}\nu_l^\omega(d\overline{y}_{n_i}).$$

It follows from (5.36), (5.37) and (5.24) that the $n = n_i \to \infty$ limit of the right-hand side of (5.40) is equal to $[\tilde{Q}_P^{(r)}(U) - \tilde{h}_{\mu_+}^{(r)}(\tau_+)]$.

Since $\frac{1}{n}\sum_{i=0}^{n-1} \tilde{U}^{\theta^i\omega}(\sigma_+^i(\omega)y) \to \int \tilde{U}^\omega(y)d\mu_+$ for μ_+-a.a. (y, ω) and ν_l is absolutely continuous with respect to μ_+ we obtain that the left-hand side of (5.40) tends to $\int \tilde{U}(\omega)(y)d\mu_+$ as $n_i \to \infty$. Hence,

$$(5.41) \qquad \int \tilde{U}^\omega(y)\mu_+(dy, d\omega) + \tilde{h}_{\mu_+}^{(r)}(\tau_+) = \tilde{Q}_P^{(r)}(U).$$

The proof of the uniqueness is also standard (see **[R]**), but slightly more technical. Consider the set of potentials $\tilde{U}^\omega(y)$ such that $\tilde{U}^\omega(y_m(\omega)) = 0$ and

$$(5.42) \qquad C_{\tilde{U}}(\omega) \equiv \sup_n \sup_{y', y'': y'_i = y''_i,\, 0 \le i < n} e^{\tilde{\alpha} n}|\tilde{U}^\omega(y') - \tilde{U}^\omega(y'')| < \infty.$$

Denote by $\mathcal{B}$ the Banach space of potentials that satisfy (5.42) with the norm

$$(5.43) \qquad |\tilde{U}\|_{\mathcal{B}} = \int (\max_y \tilde{U}^\omega(y) - \min_y \tilde{U}^\omega(y))dP(\omega) + \int \log(1 + C_{\tilde{U}}(\omega))dP(\omega).$$

For any potential $\tilde{U}^{\omega} \in \mathcal{B}$, the pressure $\tilde{Q}_P^{(r)}(\tilde{U})$ is defined (see (5.34)). It is easy to show that $\tilde{Q}_P^{(r)}(\tilde{U})$ is a continuous convex function on $\mathcal{B}$. Continuity follows from the variational principle (5.35). To prove the convexity one should consider potentials $\tilde{U}_\alpha = \alpha\tilde{U}_1 + (1-\alpha)\tilde{U}_2$ and take the limit of $\frac{1}{n_i}\log Z_{n_i}^{\omega}(\overline{y}_{n_i})$ as $i \to \infty$, where Z is the partition function corresponding to $\tilde{U}_\alpha$ and n_i are the "nice" moments for both of the potentials $\tilde{U}_1^{\omega}$ and $\tilde{U}_2^{\omega}$ simultaneously.

Suppose that for τ_+-invariant measure $\overline{\chi}$ on $\mathcal{P}_P(\mathcal{Y}(A))$ we have

$$\int \tilde{U}\,d\overline{\chi} + \tilde{h}_{\overline{\chi}}^{(r)}(\tau_+) = \sup_{\chi \in \mathcal{P}_P(\mathcal{Y}(A)),\tau_+\chi=\chi} \left(\int \tilde{U}\,d\chi + \tilde{h}_{\chi}^{(r)}(\tau_+) \right). \tag{5.44}$$

It is easy to check (see **[R]**) that for any $\tilde{U}' \in \mathcal{B}$

$$\tilde{Q}_P^{(r)}(\tilde{U} + \tilde{U}') \geq Q_P^{(r)}(\tilde{U}) + \int \tilde{U}'\,d\overline{\chi}. \tag{5.45}$$

Condition (5.45) means that $\overline{\chi}$ is tangent to $\tilde{Q}_P^{(r)}$ at the point $\tilde{U}$. Thus, it follows from the theorem of Lanford-Robinson (see **[R]**) that there exists a sequence of potentials $\tilde{U}_n \in \mathcal{B}$ such that $\|\tilde{U} - \tilde{U}_n\|_{\mathcal{B}} \to 0$ as $n \to \infty$, for each of $\tilde{U}_n$ there exists a unique measure $\chi(n)$ that satisfies (5.45) and $\chi(n) \to \overline{\chi}$ in the weak topology as $n \to \infty$. Obviously, $\chi(n) = \mu_+(n)$, where $\mu_+(n)$ is the $\tilde{U}_n$-Gibbs state. Since $\overline{\chi}$ is a limit of $\tilde{U}_n$-Gibbs states and the potentials $\tilde{U}_n$ tend to U, it is easy to show that $\overline{\chi}$ is the U-Gibbs state, which implies uniqueness. □

References

[B1] T. Bogenschütz, *Entropy, pressure, and a variational principle for random dynamical systems*, Random and Computational Dynamics **1** (1992), 99–116.

[B2] ———, *Equilibrium States for Random Dynamical Systems*, Ph.D. thesis, Universität Bremen, 1993.

[BG1] T. Bogenschütz and V. M Gundlach, *Symbolic dynamics for expanding random dynamical systems,*, Random and Computational Dynamics **1** (1992), 219–227.

[BG2] ———, *Ruelle's transfer operator for random subshifts of finite type*, Preprint, Universität Bremen, 1993.

[C] R. Cogburn, *On the central limit theorem for Markov chains in random environments*, Ann. Probab **19** (1991), 587–604.

[D] R. L. Dobrushin, *Conditions for the absence of phase transitions in one-dimensional classical systems*, Math. USSR Sbornik **22** (1974), 28–48.

[Ke] M. Keane, *Strongly mixing g-measures*, Invent. Math. **16** (1972), 309–324.

[K1] Y. Kifer, *Ergodic theory of random transformations*, Birkhäuser, Boston, 1986.

[K2] ———, *Equilibrium states for random expanding transformations*, Random and Computational Dynamics **1** (1992), 1–31.

[P] R. R. Phelps, *Lectures on Choquet's theorem*, Van Nostrand, London, 1966.

[R] D. Ruelle, *Thermodynamic formalism*, Addison-Wesley, Reading MA, 1978.

[S] Ya. G. Sinai, *Gibbs measures in ergodic theory*, Russian Math. Surveys **27** (1972), 21–69.

[W] P. Walters, *Invariant measures and equilibrium states for some mappings which expand distances*, Trans. Amer. Math. Soc. **236** (1978), 121–153.

INSTITUTE OF MATHEMATICS, HEBREW UNIVERSITY OF JERUSALEM, GIVAT RAM, JERUSALEM 91904, ISRAEL

DEPARTMENT OF MATHEMATICS, PRINCETON UNIVERSITY, PRINCETON, NJ 08544

Amer. Math. Soc. Transl.
(2) Vol. **171**, 1996

Bounded Orbits of Nonquasiunipotent Flows on Homogeneous Spaces

D. Y. Kleinbock and G. A. Margulis

ABSTRACT. Let $\{g_t\}$ be a nonquasiunipotent one-parameter subgroup of a connected semisimple Lie group G without compact factors; we prove that the set of points in a homogeneous space G/Γ (Γ an irreducible lattice in G) with bounded $\{g_t\}$-trajectories has full Hausdorff dimension. Using this we give necessary and sufficient conditions for this property to hold for any Lie group G and any lattice Γ in G.

Introduction

Let G be a Lie group, Γ a lattice in G, $F = \{g_t \mid t \in \mathbb{R}\}$ a one-parameter subgroup of G. Then the action of F on the homogeneous space $\Omega = G/\Gamma$ by left translations defines a flow. Assume that g_1 is not *quasiunipotent*, that is, $\operatorname{Ad} g_1$ has an eigenvalue with modulus different from 1. The results of S. G. Dani **[Dan2, Dan3]** suggest the following

CONJECTURE (A) **[Mrg2]**. *For any nonempty open subset W of Ω the set*

$$\{x \in W \mid \text{the } F\text{-orbit of } x \text{ is bounded}\}$$

is of Hausdorff dimension equal to the dimension of G.

This conjecture was proved by Dani in the two following cases:

(i) (see **[Dan2]**) $G = SL_n(\mathbb{R})$, $\Gamma = SL_n(\mathbb{Z})$, and

$$g_t = \operatorname{diag}(e^{-t}, \dots, e^{-t}, e^{\lambda t}, \dots, e^{\lambda t}), \tag{1}$$

where λ is such that the determinant of g_t is 1;

(ii) (see **[Dan3]**) G is a connected semisimple Lie group of $\mathbb{R}$-rank 1.

In the present paper we prove Conjecture (A) in the case

(iii) G is a connected semisimple Lie group without compact factors and Γ is an irreducible lattice in G (Theorem 1.1).

In fact, the statement of this theorem is stronger: we consider orbits which are bounded and stay away from a given closed $\{g_t\}$-invariant subset Z of Ω of Haar measure 0.

1991 *Mathematics Subject Classification*. Primary 22E99; Secondary 28D15.

The work of the second author was supported in part by NSF Grant DMS-9204270.

In the general case we give necessary and sufficient conditions for Conjecture (A) to hold (Theorem 5.2 and Corollary 5.5), based on the reduction to the case (iii). In particular, these conditions are satisfied when F consists of semisimple elements.

The main idea of the proof is similar to that of Dani **[Dan2]**. He considered the *horospherical* (relative to g_{-1}) subgroup H – the abelian subgroup of matrices of the form

$$h = \begin{pmatrix} I & 0 \\ L & I \end{pmatrix}. \tag{2}$$

Then he showed that the trajectory $\{g_t h\mathbb{Z}^n\}$ in the space $SL_n(\mathbb{R})/SL_n(\mathbb{Z})$ of lattices in $\mathbb{R}^n$, with g_t as in (1) and h as in (2), is bounded if and only if the set of linear forms corresponding to the matrix L, is badly approximable in the sense of Schmidt **[S2]**. The result of the latter implies that the set of points with bounded trajectories has full Hausdorff dimension.

In fact, what is proved in **[Dan2]** (as well as in **[Dan3]**) is an apparently stronger result that the above set is winning in Schmidt's sense (cf. **[S1]**). In the general case the horospherical subgroup H is a connected simply connected nilpotent Lie group admitting a one-parametric semigroup of expanding automorphisms Φ_t-conjugations by g_t for positive t. The main reason why the method of **[Dan2]** cannot be used (at least directly) in this generality is that the restriction of Φ_t to H does not have to be conformal with respect to a Riemannian metric on H. Still, the fact that Φ_t, $t > 0$, is expanding on H is crucial for our approach. In §1 we reduce the problem to studying one-sided trajectories F^+x of points $x \in \Omega$ (here $F^+ = \{g_t \mid t \geq 0\}$); this makes it possible to imbed a sequence of sets with easy to estimate increasing dimensions into the set of points with bounded orbits.

Indeed, fix a large compact set K in Ω and some neighborhood V of identity in H. Our first purpose is to prove that, roughly speaking, for sufficiently large T and for any $x \in K$, there exists $t(x) \in [T, 2T]$ such that the major part of the set $g_{t(x)}Vx$ lies in a smaller set $K' \subset K$, the quantitative meaning of "the most" being uniform in $x \in K$ (see Proposition 2.5). After that we divide H (up to a set of measure 0) into pieces which are right translations of V by elements γ of H (this can be done for suitable V, see §3). For any $x \in K$ we choose all the translations γ such that $V\gamma \subset \Phi_{t(x)}(V)$ and $V\gamma g_{t(x)}x \subset K$, and define a compact subset $\mathbf{A}_1(x)$ of $\overline{V}$ to be the union of $\Phi_{-t(x)}(\overline{V}\gamma)$ over all γ as above. By iteration of this procedure, a sequence of compact sets $\overline{V}x \supset \mathbf{A}_1(x) \supset \mathbf{A}_2(x) \supset \cdots \supset \mathbf{A}_j(x) \supset \cdots \supset$ ($\mathbf{A}_{j+1}(x)$ being the union of $\Phi_{-t(x)}(\mathbf{A}_j(\gamma g_{t(x)}x)\gamma)$ over all γ as above) is constructed. The *limit set* $\mathbf{A}_\infty(x) = \bigcap_{i=1}^\infty \mathbf{A}_j(x)$ then consists of elements h such that the F^+-orbit of hx lies in a certain compact subset of Ω. The detailed description of this construction is given in §4; a result of C. McMullen and M. Urbanski allows one to derive a lower estimate on the Hausdorff dimension of $\mathbf{A}_\infty$ from the information about relative measure of the union of pieces $V\gamma g_{t(y)}y$ chosen at each stage for each $y \in K$.

To prove Proposition 2.5 mentioned above, we consider three cases.

Case 1. $\{g_t\}$ consists of semisimple elements (see Section 2.2; in this case we will say that the flow $(\Omega, \{g_t\})$ is *semisimple*).

Case 2. All the essential simple factors of G have $\mathbb{R}$-rank 1 (a factor G' of G is said to be *essential* if the projection of the semisimple part $\{a_t\}$ of $\{g_t\}$ onto G' is not relatively compact). This is a special case of *essentially semisimple* flows, see Section 2.3.

Case 3. The flow (Ω, F) *has property* (EM), by which we mean that the representation of the product G^e of all the essential factors of G on the space $L_2^0(\Omega) \stackrel{\text{def}}{=} \{f \in L^2(\Omega) \mid \int_\Omega f = 0\}$ is isolated from the trivial representation (see Section 2.4). In view of Kazhdan's results on property (T) (cf. Lemma 2.4.1), it covers the case when G^e contains at least one factor of $\mathbb{R}$-rank greater than 1.

In Case 1 we use mixing properties of the action of F on Ω. In Case 2 we combine the results of Case 1 with facts about nondivergence of orbits of unipotent flows. In Case 3 instead of mixing we employ stronger results on the decay of matrix coefficients of smooth functions (Section 2.4; see also Appendix for similar treatment of Hölder functions). Note that in Cases 1 and 3 one can choose $t(x)$ to be equal to T for all $x \in K$. Let us also note that it is probably sufficient to consider only Case 3, because the representation of any simple factor of G in $L_0^2(\Omega)$ seems to be always isolated from the trivial representation. However, we were unable to prove it or find a reference in the literature.

Several generalizations of the main result, as well as some open questions, are considered in §5. An estimate on exponential decay of spherical averages of Hölder continuous functions on Ω (Corollary A.8) is deduced in the Appendix from mixing-related results obtained in §2.

In what follows, $\dim(A)$ will denote the Hausdorff dimension of a metric space A. If F is a one-parameter ($F = \{g_t \mid t \in \mathbb{R}\}$) or cyclic ($F = \{g^i \mid i \in \mathbb{Z}\}$) subgroup of a Lie group G, we will use the notation F^+ and F^- for the subsemigroups of F corresponding to nonnegative (resp. nonpositive) values of t (resp. i). A subgroup F of G will be called *nonquasiunipotent* if it contains an element which is not quasiunipotent.

§1. Reduction to the subgroup H

1.1. Our principal goal is to prove the following result.

THEOREM. *Let G be a connected semisimple Lie group of dimension n without compact factors, Γ an irreducible lattice in G, F a (one-parameter or cyclic) nonquasiunipotent subgroup of G, and $Z \subset \Omega \stackrel{\text{def}}{=} G/\Gamma$ a closed set of (Haar) measure 0, which is invariant under either F^+ or F^-. Then for any nonempty open subset W of Ω,*

$$\dim\big(\{x \in W \mid Fx \text{ is bounded and } \overline{Fx} \cap Z = \varnothing\}\big) = n\,.$$

This theorem, with $Z = \varnothing$, clearly implies Conjecture (A) for the case (iii). The proof of this theorem will occupy §§1–4; unless otherwise specified, G, Γ and Ω will be as in Theorem 1.1.

1.2. Remark. Without loss of generality one can assume in the above theorem that the center $Z(G)$ of G is trivial. Indeed, let us denote the quotient group $G/Z(G)$ by G', the homomorphism $G \to G'$ by p, and the induced map $\Omega \to \Omega' \stackrel{\text{def}}{=} G'/p(\Gamma)$ by $\bar{p}$. Since $\Gamma Z(G)$ is discrete [**Rag,** Corollary 5.17], $p(\Gamma)$ is also discrete, hence $Z(G)/\big(\Gamma \cap Z(G)\big)$ is finite. This means that $(\Omega, \bar{p})$ is a finite covering of Ω', therefore for any closed null subset Z of Ω which is invariant under either F^+ or F^-, $\bar{p}(Z)$ is a closed null subset of Ω' invariant under either $p(F)^+$ or $p(F)^-$. Let $W \subset \Omega$ be an open set small enough for $\bar{p}$ to be injective on W, and denote

$$A = \{x \in \bar{p}(W) \mid p(F)x \text{ is bounded and } \overline{p(F)x} \cap \bar{p}(Z) = \varnothing\}\,.$$

Then for any $y \in \bar{p}^{-1}(A)$ the trajectory Fy is bounded and $\overline{Fy} \cap Z = \varnothing$; since $\bar{p}$ is a local isometry, the set $W \cap \bar{p}^{-1}(A)$ has dimension equal to the dimension of A.

In particular, the above statement shows that it is sufficient to prove Theorem 1.1 in the case when F is a one-parameter subgroup. Indeed, if G is centerfree, any of its cyclic subgroup contains a subgroup of finite index which can be imbedded in a one-parameter subgroup of G. On the other hand, the statement of Theorem 1.1 is stable with respect to passing to a cocompact subgroup of F (see Lemma 5.1(a); part (b) of the same lemma shows stability with respect to passing to a finite covering of Ω, which was implicitly used in the first part of this remark).

1.3. Let $\mathfrak{g}$ be a Lie algebra of G, $\mathfrak{g}_{\mathbb{C}}$ its complexification, and for $\lambda \in \mathbb{C}$, let E_λ be a *generalized eigenspace* of $\operatorname{Ad} g_1$:

$$E_\lambda = \{X \in \mathfrak{g}_{\mathbb{C}} \mid (\operatorname{Ad} g_1 - \lambda I)^j X = 0 \text{ for some } j\}.$$

Let $\mathfrak{h}$, $\mathfrak{h}^0$, $\mathfrak{h}^-$ be the subalgebras of $\mathfrak{g}$ with complexifications

$$\mathfrak{h}_{\mathbb{C}} = \operatorname{span}(E_\lambda \mid |\lambda| > 1),\ \mathfrak{h}^0_{\mathbb{C}} = \operatorname{span}(E_\lambda \mid |\lambda| = 1),\ \mathfrak{h}^-_{\mathbb{C}} = \operatorname{span}(E_\lambda \mid |\lambda| < 1).$$

The fact that g_1 is not quasiunipotent implies that $\mathfrak{h} \neq \{0\}$ and $\mathfrak{h}^- \neq \{0\}$; denote by k the common dimension of $\mathfrak{h}$ and $\mathfrak{h}^-$. Let H, H^0, H^- be the corresponding subgroups of G. Note that H^- is a horospherical subgroup with respect to g_1, while H is horospherical with respect to g_{-1}. (Recall that a subgroup

$$U(g) = \{h \in G \mid g^l h g^{-l} \to e \text{ as } l \to +\infty\}$$

of G is called *horospherical* with respect to $g \in G$.)

Clearly the subalgebras $\mathfrak{h}$, $\mathfrak{h}^0$, $\mathfrak{h}^-$ are invariant under $\operatorname{Ad} g_t$, which implies that the subgroups H, H^0, H^- are normalized by F. Moreover, it is easy to show that the inner automorphism $\Phi_t : G \to G, g \to g_t g g_{-t}$, $t > 0$, defines an *expanding* automorphism of H:

for any compact $K \subset H$, any open $V \subset H$, and any $e \in V$,
there exists t_0 such that $t > t_0 \Rightarrow K \subset \Phi_t(V)$.

Similarly, the automorphism Φ_t, $t > 0$, is *contracting* on the subgroup H^- (that is, $\Phi_t^{-1}|_{H^-}$ is expanding).

Since the group G can be assumed to be centerfree, it makes sense to consider a Jordan decomposition $g_t = a_t u_t = u_t a_t$, a_t and u_t being semisimple and unipotent parts of g_t respectively. Say that the flow (Ω, F) is *semisimple* if $\{u_t\}$ is trivial; say also that (Ω, F) is *essentially semisimple* if $\{u_t\}$ is trivial modulo the centralizer $Z(H)$ of H in G (that is, if $\{u_t\}$ commutes with H). Let us note that these are in fact properties of the subgroup F, not of its action on Ω.

In the flow (Ω, F) is semisimple, the restriction of Φ_t to the subgroup $\tilde{H} \stackrel{\text{def}}{=} H^- H^0$ is (locally) *nonexpanding* (cf. **[Bow, EP]**):

$$\text{(1.1)} \qquad \exists c > 1 \text{ such that } \forall r < 1\ \forall t > 0, \quad \Phi_t(\tilde{B}(r)) \subset \tilde{B}(cr)$$

(here and hereafter $\tilde{B}(r)$ stands for the open ball of radius r in $\tilde{H}$ centered at e).

If $\{u_t\}$ is not trivial, (1.1) need not be true. However, Φ_t is still contracting on H^-, while the restriction of $\operatorname{Ad} g_t$ to $\mathfrak{h}^0$ lies in the product of a compact and a unipotent subgroup of $GL(\mathfrak{h}^0)$. This means that Φ_t defines (locally) *at most polynomially expanding* automorphism of $\tilde{H}$:

$$\text{(1.2)} \qquad \exists c > 1,\ \varkappa \in \mathbb{N} \text{ such that } \forall t > 1\ \forall r < t^{-\varkappa}, \quad \Phi_t(\tilde{B}(r)) \subset \tilde{B}(ct^{\varkappa} r).$$

1.4. An important property of Hausdorff dimension that we will use is the following

LEMMA (Marstrand Slicing Theorem, [**Mrs**] or [**F,** Theorem 5.8]). *Let A and B be metric spaces, and let C be a subset of the direct product $A \times B$. Assume that either*

(a) *A has positive α-dimensional Hausdorff measure $\mathcal{H}_\alpha$, and*

$$\dim\big(C \cap (x \times B)\big) \geq \beta \tag{1.3}$$

for $\mathcal{H}_\alpha$-almost every $x \in A$, or

(b) *A is a Borel set in a manifold,* $\dim(A) \geq \alpha$, *and the condition* (1.3) *is satisfied for all $x \in A$.*

Then

$$\dim(C) \geq \alpha + \beta\,. \tag{1.4}$$

PROOF. The standard version, as in [**Mrs**] or [**F**], assumes (a); to derive (1.4) from (b), note that by [**Dav**] (see also [**F,** Theorem 5.6]), A contains a subset of positive α'-dimensional Hausdorff measure for any $\alpha' < \dim(A)$. □

1.5. We will repeatedly use the fact that locally (in the neighborhood of identity) G is bi-Lipschitz equivalent to the direct product of the subgroups H, H^0 and H^-. In particular, it makes it possible to reduce Theorem 1.1 to the following statement:

THEOREM. *For any $x \in \Omega \smallsetminus Z$ and for any neighborhood V of identity in H*

$$\dim\left(\{h \in V \mid F^+hx \text{ is bounded and } \overline{F^+hx} \cap Z = \varnothing\}\right) = k\,.$$

REDUCTION OF THEOREM 1.1 TO THEOREM 1.5. Given a nonempty open $W \subset \Omega$, choose a point $x \in W$ not contained in Z. After that take $U \subset G$ of the form V^-VV^0, where V^-, V and V^0 are neighborhoods of identity in H^-, H and H^0 respectively, such that the multiplication map $V^- \times V \times V^0 \to U$, $(h^-, h, h_0) \to h^-hh^0$ is bi-Lipschitz, the quotient map $\pi_x : G \to \Omega$, $g \to gx$ is injective on U, and Ux is inside $W \smallsetminus Z$. Then it is sufficient to show that

$$\dim\left(\{g \in U \mid Fgx \text{ is bounded and } \overline{Fgx} \cap Z = \varnothing\}\right) = n\,. \tag{1.5}$$

Let C be the set $\{h \in U \mid F^+hx \text{ is bounded and } \overline{F^+hx} \cap Z = \varnothing\}$. From Theorem 1.5 it follows that for any $h^0 \in V^0$, $\dim(C \cap Vh^0) = k$. In view of Lemma 1.4, this implies that the set $C \cap VV^0$ has dimension equal to $n - k$.

CLAIM. *For all $h \in C$ there exists a neighborhood $V^-(h)$ of identity in H^- such that $V^-(h)h \subset C$.*

PROOF. For $h \in C$, let $\varepsilon(h)$ be the distance between disjoint closed sets Z and $\overline{F^+hx}$. Since the map Φ_t is contracting on H^- for $t > 0$, one can find an open neighborhood $V^-(h)$ of identity in H^- such that $V^-(h)h \subset U$ and $\operatorname{diam}\big(\Phi_t(V^-(h))\big) \leq \varepsilon(h)/2$ for any $t \geq 0$. Then $g_tV^-(h)hx = \Phi_t\big(V^-(h)\big)g_thx$ is disjoint from Z for any $t \geq 0$, and clearly $\operatorname{diam}\big(F^+V^-(h)hx\big) < \varepsilon(h) + \operatorname{diam}(F^+hx)$. Therefore, the trajectory of any point from $V^-(h)hx$ is bounded and stays away from Z. □

Applying Theorem 1.5 (with F^- in place of F^+ and H^- in place of H) to each of the sets $V^-(h)$, $h \in C \cap VV^0$, one gets that

$$\forall\, h \in C \cap VV^0 \quad \dim\left(\{h^- \in V^- \mid Fh^-hx \text{ is bounded and } \overline{Fh^-hx} \cap Z = \varnothing\}\right) = k\,,$$

and another application of Lemma 1.4 yields the equality (1.5). □

1.6. Since Hausdorff dimension of the union of a countable family of sets is equal to the supremum of dimensions of these sets, Theorem 1.5 is equivalent to the following statement, which is in fact what we will be proving in the course of the paper:

THEOREM. *For any $x \in \Omega$, $x \notin Z$, there exist a sequence of neighborhoods V_s of identity in H and a sequence of compact subsets C_s of $\Omega \setminus Z$, $s \in \mathbb{N}$, such that*

$$\operatorname{diam}(V_s) \to 0 \text{ as } s \to \infty\,, \tag{1.6}$$

and

$$\dim\left(\{h \in \overline{V_s} \mid F^+hx \subset C_s\}\right) \to k \text{ as } s \to \infty\,. \tag{1.7}$$

§2. Mixing and its consequences

2.1. Preliminaries.

2.1.1. Let μ, $\bar{\mu}$, m, m^0, m^- be Haar measures on G, Ω, H, H^0, H^- respectively, normalized so that $\bar{\mu}(\Omega) = 1 = \mu$ (any fundamental domain of Γ-action on G), and μ is locally almost the product of m^-, m^0, and m. The latter, in view of [**Bou,** Ch. VII, §9, Proposition 13], means that μ can be expressed via m, m^0 and m^- in the following way: for any $\varphi \in L^1(G)$
(2.1.1)

$$\int_{H^-H^0H} \varphi(g)\,d\mu(g) = \int_{H^-\times H^0\times H} \varphi(h^-h^0h)\Delta_{\tilde{H}}(h^0)\,dm^-(h^-)\,dm^0(h^0)\,dm(h)\,,$$

where $\Delta_{\tilde{H}}$ is the modular function of the group $\tilde{H} = H^-H^0$.

Observe that the measures μ and m^0 are preserved by the automorphism Φ_t, while for m and m^- one has $\Phi_t(m^-) = e^{-\chi t}m^-$, and

$$\Phi_t(m) = e^{\chi t}m\,, \tag{2.1.2}$$

where χ is a positive number (equal to $\operatorname{Tr}\operatorname{ad} Y|_{\mathfrak{h}}$, if $Y \in \mathfrak{g}$ is such that $g_1 = \exp(Y)$).

2.1.2. It is well known (cf. [**Mo1**]) that the action of F on Ω is *mixing*. We will use the following generalized version:

THEOREM. *Let Φ, $\Psi \subset L^2(\Omega)$ be two compact (in the topology of $L^2(\Omega)$) families of functions. Then for any $\varepsilon > 0$ there exists $T > 0$ such that*

$$t \geq T \Rightarrow \forall \varphi \in \Phi\, \forall \psi \in \Psi\, \left|(g_t\varphi, \psi) - \int_\Omega \varphi\,d\bar{\mu}\int_\Omega \psi\,d\bar{\mu}\right| \leq \varepsilon$$

(here $(\cdot,\cdot)$ means the inner product in $L^2(\Omega)$).

PROOF. Since F is not quasiunipotent, $\operatorname{Ad} F$ is not relatively compact, so by [**Mo1**] the statement is true for one-element sets $\Phi = \{\varphi\}$ and $\Psi = \{\psi\}$, φ, $\psi \in L^2(\Omega)$. The general case then follows from the unitarity of the regular representation of G on $L^2(\Omega)$. □

2.1.3. Our purpose now is to derive an analogue of the mixing property for functions ϕ supported on certain proper submanifolds of Ω. Denote by v the volume form on H corresponding to the Haar measure m. Since $\mathfrak{g} = \mathrm{Lie}(G)$ is a direct sum of $\mathfrak{h} = \mathrm{Lie}(H)$ and $\tilde{\mathfrak{h}} = \mathrm{Lie}(\tilde{H})$, the projection $\mathfrak{g} \to \mathfrak{h}$ is well defined and induces the map $p_x : T_x\Omega \to T_x(Hx)$ for any $x \in \Omega$. Thus the form v induces a k-form ω_H on Ω given by $(\omega_H)_x = v \circ d\pi_x^{-1} \circ p_x$ (here π_x as before stands for the quotient map $G \to \Omega$, $g \to gx$).

Let M be a smooth k-dimensional manifold, and let π be a C^∞ immersion $M \to \Omega$ such that $\pi(M)$ is transversal to the orbit $\tilde{H}\pi(y)$ for all $y \in M$. Then one can pull the form ω_H back to get a k-form ω_π on M. Further, let f be a function on M and ψ a function on Ω. Then one can integrate the product $f(y)\psi\big(\pi(y)\big)$ with respect to ω_π and, furthermore, look at the asymptotics of $\int_M f(y)\psi\big(g_t\pi(y)\big)\,\omega_\pi$ (which is by (2.1.2) equal to $e^{-\chi t}\int_M f(y)\psi\big(g_t\pi(y)\big)\,\omega_{g_t\circ\pi}$) as $t \to \infty$. We will return to this level of generality in Appendix (Section A.7), while for the proof of Theorem 1.6 it suffices to consider a special case $M = H$ and $\pi = \pi_x$, where $x \in \Omega$.

If $f \in L^2(H)$ and $x \in \Omega$, the integral $\int_H f(h)\psi\big(g_t\pi_x(h)\big)\,\omega_{\pi_x}$ can be written as $\int_H f(h)\psi(g_t hx)\,dm(h)$. Using (2.1.2), one gets

$$\int_H f(h)\psi(g_t hx)\,dm(h) = e^{-\chi t}\int_H f(g_t h g_{-t})\psi(hg_t x)\,dm(h), \tag{2.1.3}$$

the last integral being equal to $\int_H f(h)\psi\big(g_t\pi_x(h)\big)\,\omega_{g_t\circ\pi_x}$.

We will describe asymptotics of integrals of type (2.1.3) for sufficiently large classes of functions f and ψ; moreover, our estimates will be uniform in x lying in a fixed compact subset of Ω. However, additional assumptions on the subgroup F will be necessary.

2.2. Semisimple flows.

2.2.1. The proof of the next proposition is based on Theorem 2.1.2 and the non-expanding property (1.1) of $\Phi_t|_{\tilde{H}}$. Similar methods were developed in **[Bow, EM, EP, Mrc, Mrg1]**.

PROPOSITION. *Assume that F consists of semisimple elements. We are given $f \in L^2(H)$ with compact support and a uniformly continuous $\psi \in L^2(\Omega)$. Then for any compact subset L of Ω and any $\varepsilon > 0$ there exists $T > 0$ such that*

$$\left|\int_H f(h)\psi(g_t hx)\,dm(h) - \int_H f\,dm\int_\Omega \psi\,d\bar{\mu}\right| \le \varepsilon \tag{2.2.1}$$

for all $x \in L$ and $t \ge T$.

PROOF. Since $\mathrm{supp}(f)$ and L are compact, and Γ is discrete, f can be written as a sum of functions f_j, $1 \le j \le N$, with π_x injective on $\mathrm{supp}(f_j)$ for all $x \in L$ and for each j. Hence one can without loss of generality assume that the maps π_x are injective on $\mathrm{supp}(f)$ for all $x \in L$.

If $f \equiv 0$ a.e., there is nothing to prove. Otherwise, denote by a the positive number $\varepsilon\big(2\int_H |f|\,dm\big)^{-1}$. Since ψ is uniformly continuous, there exists r, $0 < r < 1$, such that

$$\forall \tilde{h} \in \tilde{B}(r)\ \forall x \in \Omega \quad \big|\psi(\tilde{h}x) - \psi(x)\big| \le a\,. \tag{2.2.2a}$$

We will choose this r small enough to ensure that

$$\forall x \in L \quad \pi_x \text{ is injective on } \tilde{B}(r)\cdot \mathrm{supp}(f)\,. \tag{2.2.2b}$$

Pick two nonnegative functions $f^- \in L^2(H^-)$, $f^0 \in L^2(H^0)$ such that

$$\int_{H^-} f^- \, dm^- = 1 = \int_{H^0} f^0 \, dm^0 , \tag{2.2.3}$$

and

$$\operatorname{supp}(f^-) \cdot \operatorname{supp}(f^0) \subset \tilde{B}(r/c) \quad (\text{with } c > 1 \text{ from } (1.1)) . \tag{2.2.4}$$

Then for any $t \in \mathbb{R}$ by (2.1.1)

(2.2.5)

$$\begin{aligned}\int_H f(h)\psi(g_t hx)\, dm(h) &= \int_{H^-} f^- \, dm^- \int_{H^0} f^0 \, dm^0 \int_H f(h)\psi(g_t hx)\, dm(h) \\ &= \int_G f^-(h^-) f^0(h^0) f(h) \psi(g_t hx) \Delta_{\tilde{H}}(h^0)^{-1} \, d\mu(h^- h^0 h) .\end{aligned}$$

Now observe that

$$\begin{gathered}\left|\psi(g_t hx) - \psi(g_t h^- h^0 hx)\right| = \left|\psi(g_t hx) - \psi\left((\Phi_t(h^- h^0)) g_t hx\right)\right| \le a \\ \text{whenever } h^- \in \operatorname{supp}(f^-) \text{ and } h^0 \in \operatorname{supp}(f^0)\end{gathered} \tag{2.2.6}$$

(this follows from (2.2.4), nonexpanding property (1.1) and (2.2.2a)). Hence

(2.2.7)

$$\begin{aligned}&\Big| \int_H f(h)\psi(g_t hx)\, dm(h) \\ &\qquad - \int_G f^-(h^-) f^0(h^0) f(h) \psi(g_t h^- h^0 hx) \Delta_{\tilde{H}}(h^0)^{-1} \, d\mu(h^- h^0 h) \Big| \\ &= \left| \int_G f^-(h^-) f^0(h^0) f(h) \big(\psi(g_t hx) - \psi(g_t h^- h^0 hx)\big) \Delta_{\tilde{H}}(h^0)^{-1} \, d\mu(h^- h^0 h) \right| \\ &\le a \int_G \left| f^-(h^-) f^0(h^0) f(h) \Delta_{\tilde{H}}(h^0)^{-1} \right| \, d\mu(h^- h^0 h) = a \int_H |f| \, dm = \frac{\varepsilon}{2} .\end{aligned}$$

Let φ be a function on $H^- H^0 H$ defined by the formula

$$\varphi(h^- h^0 h) = f^-(h^-) f^0(h^0) f(h) \Delta_{\tilde{H}}(h^0)^{-1} .$$

Denote by φ_x the function $\varphi \circ \pi_x^{-1}$. Then, by (2.2.2b) and (2.2.4), φ_x is well defined for all $x \in L$, and

$$\int_\Omega \varphi_x \, d\bar{\mu} = \int_G \varphi \, d\mu = \int_H f \, dm \tag{2.2.8}$$

(the last equality follows from (2.1.1) and (2.2.3)). Moreover,

$$\begin{aligned}&\int_G f^-(h^-) f^0(h^0) f(h) \psi(g_t h^- h^0 hx) \Delta_{\tilde{H}}(h^0)^{-1} \, d\mu(h^- h^0 h) \\ &\qquad = (\varphi_x, g_{-t}\psi) = (g_t \varphi_x, \psi) .\end{aligned} \tag{2.2.9}$$

Since the family $\{\varphi_x \mid x \in L\}$ is compact in $L^2(\Omega)$, by Theorem 2.1.2 there exists $T > 0$ such that for $t \ge T$ and for all $x \in L$

$$\left| (g_t \varphi_x, \psi) - \int_H f \, dm \int_\Omega \psi \, d\bar{\mu} \right| \le \frac{\varepsilon}{2} ,$$

and the claim follows from (2.2.7), (2.2.9) and the above inequality. □

2.2.2. The assumption of the uniform continuity of the function ψ was essential in the above proof. However, it is not hard to see that Proposition 2.2.1 also holds for functions ψ which can be approximated by uniformly continuous functions in a suitable way. To better describe it, we make the following definition: for any property $\mathcal{P}$ of measurable functions on a measure space X, say that a function φ *almost has* property $\mathcal{P}$ if there exist two sequences of functions $\{\varphi_j\}$ and $\{\varphi_j'\}$ having property $\mathcal{P}$, such that one of them is nondecreasing ($\varphi_{j+1} \geq \varphi_j$ a.e.), the other one nonincreasing ($\varphi_{j+1}' \leq \varphi_j'$ a.e.), and both converge to φ almost everywhere. The main example: characteristic functions of sets with null boundary in a metric space with Borel measure almost have many nice properties, such as continuity or infinite differentiability.

From Proposition 2.2.1 we now deduce the following

COROLLARY. *Let F and f be as in Proposition* 2.2.1, *and let $\psi \in L^2(\Omega)$ be almost uniformly continuous. Then for any compact $L \subset \Omega$ and any $\varepsilon > 0$, there exists $T = T(L, \varepsilon) > 0$ such that* (2.2.1) *holds for all $x \in L$ and $t \geq T$.*

PROOF. Without loss of generality one can assume that the function f is a.e. nonnegative. Indeed, if $f = f_1 - f_2$ is a difference of a.e. nonnegative functions, take T such that

$$t \geq T \Rightarrow \forall x \in L \quad \left| \int_H f_i(h)\psi(g_t hx)\, dm(h) - \int_H f_i\, dm \int_\Omega \psi\, d\bar{\mu} \right| \leq \frac{\varepsilon}{2}$$

for $i = 1, 2$.

Let $\{\psi_j\}$ be a nondecreasing sequence of uniformly continuous functions converging to ψ. For $\varepsilon > 0$, find j such that

$$\int_H f\, dm \left| \int_\Omega \psi_j\, d\bar{\mu} - \int_\Omega \psi\, d\bar{\mu} \right| \leq \frac{\varepsilon}{2}.$$

Then, using Proposition 2.2.1, find $T' > 0$ such that

$$\left| \int_H f(h)\psi_j(g_t hx)\, dm(h) - \int_H f\, dm \int_\Omega \psi_j\, d\bar{\mu} \right| \leq \frac{\varepsilon}{2}$$

for any $t \geq T'$ and for all $x \in L$.

Since $f \geq 0$ and $\psi \geq \psi_j$ by construction, $\int_H f(h)\psi(g_t hx)\, dm(h)$ is not less than $\int_H f(h)\psi_j(g_t hx)\, dm(h)$. Thus from the above inequalities it follows that

$$\int_H f(h)\psi(g_t hx)\, dm(h) \geq \int_H f\, dm \int_\Omega \psi\, d\bar{\mu} - \varepsilon$$

for any $t \geq T'$ and for all $x \in L$. Repeating all the above arguments with a nonincreasing sequence of uniformly continuous functions approximating ψ, one gets $T'' > 0$ such that

$$\int_H f(h)\psi(g_t hx)\, dm(h) \leq \int_H f\, dm \int_\Omega \psi\, d\bar{\mu} + \varepsilon$$

for any $t \geq T''$ and for all $x \in L$, and it suffices to take $T = \max(T', T'')$. □

2.3. Essentially semisimple flows.

2.3.1. Recall that if the unipotent part $\{u_t\}$ of $\{g_t\}$ is nontrivial, the restriction $\Phi_t|_{\tilde{H}}$ does not have to be locally nonexpanding, so the above proof of Proposition 2.2.1 cannot be carried out. However, it turns out that if the subgroup $\{u_t\}$ is small enough, one can prove a weaker (but still good enough for our purposes) version of the estimates from the preceding section. Here "small enough" means "lies in the centralizer $Z(H)$ of H in G", that is, the flow (Ω, F) is essentially semisimple as defined in Section 1.3.

The following lemma gives a large class of nontrivial examples of essentially semisimple flows:

LEMMA. *Let G be a direct product of G' and G'' such that*
(a) *the projection of $\{a_t\}$ onto G' is relatively compact, and*
(b) *the projection of $\{u_t\}$ onto G'' is trivial.*
Then the flow (Ω, F) is essentially semisimple.

PROOF. Indeed, (a) implies that $H \subset G''$, while $\{u_t\} \subset G'$ by (b). □

2.3.2. To apply the above lemma, decompose the group G (assumed to be center-free) into a direct product of simple groups. Then say that a factor G_i of G is *essential* with respect to F if the projection of the semisimple part $\{a_t\}$ of F onto G_i is not relatively compact. Denote by G^{e} the product of all the essential factors, and by G^{i} the product of all other factors; g_t^{e} and g_t^{i} will mean projections of g_t on G^{e} and G^{i} respectively.

COROLLARY. *Assume that all the essential factors of G are of $\mathbb{R}$-rank* 1. *Then the flow is essentially semisimple.*

PROOF. Indeed, the projection of u_t onto G^{e} is a product of unipotent elements of groups of rank 1, and each of these elements centralizes a noncompact torus in the corresponding factor. Since the centralizer of a maximal torus for groups of rank 1 coincides with the torus itself, the projection of u_t on each factor of G^{e} must be semisimple, hence trivial. On the other hand, the projection of $\{a_t\}$ onto G^{i} is by definition relatively compact, and the claim follows from Lemma 2.3.1. □

2.3.3. Observe now that if $\{u_t\}$ commutes with H, the integral (2.1.3) can be written in the form

$$\int_H f(h)\psi\big(g_t\pi_x(h)\big)\,\omega_{\pi_x} = \int_H f(h)\psi(a_t u_t hx)\,dm(h)$$
$$= \int_H f(h)\psi(a_t h u_t x)\,dm(h) = \int_H f(h)\psi\big(a_t\pi_{u_t x}(h)\big)\,\omega_{\pi_{u_t x}}\,.$$

This suggests that in order to estimate the asymptotics of the above integral, one may wish to apply Proposition 2.2.1 to the semisimple part $\{a_t\}$ of $\{g_t\}$. However, to get a uniform estimate, one has to be sure that the points $u_t x$ lie in a fixed compact set. This is made possible by the following theorem, describing nondivergence of orbits of unipotent flows $(\Omega, \{u_t\})$:

THEOREM **[Dan4, DM]**. *Let L be a compact subset of Ω and let $\sigma > 0$ be given. Then there exists a compact subset $Q = Q(L,\sigma)$ of Ω such that for any unipotent one-parameter subgroup $\{u_t\}$ of G, any $x \in L$ and any $T \geq 0$,*

$$l\big(\{t \in [0,T] \mid u_t x \in Q\}\big) \geq (1-\sigma)T\,,$$

where l denotes the Lebesgue measure on $\mathbb{R}$.

2.3.4. PROPOSITION. *Assume that the flow (Ω, F) is essentially semisimple. Then for any compactly supported $f \in L^2(H)$, almost uniformly continuous $\psi \in L^2(\Omega)$, a compact set $L \subset \Omega$ and positive ε and σ, there exists $T_0 > 0$ such that for all $x \in L$ and $T \geq T_0$,*

$$l\big(\{t \in [T, 2T] \mid (2.2.1) \text{ holds}\}\big) \geq (1 - \sigma)T\,.$$

PROOF. Given $\sigma > 0$ and a compact set $L \subset \Omega$, take $Q = Q(L, \sigma/2)$ from Theorem 2.3.3. After that for any $\varepsilon > 0$ put T_0 to be equal to $T(Q, \varepsilon)$ from Corollary 2.2.2. Then it follows from the latter corollary that (2.2.1) is satisfied whenever $u_t x \in Q$. Therefore for any $x \in L$ and $T \geq T_0$, one has

$$\begin{aligned} l\big(\{t \in [T, 2T] \,|(2.2.1) \text{ holds}\}\big) &\geq l\big(\{t \in [T, 2T] \mid u_t x \in Q\}\big) \\ &\geq l\big(\{t \in [0, 2T] \mid u_t x \in Q\}\big) - T \geq 2T(1 - \sigma/2) - T \end{aligned}$$

which finishes the proof. □

2.4. Flows with property (EM).

2.4.1. We will show in this section that certain properties of representations of semisimple groups allow one to derive an analogue of Proposition 2.2.1, valid, in particular, for flows which are not essentially semisimple.

Denote by ρ_0 the regular representation of G on the subspace $L^2_0(\Omega)$ of $L^2(\Omega)$ orthogonal to constant functions, and say that the flow (Ω, F) *has property (EM)* (an abbreviation for exponential mixing, see Theorem 2.4.5 below for justification) if

(EM) the restriction of ρ_0 to G^{e} is isolated (in the Fell topology) from the trivial representation.

Note that this condition, in particular, means that G^{e} is not trivial (there exists at least one essential factor), hence F is not quasiunipotent.

LEMMA. *Assume that there exists an essential factor G' of G not locally isomorphic to $SO(m, 1)$ or $SU(m, 1)$, $m \in \mathbb{N}$. Then the flow (Ω, F) has property* (EM).

PROOF. Write $G^{\mathrm{e}} = G' \times G''$ (G'' being the product of all the essential factors of G except G'), and consider a direct integral decomposition

$$\rho_0|_{G^{\mathrm{e}}} = \int_X \rho'_x \otimes \rho''_x \, d\mu_X\,, \tag{2.4.1}$$

where ρ'_x and ρ''_x are irreducible representations of G' and G'' respectively. From the irreducibility of the lattice Γ it follows that μ_X-almost all the representations ρ'_x are nontrivial. Since G' has Kazhdan's property (T) (see e.g. [**Cow**, Section 2.3]), all these representations are separated from the trivial representation of G', which, together with the decomposition (2.4.1), implies (EM). □

Note that, in view of Corollary 2.3.2, this lemma implies that the flow (Ω, F) has property (EM) whenever it is not essentially semisimple.

2.4.2. REMARK. It is well known (although we failed to identify the exact reference) that ρ_0 itself is isolated from the trivial representation. This means that (EM) is satisfied whenever $G = G^{\mathrm{e}}$ (all the simple factors of G are essential with respect to F). In fact, it seems likely that the condition (EM) is always satisfied, though we are not aware of the proof, if it exists.

2.4.3. It turns out that for the flows with property (EM) one can formulate a refinement of the mixing property. Indeed, the condition (EM) allows one to make use of the results on the decay of matrix coefficients of semisimple Lie groups. The following exponential estimate is a modification of Theorem 3.1 from **[KS]**.

Fix a maximal compact subgroup $\mathcal{K}$ of a connected semisimple Lie group G, and denote by $\mathfrak{k}$ its Lie algebra. Take an orthonormal basis $\{Y_j\}$ of $\mathfrak{k}$, and set $\Upsilon = 1 - \sum Y_j^2$. Then Υ belongs to the center of the universal enveloping algebra of $\mathfrak{k}$ and acts on smooth vectors of any representation space of G.

Let $\mathfrak{a}$ be a maximal split Cartan subalgebra of the Lie algebra $\mathfrak{g}$ of G. Fix an order on the roots, and let $\mathfrak{c}$ be a positive Weyl chamber; denote by ϑ half the sum of the positive roots on $\mathfrak{c}$.

THEOREM. *Let G be a connected semisimple Lie group with finite center, and let Π be a family of unitary representations of G with discrete kernel, which is isolated from the trivial representation. Then there exists a universal constant $B > 0$, a positive integer l (depending only on G), and a positive integer p such that for any $\rho \in \Pi$, any C^∞-vectors v, w in a representation space of ρ, all $Y \in \bar{\mathfrak{c}}$ and $t \geq 0$, we have*

$$\big|\big(\rho\big(\exp(tY)\big)v, w\big)\big| \leq Be^{-\frac{t}{2p}\vartheta(Y)}\|\Upsilon^l(v)\|\|\Upsilon^l(w)\|. \tag{2.4.2}$$

PROOF. Decomposing any $\rho \in \Pi$ into a direct integral of irreducible representations and using Cauchy-Schwartz inequality, one can without loss of generality assume that Π consists of irreducible representations. A. Katok and R. Spatzier, using R. Howe's estimates **[H]** for matrix coefficients of $\mathcal{K}$-finite vectors, proved in **[KS]** that (2.4.2) holds whenever an irreducible representation ρ of G with discrete kernel is strongly L^p. Then from M. Cowling's results **[Cow]** they deduced that p depends only on G if the latter contains no factors locally isomorphic to $SO(m, 1)$ or $SU(m, 1)$, $m \in \mathbb{N}$. However, using the argument from **[Cow,** Section 3.1], one can show that all irreducible representations of G outside a fixed neighborhood of the trivial representation are strongly L^p for some p. □

2.4.4. Let us fix a Riemannian metric $\operatorname{dist}(\cdot,\cdot)$ on G bi-invariant with respect to $\mathcal{K}$.

COROLLARY. *Let G, Π, ρ, v and w be as in Theorem 2.4.3. Then there exist constants $E > 0$, $l \in \mathbb{N}$ (depending only on G) and $\alpha > 0$ (depending only on G and Π) such that for any $g \in G$,*

$$\big|\big(\rho(g)v, w\big)\big| \leq Ee^{-\alpha \operatorname{dist}(e,g)}\|\Upsilon^l(v)\|\|\Upsilon^l(w)\|.$$

PROOF. If $k_1 \exp(tY)k_2$, with $\vartheta(Y) = 1$, is the Cartan decomposition of g, one can apply Theorem 2.4.3 to the subgroup $\exp(tY)$. □

2.4.5. We now return to the original problem and apply the above corollary to the subgroup G^{e} of G.

THEOREM. *Let the condition* (EM) *be satisfied. Then there exist constants $\gamma > 0$, $E > 0$, $l \in \mathbb{N}$ such that for any two functions φ, $\psi \in C^\infty_{\mathrm{comp}}(\Omega) \cap L^2_0(\Omega)$ and for any $t \geq 0$,*

$$|(g_t\varphi, \psi)| \leq Ee^{-\gamma t}\|\varphi\|_l\|\psi\|_l, \tag{2.4.3}$$

where $\|\cdot\|_l$ *means the norm in the Sobolev space* $W_l^2(\Omega)$.

In other words, if the flow has property (EM), the rate of mixing for smooth functions is at least exponential.

PROOF. From the definition of G^{e} it follows that $\operatorname{dist}(e, g_t^{\mathrm{e}})$ is for any $t \geq 0$ bounded from below by βt for some $\beta > 0$. Hence by Corollary 2.4.4, for any irreducible component ρ of $\rho_0|_{G^{\mathrm{e}}}$, and for any two C^∞-vectors v and w of ρ,

$$\left|\left(\rho(g_t^{\mathrm{e}})v, w\right)\right| \leq E e^{-\alpha\beta t}\|\Upsilon^l(v)\|\|\Upsilon^l(w)\|. \tag{2.4.4}$$

Now restrict ρ_0 to the group $G^{\mathrm{e}} \times \{g_t^{\mathrm{i}}\}$ and then decompose this restriction into a direct integral of irreducible representations. Each of them will be a tensor product of an irreducible representation of G^{e} and a one-dimensional representation of $\{g_t^{\mathrm{i}} \mid t \in \mathbb{R}\}$. Therefore the estimate (2.4.3) for matrix coefficients $(g_t\varphi, \psi)$ can be obtained by integration of inequalities of type (2.4.4) (by irreducibility of the lattice Γ, almost all irreducible components of ρ_0 have discrete kernel). □

Note that for Theorem 2.4.3 (resp. Theorem 2.4.5) to hold, it clearly suffices for v and w (resp. φ and ψ) to be C^l for sufficiently large l. Moreover, similar estimates can be proved for Hölder vectors – see Appendix.

2.4.6. COROLLARY. *With the assumptions of Theorem* 2.4.5, *for any two functions* $\varphi, \psi \in C^\infty_{\mathrm{comp}}(\Omega)$ *and for any* $t \geq 0$,

$$\left|(g_t\varphi, \psi) - \int_\Omega \varphi\, d\bar{\mu} \int_\Omega \psi\, d\bar{\mu}\right| \leq E e^{-\gamma t}\|\varphi\|_l\|\psi\|_l.$$

PROOF. Apply (2.4.3) to the functions $\varphi - \int_\Omega \varphi\, d\bar{\mu}$ and $\psi - \int_\Omega \psi\, d\bar{\mu}$. □

2.4.7. To make use of the above facts, let us list simple properties of the Sobolev norm $\|\cdot\|_l$ we will need later.

LEMMA. (a) *Let* X, Y *be Riemannian manifolds,* $\varphi \in C^\infty_{\mathrm{comp}}(X)$, $\psi \in C^\infty_{\mathrm{comp}}(Y)$; *consider* $\varphi \cdot \psi$ *as a function on* $X \times Y$. *Then* $\|\varphi \cdot \psi\|_l \leq c(X, Y)\|\varphi\|_l\|\psi\|_l$, *where* $c(X, Y)$ *is a constant independent on the functions* φ *and* ψ.

(b) *Let* X *be a Riemannian manifold of dimension* N, x *a point in* X. *Then for any* r, $0 < r < 1$, *there exists a nonnegative function* $f \in C^\infty_{\mathrm{comp}}(X)$ *such that* $\operatorname{supp}(f)$ *is inside the ball of radius* r *centered at* x, $\int_X f = 1$, *and* $\|f\|_l \leq c(X, x)r^{-(l+N/2)}$, *where* $c(X, x)$ *is a constant independent on* r.

2.4.8. PROPOSITION. *Let the condition* (EM) *be satisfied. Then for any* $f \in C^\infty_{\mathrm{comp}}(H)$, *for any* $\psi \in C^\infty_{\mathrm{comp}}(\Omega)$ *and for any compact subset* L *of* Ω *there exists a constant* $C = C(f, \psi, L)$ *such that for all* $x \in L$ *and for any* $t \geq 0$,

$$\left|\int_H f(h)\psi(g_t hx)\, dm(h) - \int_H f\, dm \int_\Omega \psi\, d\bar{\mu}\right| \leq C t^q e^{-\lambda t}, \tag{2.4.5}$$

where $\lambda = \gamma/(2l + 1 + \frac{n-k}{2})$ (γ *and* l *from Theorem* 2.4.5), *and* $q = \varkappa/(2l + 1 + \frac{n-k}{2})$ ($\varkappa$ *from* (1.2)).

PROOF. The proof will basically go along the same lines as that of Proposition 2.2.1. It clearly suffices to prove (2.4.5) for sufficiently large t and for f such that the maps π_x are injective on $\operatorname{supp}(f)$ for all $x \in L$; then it can be extended to hold arbitrarily for all $f \in C^\infty_{\text{comp}}(H)$ and $t \geq 0$ by appropriate variation of the constant $C(f, \psi, L)$.

If either $f \equiv 0$ or $\psi \equiv 0$, there is nothing to prove. Otherwise, pick $t > 1$, denote by a the positive number $t^q e^{-\lambda t} \cdot \left(\int_H |f| \, dm \right)^{-1}$, and by r the positive number $a \cdot d(\psi)^{-1}$, where $d(\psi) \stackrel{\text{def}}{=} \max_{x \in \Omega} |\nabla \psi(x)|$. Then (2.2.2a) is satisfied, and making t large enough one can satisfy (2.2.2b) as well.

Using Lemma 2.4.7, choose nonnegative functions $f^- \in C^\infty_{\text{comp}}(H^-)$, $f^0 \in C^\infty_{\text{comp}}(H^0)$ satisfying (2.2.3) such that

$$\operatorname{supp}(f^-) \cdot \operatorname{supp}(f^0) \subset \tilde{B}(r/ct^\varkappa) \quad (c \text{ and } \varkappa \text{ from } (1.2)), \tag{2.4.6}$$

and at the same time[1]

$$\|\tilde{f}\|_l \leq \text{const} \cdot (r/ct^\varkappa)^{(2l+(n-k)/2)}, \tag{2.4.7}$$

where $\tilde{f} \in C^\infty_{\text{comp}}(\tilde{H})$ is defined by $\tilde{f}(h^- h^0) = f^-(h^-) f^0(h^0) \Delta_{\tilde{H}}(h^0)^{-1}$. Then the formula (2.1.1) implies (2.2.5), while (2.2.6) follows from (2.2.2a), (2.4.6) and at most polynomial expanding property (1.2). Hence

(2.4.8)

$$\begin{aligned} &\Big| \int_H f(h) \psi(g_t h x) \, dm(h) \\ &\qquad - \int_G f^-(h^-) f^0(h^0) f(h) \psi(g_t h^- h^0 h x) \Delta_{\tilde{H}}(h^0)^{-1} \, d\mu(h^- h^0 h) \Big| \\ &= \left| \int_G f^-(h^-) f^0(h^0) f(h) \big(\psi(g_t h x) - \psi(g_t h^- h^0 h x) \big) \Delta_{\tilde{H}}(h^0)^{-1} \, d\mu(h^- h^0 h) \right| \\ &\leq a \int_G |f^-(h^-) f^0(h^0) f(h) \Delta_{\tilde{H}}(h^0)^{-1}| \, d\mu(h^- h^0 h) = a \int_H |f| \, dm = t^q e^{-\lambda t} . \end{aligned}$$

Define the functions φ and φ_x as in the proof of Proposition 2.2.1; then (2.2.8) and (2.2.9) are satisfied, and by Corollary 2.4.6

$$\left| (g_t \varphi_x, \psi) - \int_H f \, dm \int_\Omega \psi \, d\bar{\mu} \right| \leq E e^{-\gamma t} \|\varphi_x\|_l \|\psi\|_l \, .$$

Observe now that

$$\begin{aligned} \|\varphi_x\|_l &= \|\tilde{f} \cdot f\|_l \\ \text{(by Lemma 2.4.7)} \quad &\leq \text{const} \cdot \|\tilde{f}\|_l \|f\|_l \\ \text{(by (2.4.7))} \quad &\leq \text{const} \cdot (t^\varkappa / r)^{2l + \frac{n-k}{2}} \|f\|_l \\ \text{(by definition of } r) \quad &\leq \text{const} \cdot \left(t^{\varkappa - q} e^{\gamma t} d(\psi) \int_H |f| \, dm \right)^{2l + \frac{n-k}{2}} \|f\|_l \\ \text{(by definition of } q \text{ and } \lambda) \quad &\leq \text{const} \cdot t^q e^{(\gamma - \lambda) t} \left(d(\psi) \int_H |f| \, dm \right)^{2l + \frac{n-k}{2}} \|f\|_l \, . \end{aligned}$$

[1]The values of constants in the proof are different and independent on t, f, and ψ.

Thus

$$\left|(g_t\varphi_x,\psi) - \int_H f\,dm \int_\Omega \psi\,d\bar\mu\right| \le \text{const}\cdot t^q e^{-\lambda t}\left(d(\psi)\int_H |f|\,dm\right)^{2l+\frac{n-k}{2}} \|f\|_l\|\psi\|_l,$$

which, together with (2.4.8) and (2.2.9), finishes the proof. □

2.4.9. COROLLARY. *Let the condition* (EM) *be satisfied. Let f be almost in $C^\infty_{\text{comp}}(H)$, ψ almost in $C^\infty_{\text{comp}}(\Omega)$, and let L be a compact subset of Ω. Then given any $\varepsilon > 0$ there exists $T > 0$ such that* (2.2.1) *holds for any $x \in L$ and $t \ge T$.*

PROOF. If $f \in C^\infty_{\text{comp}}(H)$ and $\psi \in C^\infty_{\text{comp}}(\Omega)$, for any $\varepsilon > 0$ the estimate (2.4.5) gives the value of T such that (2.2.1) is satisfied for all $x \in L$ and $t \ge T$. After that one can argue as in the proof of Corollary 2.2.2. □

2.5. An application.

We now unify the results of three preceding sections in the form which will be used in the proof of Theorem 1.6. Let V be a subset[2] of H, K a subset of Ω, $x \in \Omega$, and $t \in \mathbb{R}$. Define the set

$$V(x,K,t) = \{h \in g_t V g_{-t} \mid hg_t x \in K\} = \Phi_t\big(V \cap \pi_x^{-1}(g_{-t}K)\big).$$

In other words, $h \in V(x,K,t)$ if and only if $h = \Phi_t(h')$, where $h' \in V$ and $g_t h' x \in K$. If f is the characteristic function of V and ψ the characteristic function of K, then the measure of this set is equal to

(2.5)
$$m\big(V(x,K,t)\big) = \int_H f(g_t h g_{-t})\psi(hg_t x)\,dm(h) = e^{\chi t}\int_H f(h)\psi(g_t h x)\,dm(h).$$

PROPOSITION. *Let V be a bounded subset of H with $m(\partial V) = 0$, K a bounded subset of Ω with $\bar\mu(\partial K) = 0$, L a compact subset of Ω. Then for any $\varepsilon > 0$ and $\sigma > 0$ there exists $T_1 = T_1(V,K,L,\varepsilon,\sigma) > 0$ such that*

$T \ge T_1 \Rightarrow$ *for all $x \in L$ we have*

$$l\left(\left\{t \in [T,2T] \,\middle|\, |e^{-\chi t} m\big(V(x,K,t)\big) - m(V)\bar\mu(K)| \le \varepsilon\right\}\right) \ge (1-\sigma)T.$$

PROOF. Let f be the characteristic function of V, ψ the characteristic function of K. Clearly f is almost in $C^\infty_{\text{comp}}(H)$, and ψ is almost in $C^\infty_{\text{comp}}(\Omega)$, in particular, almost uniformly continuous. Thus the proposition follows from Proposition 2.3.4 in the essentially semisimple case or from Corollary 2.4.9 in the (EM) case, once the substitution (2.5) is made. □

Note that if the flow is either semisimple or with property (EM), Corollaries 2.2.2 and 2.4.9 imply the validity of the above proposition for $\sigma = 0$ as well.

[2]All the sets considered are Borel subsets of the corresponding measure spaces.

§3. Tesselations in nilpotent Lie groups

3.1. Let X be a (locally compact separable) topological space, m a Borel measure on X, G a group of measure-preserving homeomorphisms of X. Say that an open subset V of X is a *tesselation domain* (cf. **[Mag]**) for G-action on X relative to a (countable) subset Λ of G if

(a) $m(\partial V) = 0$,

(b) $\gamma_1(V) \cap \gamma_2(V) = \varnothing$ for different $\gamma_1, \gamma_2 \in \Lambda$, and

(c) $X = \bigcup_{\gamma \in \Lambda} \gamma(\overline{V})$.

For brevity, we will refer to the pair (V, Λ) as to the *tesselation* of X. The following fact follows easily from the definition:

LEMMA. *If (V, Λ_V) is a tesselation of X, $U \subset V$, and (U, Λ_U) is a tesselation of V (with the topology and the measure coming from X), then $(U, \Lambda_V \cdot \Lambda_U)$ is a tesselation of X.*

3.2. EXAMPLE. The interior of a fundamental domain for a discrete group Λ acting on X properly discontinuously is a tesselation domain relative to Λ. In particular, the set $I_k = \{(x_1, \dots, x_k) \mid |x_j| < 1/2\} \subset \mathbb{R}^k$ is a tesselation domain relative to $\mathbb{Z}^k \subset G$, where G is any subgroup of isometries of $\mathbb{R}^k$ containing translations. Or, more generally, for any $R > 0$ the set $\frac{1}{R} I_k \subset \mathbb{R}^k$ is a tesselation domain relative to $\frac{1}{R}\mathbb{Z}^k$.

3.3. Now let H denote a connected simply connected nilpotent Lie group of dimension k, acting on itself by right translations. The purpose of the next proposition is to give an explicit construction of tesselation domains for this action. Note that any nilpotent Lie group considered below will be understood to be equipped with (left = right) Haar measure and right invariant Riemannian metric "dist".

Let $\mathfrak{h}$ be the Lie algebra of H. Pick a basis $\{X_1, \dots, X_k\}$ of $\mathfrak{h}$ such that $\mathbb{R}$-spans $\mathfrak{h}_j$ of $\{X_1, \dots, X_j\}$, $j = 1, \dots, k$, are ideals in $\mathfrak{h}$ satisfying $[\mathfrak{h}, \mathfrak{h}_j] \subseteq \mathfrak{h}_{j-1}$, that is, a *strong Malcev basis* of $\mathfrak{h}$ (cf. **[CG]**). Here $\mathfrak{h}_0 = \{0\}$, and (at least) $\mathfrak{h}_1$ lies in the center of $\mathfrak{h}$. Note also that for any j, $0 \le j \le k$, $\{p(X_{j+1}), \dots, p(X_k)\}$ is a strong Malcev basis for $\mathfrak{h}/\mathfrak{h}_j$, where p is the quotient map $\mathfrak{h} \to \mathfrak{h}/\mathfrak{h}_j$.

Let $(e) = H_0 \subset H_1 \subset \cdots \subset H_{k-1} \subset H_k = H$ be the central series of H corresponding to the ideals $\mathfrak{h}_j$, so that $H_j = \exp(\mathfrak{h}_j)$, $j = 0, \dots, k$.

PROPOSITION. *For any $r > 0$ there exists a neighborhood V of identity in H, $\operatorname{diam}(V) < r$, which is a tesselation domain for the right action of H on itself.*

PROOF. As in the above example, let $I_k = \left\{\sum_{j=1}^k x_j X_j \mid |x_j| < 1/2\right\}$ be the unit cube in the Lie algebra $\mathfrak{h}$ of H. We will put $V = \exp(\frac{1}{R} I_k)$, where R is large enough to ensure that $\operatorname{diam}(V) < r$, and then use induction on the dimension k of H to prove that V is a tesselation domain for the right action of H on itself.

The basis of the induction is treated in the Example 3.2. Suppose that the statement is true for groups of dimension less than k. Let $\hat{H}$ be the quotient group H/H_1, $\hat{\mathfrak{h}}$ its Lie algebra (isomorphic to the span of $X_2, \dots, X_k$ as a linear space), p the quotient map $\mathfrak{h} \to \hat{\mathfrak{h}}$, π the quotient map $H \to \hat{H}$. Then the set $\hat{I} = p(\frac{1}{R} I_k)$ is the cube of sidelength $\frac{1}{R}$ in $\hat{\mathfrak{h}}$. Clearly $p \circ \exp = \exp \circ \pi$, hence by the induction assumption $\pi(V)$ is a tesselation domain for the right action of $\hat{H}$ on itself; denote by $\hat{\Lambda} \subset \hat{H}$ the corresponding set of translations.

Let $\Lambda \subset H$ be any section for $\hat{\Lambda}$. Then it is clear that $\hat{V} = \pi^{-1}(\pi(V))$ is a tesselation domain of H relative to Λ. But the group $H_1 = \exp(\mathfrak{h}_1)$ acts on $\hat{V}$ by left

(= right, since it is in the center of G) translations, and one can easily show that V is a tesselation domain for this action relative to, say, $\Lambda_1 = \exp(\frac{1}{R}\mathbb{Z}X_1)$. By Lemma 3.1, $(V, \Lambda_1 \cdot \Lambda)$ is a tesselation of H. □

3.4. Let now (V, Λ) be any tesselation of H, Φ an expanding automorphism of H, and let c be the *contraction bound* of Φ^{-1} on V, that is,

$$c = \sup_{g,h \in V,\, g \neq h} \frac{\operatorname{dist}(\Phi^{-1}(g), \Phi^{-1}(h))}{\operatorname{dist}(g,h)}.$$

Define a function

$$f_V(r) = m(\{h \in H \mid \operatorname{dist}(h, \partial V) \leq r\});$$

since $m(\partial V) = 0$, $f_V(r) \to 0$ as $r \to 0$. The next inequality helps one to estimate the number of sets $V\gamma, \gamma \in \Lambda$, which lie entirely inside the expansion of V by the map Φ.

PROPOSITION.

$$\#\{\gamma \in \Lambda \mid V\gamma \subset \Phi(V)\} \geq \frac{m(\Phi(V))}{m(V)}\left(1 - \frac{f_V(c \cdot \operatorname{diam}(V))}{m(V)}\right). \tag{3.1}$$

PROOF. One has

$$\begin{aligned}\#\{\gamma \in \Lambda \mid V\gamma \subset \Phi(V)\} &= \#\{\gamma \in \Lambda \mid \Phi^{-1}(V\gamma) \subset V\} \\ &= \#\{\gamma \in \Lambda \mid \Phi^{-1}(V\gamma) \cap V \neq \varnothing\} - \#\{\gamma \in \Lambda \mid \Phi^{-1}(V\gamma) \cap \partial V \neq \varnothing\}.\end{aligned}$$

Since (V, Λ) is a tesselation of H, the minuend is not less than $m(\Phi(V))/m(V)$, while the substrahend is not greater than

$$\frac{m\left(\{h \in H \mid \operatorname{dist}(h, \partial V) \leq \operatorname{diam}(\Phi^{-1}(V))\}\right)}{m(\Phi^{-1}(V))} \leq \frac{f_V(c \cdot \operatorname{diam}(V)) \cdot m(\Phi(V))}{m(V)^2},$$

and (3.1) follows. □

3.5. COROLLARY. *Let $\{\Phi_t,\ t \geq 0\}$ be a one-parameter semigroup of expanding automorphisms of H. Then for any tesselation (V, Λ) of H and any $\varepsilon > 0$, there exists $T_2 = T_2(V, \Lambda, \varepsilon) > 0$ such that*

$$t \geq T_2 \Rightarrow \#\{\gamma \in \Lambda \mid V\gamma \subset \Phi_t(V)\} \geq \frac{m(\Phi_t(V))}{m(V)}(1 - \varepsilon).$$

PROOF. Indeed, since the contraction bounds c_t of Φ_t tend to 0 as $t \to \infty$, one has $f_V(c_t \cdot \operatorname{diam}(V)) \to 0$ as $t \to \infty$. Thus one can find T_2 such that $f_V(c_t \cdot \operatorname{diam}(V)) \leq \varepsilon m(V)$ for $t \geq T_2$. □

3.6. We return now to the setting of §1, i.e., take H to be the horospherical subgroup of $\{g_t\}$, and Φ_t to be the restriction of the inner automorphism $g \to g_t g g_{-t}$ to H. Let (V, Λ) be any tesselation of H, K a subset of Ω, $x \in \Omega$, $t \in \mathbb{R}$. Say that a translation $\gamma \in \Lambda$ is (x, K, t)-*marked* (or just *marked*, if the prefix is clear from the context) if $V\gamma$ lies entirely inside $V(x, K, t)$, in other words, $V\gamma \subset \Phi_t(V)$ and $V\gamma g_t x = g_t \Phi_t^{-1}(V\gamma)x \subset K$. We will also call *marked* the set $V\gamma$ or $\overline{V}\gamma$ for a marked γ, and denote the number of (x, K, t)-marked sets by $N(x, K, t)$.

The above results now allow one to deduce the following fact about tesselations from the measure-theoretical estimate of Proposition 2.5:

PROPOSITION. *Let K be a subset of Ω with $\bar{\mu}(\partial K) = 0$, L a compact subset of Ω. Then for any $r > 0$ there exists a tesselation (V, Λ) of H such that*

(a) $\operatorname{diam}(V) < r$,

(b) *for any $\varepsilon > 0$ and $\sigma > 0$ there exists $T_0 > 0$ such that*

(3.2)
$$T \geq T_0 \Rightarrow \forall\, x \in L, \quad l\left(\left\{t \in [T, 2T] \mid N(x, K, t) \geq e^{\chi t}\left(\bar{\mu}(K) - \varepsilon\right)\right\}\right) \geq (1 - \sigma)T,$$

and

(c) *assuming $\sigma < 1$ and $T \geq T_0$,*

(3.3) $$\forall\, x \in L \quad \exists\, t(x) \in [T, (1+\sigma)T] \quad \text{such that} \quad N\left(x, K, t(x)\right) \geq e^{\chi t}\left(\bar{\mu}(K) - \varepsilon\right).$$

PROOF. If $\bar{\mu}(K) = 0$, there is nothing to prove. Otherwise, pick a compact subset K' of K with $\bar{\mu}(\partial K') = 0$, which satisfies $\bar{\mu}(K') \geq \bar{\mu}(K) - \varepsilon/3$ and lies at a positive distance from the complement of K. Using Proposition 3.3, find a tesselation domain $V \subset H$ such that $VV^{-1}K' \subset K$ and $\operatorname{diam}(V) < r$. Then (a) is satisfied; moreover, for any $t \in \mathbb{R}$ and $x \in \Omega$,

$$\begin{aligned} N(x, K, t) &\geq N(x, VV^{-1}K', t) \\ &= \#\{\gamma \in \Lambda \mid V\gamma \subset \Phi_t(V)\ \&\ V\gamma g_t x \subset VV^{-1}K'\} \\ &\geq \#\{\gamma \in \Lambda \mid V\gamma \subset \Phi_t(V)\ \&\ \gamma g_t x \subset V^{-1}K'\} \\ &\geq \#\{\gamma \in \Lambda \mid V\gamma \subset \Phi_t(V)\ \&\ V\gamma g_t x \cap K' \neq \varnothing\} \\ &= \#\{\gamma \in \Lambda \mid V\gamma \subset \Phi_t(V)\} - \#\{\gamma \in \Lambda \mid V\gamma g_t x \cap K' = \varnothing\}. \end{aligned}$$

Now take

$$\begin{aligned} T_0 = \max\big\{ &T_1\left(V, K', L, \varepsilon m(V)/3, \sigma\right) \text{ from Proposition 2.5}, \\ &T_2(V, \Lambda, \varepsilon/3) \text{ from Corollary 3.5}\big\}. \end{aligned}$$

Then the minuend is for all $x \in L$ and $t \geq T_0$ not less than $e^{\chi t}(1 - \varepsilon/3)$ by Corollary 3.5. On the other hand, for any $x \in L$ and $T \geq T_0$, the substrahend is by Proposition 2.5 not greater than

$$\begin{aligned} \frac{m\left(\Phi_t(V) \smallsetminus V(x, K', t)\right)}{m(V)} &\leq e^{\chi t}\left(1 - \frac{1}{m(V)} e^{-\chi t} m\left(V(x, K', t)\right)\right) \\ \leq e^{\chi t}\left(1 - \frac{1}{m(V)}\left(m(V)\bar{\mu}(K') - \varepsilon m(V)/3\right)\right) &\leq e^{\chi t}\left(1 - \bar{\mu}(K) + 2\varepsilon/3\right) \end{aligned}$$

for a set of values of $t \in [T, 2T]$ with measure not less than $(1 - \sigma)T$, and part (b) follows; the statement (3.3) in (c) is clearly a direct consequence of (b). □

As before, one can remark that the above statements are true for $\sigma = 0$ if the flow is either semisimple or with property (EM); in particular, one can then take $t(x) = T$ for all $x \in L$.

3.7. In order to deduce a dimension estimate from the above proposition, it will be convenient to use the *density* (relative measure) $\delta(x, K, t)$ of all the marked translates instead of the number $N(x, K, t)$:

(3.4) $$\delta(x, K, t) = \frac{m\left(\bigcup_{\gamma \text{ is } (x,K,t)\text{-marked}} V\gamma\right)}{m\left(\Phi_t(V)\right)} = \frac{m\left(\bigcup_{\gamma \text{ is } (x,K,t)\text{-marked}} \Phi_t^{-1}(V\gamma)\right)}{m(V)};$$

this notation allows one to rewrite (3.3) as follows:
(3.5)
$$t \geq T \Rightarrow \forall x \in L \quad \exists t(x) \in [T, (1+\sigma)T] \quad \text{such that} \quad \delta\big(x, K, t(x)\big) \geq \bar{\mu}(K) - \varepsilon\,.$$

§4. The main construction

4.1. We start with the formal description of a construction (cf. [**F, Mc, U, PW**]) that served as a basic source for producing fractal sets since the times of Cantor and Hausdorff. In what follows, $\mathbb{N}_0$ will stand for the set of nonnegative integers. Let X be a Riemannian manifold, m a Borel measure on X, A_0 a compact subset of X. Say that a countable collection $\mathcal{A}$ of compact subsets of A_0 is *tree-like* relative to m if $\mathcal{A}$ is the union of finite nonempty subcollections $\mathcal{A}_j$, $j \in \mathbb{N}_0$, such that $\mathcal{A}_0 = \{A_0\}$ and the following two conditions are satisfied:

(TL1) $$\forall j \in \mathbb{N} \quad \forall A, B \in \mathcal{A}_j \quad \text{either } A = B \quad \text{or} \quad m(A \cap B) = 0\,;$$

(TL2) $$\forall j \in \mathbb{N} \quad \forall B \in \mathcal{A}_j \quad \exists A \in \mathcal{A}_{j-1} \quad \text{such that} \quad B \subset A\,.$$

Say also that $\mathcal{A}$ is *strongly tree-like* if it is tree-like and in addition

(STL) $$d_j(\mathcal{A}) \stackrel{\text{def}}{=} \sup_{A \in \mathcal{A}_j} \operatorname{diam}(A) \to 0 \quad \text{as} \quad j \to \infty\,.$$

Let $\mathcal{A}$ be a tree-like collection of sets. For each $j \in \mathbb{N}_0$, let $\mathbf{A}_j = \bigcup_{A \in \mathcal{A}_j} A$. These are nonempty compact sets, and from (TL2) it follows that $\mathbf{A}_j \subset \mathbf{A}_{j-1}$ for any $j \in \mathbb{N}$. Therefore one can define the (nonempty) *limit set* of $\mathcal{A}$ to be

$$\mathbf{A}_\infty = \bigcap_{j \in \mathbb{N}_0} \mathbf{A}_l\,.$$

Further, for any subset B of A_0 and $j \in \mathbb{N}$, define the *jth stage density* $\delta_j(B, \mathcal{A})$ of B in $\mathcal{A}$ by

$$\delta_j(B, \mathcal{A}) = \begin{cases} 0, \text{ if } m(B) = 0, \\ \dfrac{m(\mathbf{A}_j \cap B)}{m(B)}, \text{ if } m(B) > 0\,; \end{cases}$$

the condition (TL1) implies that $\delta_j(B, \mathcal{A}) \leq 1$ for any $B \subset A_0$ and $j \in \mathbb{N}$. Then for any $j \in \mathbb{N}_0$ define the *jth stage density* $\Delta_j(\mathcal{A})$ of $\mathcal{A}$ by

$$\Delta_j(\mathcal{A}) = \inf_{B \in \mathcal{A}_j} \delta_{j+1}(B, \mathcal{A})\,.$$

The following estimate, based on an application of Frostman's Lemma, is essentially proved in [**Mc**] and [**U**]:

LEMMA. *Assume that there exist constants $D > 0$ and $k > 0$ such that*

(4.1) $$m\big(B(x, r)\big) \leq Dr^k$$

for any $x \in A_0$ ($B(x, r)$ being a ball of radius r centered at x). Then for any strongly tree-like (relative to m) collection $\mathcal{A}$ of subsets of A_0,

$$\dim(\mathbf{A}_\infty) \geq k - \limsup_{j \to \infty} \frac{\sum_{i=0}^{j-1} \log(\frac{1}{\Delta_i(\mathcal{A})})}{\log(\frac{1}{d_j(\mathcal{A})})}\,.$$

4.2. We are now ready to prove Theorem 1.6.

PROOF OF THEOREM 1.6. Recall that a point $x \in \Omega \smallsetminus Z$ is given. Pick a compact set $K \subset \Omega \smallsetminus Z$ with $\bar{\mu}(\partial K) = 0$ containing some neighborhood of x, and choose a sequence $r_s \to 0$, $r_s < \mathrm{dist}(x, \Omega \smallsetminus K)$. Then, using Proposition 3.6 with $r = r_s$ and $L = K$, find a corresponding tesselation domain V, which will play the role of V_s in Theorem 1.6. By the choice of the sequence $\{r_s\}$, the set Vx is contained in K and (1.6) holds.

Choose ε and σ such that $0 < \varepsilon < \bar{\mu}(K)$ and $0 < \sigma < 1$. Using part (b) of Proposition 3.6, find $T_0 > 0$ such that (3.2) holds. Choose $T = T_s \geq T_0$ such that $T_s \to \infty$ as $s \to \infty$, and denote $(1+\sigma)T$ by T'. Then by (3.5), for any $y \in K$ there exists $t(y) \in [T, T']$ such that

$$\delta(y, K, t(y)) \geq \bar{\mu}(K) - \varepsilon\,. \tag{4.2}$$

Now for all $y \in K$ define strongly tree-like (relative to the Haar measure m on H) collections $\mathcal{A}(y)$ inductively as follows. We first let $\mathcal{A}_0(y) = \{\overline{V}\}$ for all $y \in K$, then define

$$\mathcal{A}_1(y) = \{\Phi_{-t(y)}(\overline{V}\gamma) \mid \gamma \text{ is } (y, K, t(y))\text{-marked}\}\,. \tag{4.3}$$

More generally, if $\mathcal{A}_i(y)$ is defined for all $y \in K$ and $i \leq j$, we let

$$\mathcal{A}_{j+1}(y) = \{\Phi_{-t(y)}(A\gamma) \mid \gamma \text{ is } (y, K, t(y))\text{-marked, } A \in \mathcal{A}_j(\gamma g_{t(y)} y)\} \tag{4.4}$$

(the sets $\mathcal{A}_j(\gamma g_{t(y)} y)$ are defined, since $\gamma g_{t(y)} y \in K$ by virtue of γ being a marked translation).

The properties (TL1) and (TL2) follow readily from the construction and V being a tesselation domain. Also, since $t(y)$ is for all $y \in K$ not less than T, from (4.4) it follows that for all $j \in \mathbb{N}_0$ and $y \in K$, the constant $d_j(\mathcal{A}(y))$ is not greater than $\mathrm{diam}(V) \cdot (c_T)^j$, where $c_T < 1$ is the contraction bound of Φ_T^{-1} on V, and therefore (STL) is satisfied. Moreover, (4.3) and the definition (3.4) of $\delta(x, K, t)$ imply that the density $\delta_1(\overline{V}, \mathcal{A}(y))$ is for all $y \in K$ exactly equal to $\delta(y, k, t(y))$. Hence using (4.4), (4.2) and the relative Φ_t-invariance (2.1.2) of m, one can show by induction that the jth density $\Delta_j(\mathcal{A}(y))$ is for all $y \in K$ and $j \in \mathbb{N}_0$ bounded from below by $\bar{\mu}(K) - \varepsilon$. Finally, the measure m clearly satisfies (4.1) with some positive D and $k = \dim(H)$, and an application of Lemma 4.1 yields that for all $y \in K$,

$$\dim(\mathbf{A}_\infty(y)) \geq k - \limsup_{j\to\infty} \frac{j \log\left(\frac{1}{\bar{\mu}(K)-\varepsilon}\right)}{\log\left(\frac{1}{\mathrm{diam}(V)}\right) + j\log(\frac{1}{c_T})} = k - \frac{\log\left(\frac{1}{\bar{\mu}(K)-\varepsilon}\right)}{\log(\frac{1}{c_T})}\,. \tag{4.5}$$

We now claim that for any h in the limit set $\mathbf{A}_\infty(x)$ of $\mathcal{A}(x)$, the trajectory F^+hx is contained in some compact subset of Ω. Indeed, for all $y \in K$ define a sequence $t_j(y)$, $j \in \mathbb{N}_0$, as follows: let $t_0(y) = 0$ for all $y \in K$, and then, if $t_i(y)$ is defined for all $y \in K$ and $i \leq j$, let

$$t_{j+1}(y) = t(y) + t_j(g_{t(y)} y)\,. \tag{4.6}$$

From (4.4) and (4.6) it immediately follows that $g_{t_j(y)}\mathbf{A}_j(y)y \subset K$ for all $y \in K$ and $j \in \mathbb{N}$. Recall now that V was chosen so that $\mathbf{A}_0(x)x = \overline{V}x \subset K$, therefore $g_{t_j(x)}\mathbf{A}_j(x)x \subset K$ for all $j \in \mathbb{N}_0$, which means that

$$\forall\, j \in \mathbb{N}_0\ \forall\, h \in \mathbf{A}_\infty(x), \quad g_{t_j(x)}hx \subset K\,. \tag{4.7}$$

Now define the set

$$C = C_s = \begin{cases} \bigcup_{t=-T'}^{0} g_t K, & \text{if } Z \text{ is } F^+\text{-invariant}, \\ \bigcup_{t=0}^{T'} g_t K, & \text{if } Z \text{ is } F^-\text{-invariant}. \end{cases}$$

Clearly C is compact and has empty intersection with Z. From (4.6) it easily follows that the difference $t_j - t_{j-1}$ is for any $j \in \mathbb{N}$ bounded from above by T', therefore

$$\forall h \in \mathbf{A}_\infty(x) \quad F^+ hx \subset C\,. \tag{4.8}$$

As s goes to ∞, $T = T_s$ also tends to infinity, thus the contraction bound c_T decreases to 0. Hence the right-hand side of (4.5) tends to k as $s \to \infty$, which, together with (4.8), finishes the proof of (1.7). □

4.3. One can notice that the invariance of the set Z to be avoided was used only once – when deducing a statement about continuous orbits F^+hx (cf. (4.8)) from the corresponding statement about the sequence $\{g_{t_j(x)}hx\}$ (cf. (4.7); see also the proof of Lemma 5.1(a) below, which is a generalization of the same argument). In other words, the construction presented above leads to the proof of the following

THEOREM. *Let G, Γ, Ω and $F = \{g_t\}$ be as in Theorem* 1.1, *and let K be a subset of Ω with $\bar\mu(K) > 0$ and $\bar\mu(\partial K) = 0$. Then given any nonempty open subset W of Ω, $\varepsilon > 0$ and $\sigma > 0$, there exists $T_0 > 0$ such that for any $T \geq T_0$*

$$\dim\left(\left\{x \in W \,\middle|\, \begin{array}{l} \text{there exists a sequence } 0 = t_0 < t_1 < \cdots < t_j < \cdots \\ \text{such that } \forall\, j \in \mathbb{N}_0,\ T \leq t_j - t_{j-1} \leq (1+\sigma)T, \text{ and } g_{t_j}x \in K \end{array}\right\}\right) > n - \varepsilon\,.$$

Moreover, if the flow is either semisimple or with property (EM), *one can take $\sigma = 0$ and $t_j = jT$. In other words, for any nonempty open $W \subset \Omega$ and any $\varepsilon > 0$, there exists $T > 0$ such that for any $t \geq T$*

$$\dim\left(\{x \in W \mid \forall\, j \in \mathbb{N}_0, \quad g_{jt}x \in K\}\right) > n - \varepsilon\,.$$

§5. Generalizations and concluding remarks

5.1. To make the results below easier to state, let us introduce the following notation. For a locally compact metric space X, let X^* denote the topological space $X \cup \{\infty\}$, with the topology defined so that the complements to all the compact sets constitute the basis of the neighborhoods of $\{\infty\}$. In other words, for $A \subset X$, the closure (in X^*) of A contains ∞ iff A is not bounded. Thus if X is compact, ∞ is an isolated point of X^*; otherwise X^* is a one-point compactification of X.

Let now F be any set of maps $X \to X$. We say that a subset Z of X^* is *F-escapable* if for any nonempty open subset W of X

$$\dim\left(\{x \in W \mid \overline{Fx} \cap Z = \varnothing\}\right) = \dim(W)\,,$$

with the closure taken in the topology of X^*.

The main result of the paper, Theorem 1.1, can now be stated as follows:

Let G be a connected semisimple Lie group without compact factors, Γ an irreducible lattice in G, F a (one-parameter or cyclic) nonquasiunipotent subgroup of G. Then for any closed null subset Z of G/Γ which is invariant under either F^+ or F^-, $Z \cup \{\infty\}$ is F-escapable.

Clearly any subset of an F-escapable set is F-escapable, and for any subset F' of F, any F-escapable set is F'-escapable. The partial converse for the latter statement, as well as another easy but useful (see e.g. Remark 1.2) stability property of escapable sets, are stated in the following

LEMMA. (a) *Let F and F' be sets of continuous transformations of X, and let F'' be a compact (in the compact-open topology) set of homeomorphisms $X \to X$; assume that $F \subset F''F'$. Then any F'-escapable subset of X^* which is invariant under $(F'')^{-1}$ (we set $f(\infty) = \infty$ for any map $X \xrightarrow{f} X$) is F-escapable.*

(b) *Let $\tilde{X} \xrightarrow{\varphi} X$ be a locally bi-Lipschitz covering of metric spaces, and F a set of maps $\tilde{X} \to \tilde{X}$ which factor through φ (so that elements of ${}_\varphi F \overset{\text{def}}{=} \varphi \circ F \circ \varphi^{-1}$ are transformations of X). Then $Z \subset X^*$ is ${}_\varphi F$-escapable if and only if $\varphi^{-1}(Z)$ (here $\varphi(\infty) = \infty$) is F-escapable.*

PROOF. For part (a), let $Z \subset X^*$ be $(F'')^{-1}$-invariant; we will show that the set $\{x \in \Omega \mid \overline{Fx} \cap Z = \varnothing\}$ contains $\{x \in \Omega \mid \overline{F'x} \cap Z = \varnothing\}$. Take x from the latter set, then for any $f \in F''$, $f(\overline{F'x}) \cap Z \subset f(\overline{F'x}) \cap f(Z) = f(\overline{F'x} \cap Z) = \varnothing$, and by compactness of F'', the closure of $F''F'x$ for x as above has empty intersection with Z.

For (b), let $Z \in X^*$ be such that $\varphi^{-1}(Z)$ is F-escapable. Take a nonempty open set $W \subset X$ small enough for $\varphi^{-1}(W)$ to consist of finitely many open sets bi-Lipschitz homeomorphic to W; let $\tilde{W}$ be one of these sets. Then φ gives a one-to-one correspondence between $\{x \in \tilde{W} \mid \overline{Fx} \cap \varphi^{-1}(Z) = \varnothing\}$ and $\{x \in W \mid \overline{{}_\varphi Fx} \cap Z = \varnothing\}$, while the dimension of the former set is equal to $\dim(\tilde{W}) = \dim(W)$ by the assumption; hence Z is ${}_\varphi F$-escapable. The proof of the converse statement goes along the same lines; see also Remark 1.2. □

5.2. Our goal now is to extend the class of groups G for which the statement similar to that of Theorem 1.1 is true. First let us consider the special case $Z = \varnothing$, the subject of Conjecture (A) from **[Mrg2]**.

Let G be a connected Lie group, Γ a lattice in G, F a subgroup of G. Say that the flow $(G/\Gamma, F)$ *has property* (Q) if for any connected normal subgroup $N \subset G$ with the quotient map $p : G \to G' \overset{\text{def}}{=} G/N$ such that G' is semisimple without compact factors and $p(\Gamma)$ is an irreducible lattice in G', at least one of the following three conditions is satisfied:

(Q1) $p(\Gamma)$ is cocompact in G';

(Q2) $\operatorname{Ad} p(F)$ is relatively compact;

(Q3) $p(F)$ is not quasiunipotent.

Note that if the flow is semisimple, then for any such p the subgroup $p(F)$ satisfies either (Q2) or (Q3); therefore any semisimple flow has property (Q).

THEOREM. *Let G be a connected Lie group, Γ a lattice in G, $\Omega = G/\Gamma$, F a (one-parameter or cyclic) subgroup of G. The following are equivalent:*

(a) *$\{\infty\}$ is F-escapable;*

(b) *(Ω, F) has property* (Q).

PROOF. First let us show that (a) implies (b). Let p be such that neither of the conditions (Q1), (Q2), and (Q3) holds, in other words, $\Omega' \stackrel{\text{def}}{=} G'/p(\Gamma)$ is not compact, $F' \stackrel{\text{def}}{=} p(F)$ is quasiunipotent, and $\operatorname{Ad} F'$ is not relatively compact. Then F' can be expressed as $F_c F_u$, where F_c and F_u are commuting (one-parameter or cyclic) subgroups such that $\operatorname{Ad} F_c$ is relatively compact and F_u is a nontrivial unipotent subgroup. Dividing by the center of G' and applying the argument similar to that of Remark 1.2 and the proof of Lemma 5.1(b), one can show that for any $x \in \Omega'$, the orbit $F'x$ is bounded if and only if $F_u x$ is bounded. Hence one can assume that F' is unipotent. The results on closures of unipotent orbits **[Rat2]**, see also **[DM,** Theorem 3] imply that the set $A' \stackrel{\text{def}}{=} \{x \in \Omega' \mid F'x \text{ is bounded}\}$ lies in a countable union of proper submanifolds of Ω'. Denote by $\bar{p}$ the induced map of homogeneous spaces $\Omega \to \Omega'$. Then the set $\{x \in \Omega \mid Fx \text{ is bounded}\}$ is contained in $\bar{p}^{-1}(A')$, and by Wegmann's Product Theorem **[Weg]**, the latter set has dimension at most $\dim(G) - 1$.

Now assume that (b) is satisfied. First consider the case when G is connected semisimple without compact factors and (see Remark 1.2) with trivial center. Let $G_1, \dots, G_l$ be connected normal subgroups of G such that $G = \prod_{i=1}^{l} G_i$, $G_i \cap G_j = \{e\}$ if $i \neq j$, $\Gamma_i = G_i \cap \Gamma$ is an irreducible lattice in G_i for each i, and $\prod_{i=1}^{l} \Gamma_i$ has finite index in Γ.

Take any $i \in \{1, \dots, l\}$. Denote by p_i the projection $G \to G_i$, and by Ω_i the homogeneous space $G_i/p_i(\Gamma)$. By assumption (b), at least one of the conditions (Q1), (Q2) or (Q3) (with p_i in place of p and G_i in place of G') is satisfied. Denote by A_i the set $\{x \in \Omega_i \mid p_i(F)x \text{ is bounded}\}$. If either (Q1) or (Q2) holds, then every orbit is bounded, so clearly for any nonempty open $W_i \subset \Omega_i$, the dimension of $W_i \cap A_i$ is equal to $\dim(G_i)$. By virtue of Theorem 1.1, this equality still holds if (Q3) is satisfied. This implies, in view of Wegmann's Theorem or Lemma 1.4, that for any nonempty open $W \subset \prod_{i=1}^{l} \Omega_i$, the dimension of the set $\{x \in W \mid Fx \text{ is bounded}\}$ is equal to the dimension of G. Then (a) follows from Lemma 5.1(b) and the fact that Ω is a finite covering of $\prod_{i=1}^{l} \Omega_i = G/\prod_{i=1}^{l} p_i(\Gamma)$.

Now we consider the general case. Denote by $R(G)$ the radical of G. Then we have $G/R(G) = G_0 \times \hat{G}$, where G_0 is compact and $\hat{G}$ is connected semisimple without compact factors. Let $\pi : G \to \hat{G}$ be the canonical projection, then (cf. **[Rag,** Chapter 9] or **[Dan1,** Lemma 5.1]) $\hat{\Gamma} \stackrel{\text{def}}{=} \pi(\Gamma)$ is a lattice in $\hat{G}$. Denote by $\hat{\Omega}$ the homogeneous space $\hat{G}/\hat{\Gamma}$, and let $\hat{A}$ be the set $\{x \in \hat{\Omega} \mid \pi(F)x \text{ is bounded}\}$. Since $\hat{G}$ is a quotient group of G, property (Q) for the flow (Ω, F) implies the same property for $(\hat{\Omega}, \pi(F))$. Therefore, by the first part of the proof, for any nonempty open $\hat{W} \subset \hat{\Omega}$ the dimension of $\hat{A} \cap \hat{W}$ is equal to $\dim(\hat{G})$.

Denote by $\bar{\pi}$ the induced map of homogeneous spaces $\Omega \to \hat{\Omega}$. Since (cf. the second part of the same lemma from **[Dan1]**) $\operatorname{Ker} \pi \cap \Gamma$ is a uniform lattice in $\operatorname{Ker} \pi$, the preimage $\bar{\pi}^{-1}(x)$ of any point $x \in \hat{\Omega}$ is compact. Therefore the F-orbit of any point from $\bar{\pi}^{-1}(\hat{A})$ is bounded, while Wegmann's Theorem implies that the intersection of $\bar{\pi}^{-1}(\hat{A})$ with any nonempty open subset of Ω has full dimension. □

5.3. The same argument as in the proof of (b)⇒(a) leads to the following

THEOREM. *Let G, Γ, Ω and F be as in Theorem 5.2, and let $Z \subset \Omega$ be invariant under either F^+ or F^-. Assume that for any connected normal subgroup $N \subset G$ with the* *quotient map* $p : G \to G' \stackrel{\text{def}}{=} G/N$ *such that* G' *is semisimple without compact factors and* $p(\Gamma)$ *is an irreducible lattice in* G',

(a) $p(F)$ *is not quasiunipotent, and*

(b) *the closure of* $\bar{p}(Z)$ (*here* $\bar{p}$ *is the induced map of homogeneous spaces* $\Omega \to G'/p(\Gamma)$) *has Haar measure* 0.

Then $Z \cup \{\infty\}$ *is* F*-escapable.*

We conjecture that this theorem is still true if the cumbersome condition (b) is replaced by "the closure of Z has measure 0" as in Theorem 1.1.

5.4. Our next objective is to treat the case of disconnected groups. Let G^0 be the connected component of identity in a Lie group G. If $\Gamma \subset G$ is a lattice, the subgroup $G^0\Gamma$ is of finite index in G, thus $\Omega = G/\Gamma$ consists of finitely many connected components $\Omega_1, \ldots, \Omega_l$, the component Ω_1 of $[e]$ being naturally identified with $\Omega^0 \stackrel{\text{def}}{=} G^0/(G^0 \cap \Gamma) \cong G^0\Gamma/\Gamma$. In this situation, the case $F = \{g^i \mid i \in \mathbb{Z}\}$, with g not contained in G^0, requires separate treatment. First let us consider subgroups F which are contained in G^0 (in particular, one-parameter subgroups of G).

THEOREM. *Let* $\mathcal{F}$ *be a class of subsets of* G^0 *which is invariant under all inner automorphisms of* G, *and for any* $F \in \mathcal{F}$, *let* $\mathcal{G}(F)$ *be another class of subsets of* G^0 *such that the correspondence* $\mathcal{G}$ *is also invariant under inner automorphisms of* G. *Then the following conditions are equivalent*:

(a) *for any* $F \in \mathcal{F}$ *and for any closed null* $Z \subset \Omega$ *which is invariant under at least one of the sets in* $\mathcal{G}(F)$, $Z \cup \{\infty\}$ *is* F*-escapable*;

(b) *for any* $F \in \mathcal{F}$ *and for any closed null* $Z^0 \subset \Omega^0$ *which is invariant under at least one of the sets in* $\mathcal{G}(F)$, $Z^0 \cup \{\infty\}$ *is* F*-escapable.*

PROOF. Since Ω^0 is one of the connected components of Ω, (a) trivially implies (b). For the converse, choose an element g_j in each coset of $G/G^0\Gamma$ ($g_1 = e$), then left multiplication by g_j gives a bijective map $\Omega^0 \to \Omega_j$. Given an open nonempty set $W \subset \Omega$ and a closed null $Z \subset \Omega$, for any j, $1 \le j \le l$, define $W_j = W \cap \Omega_j$ and $Z_j = Z \cap \Omega_j$.

Take any $F \in \mathcal{F}$; since $F \subset G^0$, it leaves each of the components Ω_j invariant. Therefore, the set $\{x \in W \mid \overline{Fx} \cap (Z \cup \{\infty\}) = \varnothing\}$ is the union of l sets

$$\begin{aligned} &\{x \in W_j \mid \overline{Fx} \cap (Z_j \cup \{\infty\}) = \varnothing\} \\ &\quad = g_j\{x \in g_j^{-1}W_j \mid \overline{g_j^{-1}Fg_jx} \cap (g_j^{-1}Z_j \cup \{\infty\}) = \varnothing\}. \end{aligned} \tag{5.1}$$

Let $F' \in \mathcal{G}(F)$ be such that Z is F'-invariant. Take j, $1 \le j \le l$, such that W_j is nonempty. Clearly the closed null set Z_j is F'-invariant, hence the closed null subset $g_j^{-1}Z_j$ of Ω^0 is invariant under $g_j^{-1}F'g_j$. Since $\mathcal{F}$ and $\mathcal{G}$ are invariant under inner automorphisms of G, $g_j^{-1}Fg_j \in \mathcal{F}$ and $g_j^{-1}F'g_j \in \mathcal{G}(F)$. Thus, by assumption (b), at least one (with j as above) of the sets (5.1) has full Hausdorff dimension. $\square$

5.5. COROLLARY. *Let* G, Ω, G^0 *and* Ω^0 *be as in Section 5.4, and let* F *be a (one-parameter or cyclic) subgroup of* G^0. *Then the following are equivalent:*

(a) $\{\infty\} \subset \Omega^*$ *is* F*-escapable*;

(b) (Ω^0, F) *has property* (Q).

PROOF. Consider the class $\mathcal{F}$ of subgroups $F \subset G^0$ such that the flow (Ω^0, F) has property (Q), with the correspondence $\mathcal{G}(F) = \{G^0\}$ for any $F \in \mathcal{F}$. Clearly both $\mathcal{F}$ and $\mathcal{G}$ are invariant under inner automorphisms of G, therefore one can combine Theorems 5.2 and 5.4 to get the desired statement. □

5.6. Before considering subgroups of the form $\{g^i \mid i \in \mathbb{Z}\}$ with g not contained in G^0, we will prove an auxillary result which seems to be of independent interest, indicating yet another direction for the generalization of Theorem 1.1.

THEOREM. *Let G, Γ and Ω be as in §1, and let g and γ be two elements of G such that g is not quasiunipotent and γ normalizes Γ. Define the map $f : \Omega \to \Omega$ by $f(h\Gamma) = gh\gamma\Gamma$, $h \in G$, and denote by F the cyclic subgroup generated by f. Then for any closed null subset Z of Ω which is invariant under either f or f^{-1}, $Z \cup \{\infty\}$ is F-escapable.*

PROOF. Denote by Γ' the normalizer of Γ in G. Then it is known [**Rag,** Corollary 5.17] that Γ' is a lattice in G containing Γ, therefore Ω is a finite covering of the homogeneous space $\Omega' \overset{\text{def}}{=} G/\Gamma'$, and the covering map φ sends the transformation f of Ω to the left multiplication by g in Ω'. In other words, ${}_{\varphi}F$ can be identified with the cyclic subgroup generated by g. For any closed null subset Z of Ω which is invariant under either f or f^{-1}, $\varphi(Z)$ will be closed, null and invariant under either g or g^{-1}, hence ${}_{\varphi}F$-escapable by Theorem 1.1. Lemma 5.1(b) now implies that the set $\varphi^{-1}(\varphi(Z))$ (which contains Z) is F-escapable. □

5.7. THEOREM. *Let G be a semisimple Lie group such that G^0 is without compact factors, Γ a lattice in G such that $\Gamma^0 \overset{\text{def}}{=} G^0 \cap \Gamma$ is irreducible, and let F be a (one-parameter or cyclic) nonquasiunipotent subgroup of G. Then for any closed null subset Z of Ω which is invariant under either F^+ or F^-, $Z \cup \{\infty\}$ is F-escapable.*

PROOF. The case $F \subset G^0$ follows from Theorems 1.1 and 5.4 (in the latter, $\mathcal{F}$ stands for the class of (one-parameter or cyclic) nonquasiunipotent subgroups of G^0, with $\mathcal{G}(F) = \{F^+, F^-\}$ for $F \in \mathcal{F}$). Now pick any (nonquasiunipotent) $g \in G$; since $G^0\Gamma$ is of finite index in G, one can find $r \in \mathbb{N}$ such that $g^r \in \bigcap_{h \in G} h^{-1}G^0\Gamma h$, then g^r leaves invariant every connected component Ω_j of Ω. Arguing as in the proof of Theorem 5.4 and using Lemma 5.1(a), one can reduce the problem to studying the action of g^r on Ω^0.

Write g^r in the form $h\gamma$, where $h \in G^0$ and $\gamma \in \Gamma$. For any $s \in \mathbb{N}$ and any $g^0 \in G^0$, multiplication by g^{rs} sends $g^0\Gamma^0$ to $g^{rs}g^0\gamma^{-s}\Gamma^0 = \tilde{h}\gamma^s g^0\gamma^{-s}\Gamma^0$, where $\tilde{h}$ is an element of G^0 depending on s. Since G^0 is a connected semisimple group assumed to be centerfree (see Remark 1.2), the group of its inner automorphisms has finite index in $\operatorname{Aut} G^0$. Thus for some $s \in \mathbb{N}$, $\gamma^s = az$, where $a \in G^0$ and z lies in the centralizer of G^0 in G. For this s, multiplication by g^{rs} sends $g^0\Gamma^0$ to $\tilde{h}ag^0a^{-1}\Gamma^0$.

Observe now that $a^{-1} = \gamma^{-s}z$ lies in the normalizer of Γ^0 in G^0, and $\tilde{h}a = g^{rs}z^{-1}$ is a nonquasiunipotent element of G^0. Therefore Theorem 5.6 implies that for any closed null subset Z of Ω^0 which is invariant under the action of either g or g^{-1}, $Z \cup \{\infty\}$ is $\{g^{rsi} \mid i \in \mathbb{Z}\}$-escapable, and another application of Lemma 5.1(a) finishes the proof. □

The above proof shows that the statement of the main result (the analogue of Theorem 5.2 and Corollary 5.5) for the general case of a Lie group G and any nonquasiunipotent cyclic subgroup F of G would be more complicated. However, one can modify the proof of Theorem 5.2 to show that $\{\infty\}$ is F-escapable for any (one-parameter or cyclic) F which consists of semisimple elements.

5.8. Let G be semisimple without compact factors, and Γ irreducible and nonuniform (that is, the homogeneous space Ω noncompact). Then from Theorem 1.1 and the ergodicity of F-action on Ω it follows that for a dense subset Ω_F of Ω the closure of the orbit Fx has Haar measure 0 for any $x \in \Omega_F$. Applying Theorem 1.1 with $Z = \bigcup_{j=1}^{l} \overline{Fx_j}$, where $x_1, \dots, x_l$ are from Ω_F, one can estimate the dimension of the set of points whose orbits are bounded and stay away from any given finite subset of this dense set.

THEOREM. *Let G, Ω, F and Z be as in Theorem* 1.1; *assume that Ω is not compact. Then there exist a set Ω_F satisfying* $\dim(W \cap \Omega_F) = n$ *for any nonempty open subset W of Ω, such that for any l points $x_1, \dots, x_l \in \Omega_F$, $l \in \mathbb{N}$, the set $Z \cup \{x_1, \dots, x_l, \infty\}$ is F-escapable.*

Conjecture (B) in **[Mrg2]** states that the above (with $Z = \varnothing$) should be true without any restrictions on G, Ω and the choice of the points x_j; this is going to be the topic of a forthcoming paper. Let us note that related problems were considered in **[Dan5]** for endomorphisms of tori and in **[U]** for Anosov flows on compact Riemannian manifolds.

5.9. A special case of the flows studied above is the geodesic flow on the unit tangent bundle SM of a Riemannian manifold M of constant negative curvature and finite Riemannian volume. Since the ambient group in this case has $\mathbb{R}$-rank 1, the fact that the set of points in SM with bounded geodesics has full Hausdorff dimension follows from the result of Dani **[Dan3]** (see also **[AL]** for a strengthening of this result). In **[Dan3]** it was asked whether an analogous statement would be true for manifolds of variable negative curvature. A certain progress in this direction has been recently obtained by D. Dolgopiat **[Do]**.

It is worthwhile to note that in **[Do]** the case of infinite volume is also considered, the goal being to prove that the set of points in SM with bounded geodesics has the same Hausdorff dimension as the set of points with geodesics returning to some compact subset of SM infinitely often. In the case when M is a rank-1 locally symmetric space this is done in **[BJ]** and **[Do]**. It would certainly be interesting to know whether a similar assertion holds for flows on homogeneous spaces of higher rank semisimple Lie groups.

Appendix

A.1. The first estimates on the decay of matrix coefficients of $\mathcal{K}$-finite vectors ($\mathcal{K}$ a maximal compact subgroup of a semisimple Lie group G) appeared in the work of Harish-Chandra, and then were refined by several people. Katok and Spatzier **[KS]** used Howe's estimates **[H]** for the $\mathcal{K}$-finite case, as well as Cowling's results **[Cow]** on strongly L^p representations, to prove their exponential estimate (cf. Theorem 2.4.3 above) for C^∞-vectors. Here we will use their treatment of C^∞-case to show exponential decay of matrix coefficients of Hölder vectors. This was done earlier for certain special cases by C. C. Moore **[Mo2]** and M. Ratner **[Rat1]**.

Let G be a Lie group with left Haar measure μ, ρ a unitary representation of G on a Hilbert space $\mathcal{H}$. Say that a vector $v \in \mathcal{H}$ is *Hölder* with exponent $\alpha > 0$ if

$$C = \sup_{g \in G - \{e\}} \frac{\|\rho(g)v - v\|}{\operatorname{dist}(e,g)^\alpha} < \infty ;$$

we will refer to the number C as the *α-Hölder coefficient* of v, and say that the vector v is *(C,α)-Hölder*, or *α-Hölder*, if the coefficient C is irrelevant.

A.2. REMARK. For $r > 0$, denote by $B(r)$ the open ball in G of radius r centered in e. The representation ρ can be extended to the Banach algebra $L^1(G)$: for $\varphi \in L^1(G)$ and $v \in \mathcal{H}$, $\rho(\varphi)v \stackrel{\text{def}}{=} \int_G \varphi(g)\rho(g)v \, d\mu$. Clearly the norm of $\rho(\varphi)$ is not greater than the L^1-norm of φ: for any $v \in \mathcal{H}$, $\|\rho(\varphi)v\| \le \|\varphi\|_{L^1} \|v\|$. It is also clear that if v is (C,α)-Hölder and φ is a.e. nonnegative with $\|\varphi\|_{L^1} = 1$ and $\operatorname{supp}(\varphi) \subset B(r)$, then $\|\rho(\varphi)v - v\| \le Cr^\alpha$.

A.3. Let $\mathfrak{k}$ be the Lie algebra of a maximal compact subgroup $\mathcal{K}$ of G, and, as in Section 2.4, denote by Υ the element $1 - \sum Y_j^2$, where $\{Y_j\}$ is an orthonormal basis of $\mathfrak{k}$. We will need the following simple lemma.

LEMMA. (cf. Lemma 2.4.7(b)). *There exists a constant A_1 depending only on G such that if $0 < r < 1$ and $l \in \mathbb{N}$, one can find a nonnegative function $\varphi \in C^\infty_{\text{comp}}(G)$ with* $\operatorname{supp}(\varphi) \subset B(r)$, $\int_G \varphi \, d\mu = 1$, *and* $\|\Upsilon^l(\varphi)\|_{L^1} \le A_1 r^{-2l}$.

A.4. THEOREM. *Let G, $\mathfrak{a}$, $\mathfrak{c}$, ϑ and Π be as in Theorem 2.4.3, and let v_i, $i = 1,2$, be (C_i,α_i)-Hölder vectors in the representation space of $\rho \in \Pi$. Then for any $Y \in \bar{\mathfrak{c}}$ and for any $t \ge 0$,*

$$|\left(\rho(\exp(tY))v_1, v_2\right)| \le \left(A_1^2 B\|v_1\|\|v_2\| + C_1\|v_2\| + C_2\|v_1\| + C_1C_2\right)e^{-\frac{t\xi}{2p}\vartheta(Y)},$$

where $\xi = \left(1 + \frac{2l}{\alpha_1} + \frac{2l}{\alpha_2}\right)^{-1}$, *$B$, p and l are from Theorem 2.4.3, and A_1 is from Lemma A.3.*

PROOF. Take $a_1 > 0$, $a_2 > 0$, $r_1 = e^{-\frac{ta_1}{2p}\vartheta(Y)}$, $r_2 = e^{-\frac{ta_2}{2p}\vartheta(Y)}$. Using Lemma A.3, find nonnegative C^∞-functions φ_1, φ_2 on G with $\int_G \varphi_i \, d\mu = 1$, $\operatorname{supp}(\varphi_i) \subset B(r_i)$ and

$$\|\Upsilon^l(\varphi_i)\|_{L^1} \le A_1 r_i^{-2l}, \quad i = 1,2. \tag{A.1}$$

Clearly $\rho(\varphi_1)v_1$ and $\rho(\varphi_2)v_2$ are C^∞-vectors, and (see Remark A.2)

$$\|\rho(\varphi_i)v_i - v_i\| \le C_i r_i^{\alpha_i}, \quad i = 1,2. \tag{A.2}$$

We now apply Theorem 2.4.3 to the vectors $\rho(\varphi_1)v_1$ and $\rho(\varphi_2)v_2$:

$$\begin{aligned}
|\left(\rho(\exp(tY))\rho(\varphi_1)v_1, \rho(\varphi_2)v_2\right)| &\le Be^{-\frac{t}{2p}\vartheta(Y)}\|\Upsilon^l(\rho(\varphi_1)v_1)\|\|\Upsilon^l(\rho(\varphi_2)v_2)\| \\
&= Be^{-\frac{t}{2p}\vartheta(Y)}\|\rho(\Upsilon^l(\varphi_1))v_1\|\|\rho(\Upsilon^l(\varphi_2))v_2\| \\
\text{(see Remark A.2)} \quad &\le Be^{-\frac{t}{2p}\vartheta(Y)}\|\Upsilon^l(\varphi_1)\|_{L^1}\|\Upsilon^l(\varphi_2)\|_{L^1}\|v_1\|\|v_2\| \\
\text{(by (A.1))} \quad &\le A_1^2 Be^{-\frac{t}{2p}\vartheta(Y)} r_1^{-2l} r_2^{-2l}\|v_1\|\|v_2\| \\
\text{(by definition of } r_1 \text{ and } r_2) \quad &\le A_1^2 Be^{-\frac{t}{2p}\vartheta(Y)(1-2la_1-2la_2)}\|v_1\|\|v_2\| .
\end{aligned}$$

From (A.2) and the unitarity of ρ it follows that

$$|\big(\rho\big(\exp(tY)\big)\big(\rho(\varphi_1)v_1 - v_1\big), \rho(\varphi_2)v_2\big)| \le C_1 r_1^{\alpha_1}\|v_2\| = C_1\|v_2\|e^{-\frac{t}{2p}\vartheta(Y)a_1\alpha_1},$$
$$|\big(\rho\big(\exp(tY)\big)v_1, \rho(\varphi_2)v_2 - v_2\big)| \le C_2 r_2^{\alpha_2}\|v_1\| = C_2\|v_1\|e^{-\frac{t}{2p}\vartheta(Y)a_2\alpha_2},$$
$$|\big(\rho\big(\exp(tY)\big)\big(\rho(\varphi_1)v_1 - v_1\big), \rho(\varphi_2)v_2 - v_2\big)| \le C_1C_2 r_1^{\alpha_1} r_2^{\alpha_2} = C_1C_2 e^{-\frac{t}{2p}\vartheta(Y)(a_1\alpha_1 + a_2\alpha_2)}.$$

Therefore

(A.3)
$$\begin{aligned}|\big(\rho\big(\exp(tY)\big)v_1, v_2\big)| &\le |\big(\rho\big(\exp(tY)\big)\rho(\varphi_1)v_1, \rho(\varphi_2)v_2\big)| \\ &\quad + |\big(\rho\big(\exp(tY)\big)v_1, \rho(\varphi_2)v_2 - v_2\big)| \\ &\quad + |\big(\rho\big(\exp(tY)\big)\big(\rho(\varphi_1)v_1 - v_1\big), \rho(\varphi_2)v_2\big)| \\ &\quad + |\big(\rho\big(\exp(tY)\big)\big(\rho(\varphi_1)v_1 - v_1\big), \rho(\varphi_2)v_2 - v_2\big)| \\ &\le A_1^2 B e^{-\frac{t}{2p}\vartheta(Y)(1-2la_1-2la_2)}\|v_1\|\|v_2\| + C_2\|v_1\|e^{-\frac{t}{2p}\vartheta(Y)a_2\alpha_2} \\ &\quad + C_1\|v_2\|e^{-\frac{t}{2p}\vartheta(Y)a_1\alpha_1} + C_1C_2e^{-\frac{t}{2p}\vartheta(Y)(a_1\alpha_1+a_2\alpha_2)}.\end{aligned}$$

Now choose a_1 and a_2 such that

$$\alpha_1 a_1 = \alpha_2 a_2 = 1 - 2la_1 - 2la_2 = \xi = \left(1 + \frac{2l}{\alpha_1} + \frac{2l}{\alpha_2}\right)^{-1}.$$

Then from (A.3) one gets

$$\begin{aligned}|\big(\rho\big(\exp(tY)\big)v_1, v_2\big)| &\le \big(A_1^2B\|v_1\|\|v_2\| + C_1\|v_2\| + C_2\|v_1\|\big)e^{-\frac{t\xi}{2p}\vartheta(Y)} \\ &\quad + C_1C_2e^{-\frac{2t\xi}{2p}\vartheta(Y)} \\ &\le \big(A_1^2B\|v_1\|\|v_2\| + C_1\|v_2\| + C_2\|v_1\| + C_1C_2\big)e^{-\frac{t\xi}{2p}\vartheta(Y)},\end{aligned}$$

which is exactly the desired statement. □

A.5. Let now G, Γ and Ω be as in Theorem 1.1, and let $\{g_t \mid t \in \mathbb{R}\}$ be a one-parameter subgroup of G such that the condition (EM) is satisfied.

THEOREM. *There exists a constant $A_2 > 0$, depending only on G, such that if $\psi_i \in L_0^2(\Omega)$, $i = 1, 2$, are (C_i, α_i)-Hölder, then for any $t \ge 0$,*

$$|(g_t\psi_1, \psi_2)| \le \big(A_2E\|\psi_1\| \cdot \|\psi_2\| + C_1\|\psi_2\| + C_2\|\psi_1\| + C_1C_2\big)e^{-\gamma\eta t},$$

where γ, E and l are from Theorem 2.4.5, *and* $\eta = \left(1 + \frac{l}{\alpha_1} + \frac{l}{\alpha_2}\right)^{-1}$.

The proof is based on the same kind of argument that was used in the proofs of Theorems 2.4.5 and A.4. (Instead of functions φ_1, φ_2 on G one should consider C^∞ functions supported in suitable neighborhoods of identity in the group G^e.)

A.6. Say that a function on a metric space X with distance "dist" is *α-Hölder*, if

$$\sup_{x,y\in X,\, x\neq y} \frac{|f(x) - f(y)|}{\operatorname{dist}(x,y)^\alpha} < \infty.$$

Clearly if Γ is a closed subgroup of a Lie group G, then any Hölder square-integrable function on G/Γ is a Hölder vector of a regular representation of G on $L^2(G/\Gamma)$.

The following Hölder analogue of Proposition 2.4.8 can be proved:

PROPOSITION. *Let the condition* (EM) *be satisfied. Then for any α_1-Hölder compactly supported function $f \in L^2(H)$, for any α_2-Hölder function $\psi \in L^2(\Omega)$ and for any compact subset L of Ω, there exists a constant $C = C(f, \psi, L)$ such that for all $x \in L$ and for any $t \geq 0$,*

$$\left| e^{-\chi t} \int_H f(g_t h g_{-t}) \psi(h g_t x)\, dm(h) - \int_H f\, dm \int_\Omega \psi\, d\bar{\mu} \right| \leq C e^{-\zeta t},$$

where

$$\zeta = \gamma \left(1 + \frac{l}{\alpha_1} + \frac{n-k+3l}{\alpha_2} + \frac{3(n-k)l}{\alpha_1 \alpha_2} + \frac{(n-k)l}{\alpha_2^2} \right)^{-1},$$

with γ and l from Theorem 2.4.5.

An indication for the proof is given in the next section, where a more general theorem is considered. Note that one of the reasons why the Hölder setting for this situation seems to be natural is that the constant $d(\psi)$ which appears in the proof of Proposition 2.4.8 is nothing but the Hölder coefficient of ψ corresponding to the exponent 1.

A.7. Finally, we note that similar methods can be applied to the situation of Section 2.1.3, where we considered integrals of type $\int_M f(y)\psi(g_t\pi(y))\, \omega_\pi(y)$ (M being a k-dimensional manifold and π an immersion $M \to \Omega$ with image transversal to the $\tilde{H}$-orbit foliation at any point). More generally, one can prove the following generalization of the above proposition:

THEOREM. *Let M be a smooth k-dimensional Riemannian manifold, and let $\{\pi_q \mid q \in Q\}$ be a compact family of C^∞ immersions $M \to \Omega$ such that $\pi_q(M)$ is transversal to the orbit $\tilde{H}\pi(y)$ for every $q \in Q$ and $y \in M$. Denote by ω_q the pullback of the form ω_H to M via π_q. Further, let $\{f_q \mid q \in Q\}$ be a family of functions on M with common compact support M_0 such that f_q are uniformly α_1-Hölder for some $\alpha_1 > 0$, that is*

$$\sup_{q \in Q,\, x, y \in M,\, x \neq y} \frac{|f_q(x) - f_q(y)|}{\operatorname{dist}(x, y)^{\alpha_1}} < \infty.$$

Assume that the condition (EM) *is satisfied. Then for any α_2-Hölder function $\psi \in L^2(\Omega)$ there exists a constant $C = C(\{\pi_q\}, \{f_q\}, \psi)$ such that for all $q \in Q$ and $t \geq 0$,*

$$\left| \int_M f_q(y) \psi(g_t \pi_q(y))\, \omega_q(y) - \int_M f_q\, \omega_q \int_\Omega \psi\, d\bar{\mu} \right| \leq C e^{-\zeta t}, \tag{A.4}$$

where ζ is as in Proposition A.6.

SKETCH OF PROOF. Using a smooth partition of unity on M_0, one can assume that all the immersions π_q are injective when restricted to M_0. Similarly to what was done in the proof of Proposition 2.4.8, for sufficiently large t choose $r = e^{-at}$ and α_1-Hölder nonnegative functions f^0 on H^0 and f^- on H^- satisfying

$$\int_{H^-} f^-\, dm^- = 1 = \int_{H^0} f^0\, dm^0 \quad \text{and} \quad \operatorname{supp}(f^-) \cdot \operatorname{supp}(f^0) \subset \tilde{B}(r^2),$$

and such that the Hölder coefficient of the function $f^- \cdot f^0$ is bounded from above by const $\cdot r^{-2(n-k+\alpha_1)}$. Then, using f^-, f^0 and f_q, build an α_1-Hölder function on Ω, and apply the averaging trick (with radii $r_1 = e^{-a_1\gamma t}$ and $r_2 = e^{-a_2\gamma t}$) from the proof of Theorem A.4 to the latter function and ψ to get an upper estimate for the left hand side of (A.4), which can be then minimized by choosing appropriate values of the parameters a, a_1 and a_2. The critical values satisfy the following system of equations:

$$\zeta = \alpha_2 a = \gamma(1 - la_1 - la_2) - (n-k)a = \gamma\alpha_2 a_2 - (n-k)a = \gamma\alpha_1 a_1 - 2(n-k+\alpha_1)a\,,$$

which yields the value of ζ as in the Proposition A.6. □

A.8. COROLLARY. *Let, as before, $\mathcal{K}$ stand for a maximal compact subgroup of G, and let ν be a normalized Haar measure on $\mathcal{K}$. Let $\{g_t\}$ be a one-parameter subgroup of a maximal split torus of G such that the flow $(\Omega, \{g_t\})$ has property* (EM). *Then for any α-Hölder function ψ in $L^2(\Omega)$ the spherical averages of ψ converge exponentially to $\int_\Omega \psi\, d\bar{\mu}$. More precisely, for any compact subset L of Ω there is a constant $C = C(\psi, L)$ such that for all $t \geq 0$,*

$$\text{(A.5)} \qquad \left| \int_{\mathcal{K}} \psi(g_t h x)\, d\nu(h) - \int_\Omega \psi\, d\bar{\mu} \right| \leq C e^{-\zeta_1 t}\,,$$

where

$$\zeta_1 = \gamma \left(1 + l + \frac{n-k+3l+3(n-k)l}{\alpha} + \frac{(n-k)l}{\alpha^2} \right)^{-1}$$

is the value of ζ from Proposition A.6 *corresponding to $\alpha_1 = 1$ and $\alpha_2 = \alpha$.*

SKETCH OF PROOF. Consider a smooth finite partition of unity $\{f_j\}$ on $(\mathcal{K} \cap H^0)\backslash\mathcal{K}$ subordinate to a covering $\{U_j\}$ such that the bundle $\mathcal{K} \to (\mathcal{K} \cap H^0)\backslash\mathcal{K}$ is trivial over each of U_j. For each j choose a smooth section $\varphi_j : U_j \to \mathcal{K}$, then for any $x \in \Omega$, the set $\varphi(U_j)x$ will be transversal to the orbit $\tilde{H}x$. Thus one can apply Theorem A.7 to f_j, the immersions $\{\pi_x \circ \varphi_j \mid x \in L\}$ and to the function obtained from ψ by averaging over the translations by elements of $\mathcal{K} \cap H^0$. □

Note that under the assumption that the flow (Ω, F) is semisimple, one can prove in the above corollary (resp. Theorem A.7) that the left hand side of (A.5) (resp. (A.4)) tends to zero as $t \to \infty$.

Acknowledgements

The first author wants to thank Yuval Peres and Dimitrios Gatzouras for helpful discussions.

References

[AL] C. S. Aravinda and F. E. Leuzinger, *Bounded geodesics in rank-1 locally symmetric spaces*, Ergodic Theory Dynamical Systems (to appear).

[Bou] N. Bourbaki, *Éléments de mathematique*, Fascicule XXIX, Livre VI: Integration, Chapitres 7 et 8, Hermann, Pàris, 1963.

[Bow] R. Bowen, *Weak mixing and unique ergodicity on homogeneous spaces*, Israel J. Math. **23** (1976), 267–273.

[BJ] C. Bishop and P. W. Jones, *Hausdorff dimension and Kleinian groups*, Acta Math. (to appear).

[Cow] M. Cowling, *Sur le coefficients des représentations unitaries des groupes de Lie simple*, Lecture Notes in Math., vol. 739, Springer-Verlag, Berlin and New York, 1979, pp. 132–178.

[CG] L. J. Corwin and F. P. Greenleaf, *Representations of nilpotent Lie groups and their applications.* Part I, Cambridge Stud. Adv. Math., vol. 18, Cambridge Univ. Press, Cambridge, 1990.

[Dan1] S. G. Dani, *Invariant measures of horospherical flows on noncompact homogeneous spaces*, Invent. Math. **47** (1978), 101–138.

[Dan2] ———, *Divergent trajectories of flows on homogeneous spaces and Diophantine approximation*, J. Reine Angew. Math. **359** (1985), 55–89.

[Dan3] ———, *Bounded orbits of flows on homogeneous spaces*, Comment. Math. Helv. **61** (1986), 636–660.

[Dan4] ———, *On orbits of unipotent flows on homogeneous spaces*, II, Ergodic Theory Dynamical Systems **6** (1986), 167–182.

[Dan5] ———, *On orbits of endomorphisms of tori and the Schmidt game*, Ergodic Theory Dynamical Systems **8** (1988), 523–529.

[Dav] R. O. Davies, *Subsets of finite measure in analytic sets*, Indag. Math. **14** (1952), 488–489.

[DM] S. G. Dani and G. A. Margulis, *Limit distributions of orbits of unipotent flows and values of quadratic forms*, Adv. in Soviet Math., vol. 16, Amer. Math. Soc., Providence, RI, 1993, pp. 91–137.

[Do] D. Dolgopiat, *Bounded orbits of Anosov flows*, submitted to Duke Math. J.

[EM] A. Eskin and C. McMullen, *Mixing, counting and equidistribution in Lie groups*, Duke Math. J. **71** (1993), 181–209.

[EP] R. Ellis and W. Perrizo, *Unique ergodicity of flows on homogeneous spaces*, Israel J. Math. **29** (1978), 276–284.

[F] K. J. Falconer, *The geometry of fractal sets*, Cambridge Tracts in Math., vol. 85, Cambridge Univ. Press, Cambridge and New York, 1986.

[H] R. Howe, *A notion of rank for unitary representations of the classical groups*, Harmonic Analysis and Group Representations, Liguori, Naples, 1982, pp. 223–331.

[KS] A. Katok and R. Spatzier, *Differential rigidity of hyperbolic abelian actions*, Inst. Hautes Études Sci. Publ. Math. **79** (1994), 131–156.

[Mag] W. Magnus, *Noneuclidean tesselations and their groups*, Academic Press, New York and London, 1974.

[Mc] C. McMullen, *Area and Hausdorff dimension of Julia sets of entire functions*, Trans. Amer. Math. Soc. **300** (1987), 329–342.

[Mo1] C. C. Moore, *Ergodicity of flows on homogeneous spaces*, Amer. J. Math. **88** (1966), 154–178.

[Mo2] ———, *Exponential decay of correlation coefficients for geodesic flows*, Group Representations, Ergodic Theory, Operator Algebras and Mathematical Physics, Math. Sci. Res. Inst. Publ., vol. 6, Springer-Verlag, Berlin and New York, 1987, pp. 163–181.

[Mrc] B. Marcus, *Unique ergodicity of the horocycle flow: variable negative curvature case*, Israel J. Math. **21** (1975), 133–144.

[Mrg1] G. A. Margulis, *On some problems in the theory of U-systems*, Ph.D. Thesis, Moscow State University, 1970. (Russian)

[Mrg2] ———, *Dynamical and ergodic properties of subgroup actions on homogeneous spaces with applications to number theory*, Proceedings of the International Congress of Mathematicians, Vol. I, II (Kyoto, 1990), Math. Soc. Japan, Tokyo, 1991, pp. 193–215.

[Mrs] J. M. Marstrand, *The dimension of Cartesian product sets*, Proc. Camb. Phil. Soc. **50** (1954), 198–202.

[PW] Ya. Pesin and H. Weiss, *On the dimension of deterministic and random Cantor-like sets*, Math. Res. Lett. **1** (1994), 519–529.

[Rag] M. S. Raghunathan, *Discrete subgroups of Lie groups*, Springer-Verlag, Berlin and New York, 1972.

[Rat1] M. Ratner, *The rate of mixing for geodesic and horocycle flows*, Ergodic Theory Dynamical Systems **7** (1987), 267–288.

[Rat2] ———, *Raghunathan's topological conjecture and distribution of unipotent flows*, Duke Math. J. **63** (1991), 235–280.

[S1] W. M. Schmidt, *On badly approximable numbers and certain games*, Trans. Amer. Math. Soc. **123** (1966), 178–199.

[S2] ———, *Badly approximable systems of linear forms*, J. Number Theory **1** (1969), 139–154.

[U] M. Urbanski, *The Hausdorff dimension of the set of points with nondense orbit under a hyperbolic dynamical system*, Nonlinearity **2** (1991), 385–397.

[Weg] H. Wegmann, *Die Hausdorff-Dimension von kartesischen Produktmengen in metrischen Räumen*, J. Reine Angew. Math. **234** (1969), 163–171.

D. Y. Kleinbock, Department of Mathematics, Yale University, New Haven, CT 06520
E-mail address: `kleinboc@math.yale.edu`

G. A. Margulis, Department of Mathematics, Yale University, New Haven, CT 06520
E-mail address: `margulis@math.yale.edu`

Added in proof (September 20, 1995)

Yehuda Shalom informed us that Theorem 2.4.3 is not correct as stated. Therefore the following changes should be made.

1. The first sentence in the statement of Theorem 2.4.3 should be

Let G be a connected semisimple Lie group with finite center, and let Π be a family of unitary representations of G such that the restriction of Π to any simple factor of G is isolated from the trivial representation.

The conclusion of the theorem, as well as the proof, remain unchanged.

2. In the proof of Theorem 2.4.5, one should apply Corollary 2.4.4 to simple factors of G^{e} rather than to the group G^{e} itself. Specifically, the proof should be modified as follows:

PROOF. Let G_j be the simple factors of G, and let $\{g_{t,j}\}$ and $\{g'_{t,j}\}$ denote the one-parameter subgroups of G which are natural projections of $\{g_t\}$ onto G_j and $\prod_{i \neq j} G_j$ respectively. From the definition of G^{e} it follows that for any $t \geq 0$ $\mathrm{dist}(e, g_{t,j})$ is bounded from below by βt for some $\beta > 0$ whenever $G_j \subset G^{\mathrm{e}}$. Hence by Corollary 2.4.4 and property (EM), for any irreducible component ρ of ρ_0 one can choose j such that $G_j \subset G^{\mathrm{e}}$ and for any two C^∞-vectors v and w of $\rho|_{G_j}$,

$$\big|(\rho(g_{t,j})v, w)\big| \leq E e^{-\alpha\beta t} \|\Upsilon^l(v)\| \|\Upsilon^l(w)\|, \tag{2.4.4}$$

where $E > 0$, $l \in \mathbb{N}$, and $\alpha > 0$ do not depend on ρ.

Now restrict ρ to the group $G_j \times \{g'_{t,j}\}$ and then decompose this restriction into a direct integral of irreducible representations. Each of them will be a tensor product of $\rho|_{G_j}$ and a one-dimensional representation of $\{g'_{t,j}\}$. Therefore, the estimate (2.4.3) for matrix coefficients $(g_t\varphi, \psi)$ can be obtained by integration of inequalities of type (2.4.4). □

Amer. Math. Soc. Transl.
(2) Vol. **171**, 1996

On Solutions of Infinite-Dimensional Systems of Ordinary Differential Equations Originating in Statistical Mechanics

T. V. Lokot′ and L. D. Pustyl′nikov

ABSTRACT. Solutions of some infinite-dimensional systems of ordinary differential equations generalizing Frenkel′-Kontorova equation are studied. These systems describe motions of atoms in multi-dimensional rigid body. The families of periodic and almost periodic solutions are constructed that have continual, countable, or arbitrary finite number of rationally independent frequencies in any neighborhood of a stationary solution.

Introduction

Let r be a natural number, $\mathbb{Z}_r$ the integer lattice in the r-dimensional real space $\mathbb{R}_r$, $\vec{n} = (n_1, \dots, n_r) \in \mathbb{Z}_r$. We consider points (or particles) $P_{\vec{n}}$ which we number by elements $\vec{n} \in \mathbb{Z}_r$.

Further, we assume that each particle $P_{\vec{n}}$ interacts only with the neighboring particles along each direction in $\mathbb{Z}_r$ in accordance with the law of elastic spring. We also assume that each particle is under an action of a field with periodic potential along each coordinate axis in $\mathbb{R}_r$. As a result the potential $\mathcal{U}$ of the interaction has the following form:

$$\mathcal{U} = \gamma \sum_{\vec{n} \in \mathbb{Z}_r} \sum_{S=1}^{r} f^{(S)}(q_{\vec{n}}^{(S)}) + \chi \sum_{(\vec{n}', \vec{n}'') \in \Gamma} |q_{\vec{n}'} - q_{\vec{n}''} - \vec{a}|^2,$$

where

$$q_{\vec{n}} = (q_{\vec{n}}^{(1)}, \dots, q_{\vec{n}}^{(r)}) \in \mathbb{R}_r, \qquad \vec{n}' = (n_1', \dots, n_r') \in \mathbb{Z}_r,$$
$$\vec{n}'' = (n_1'', \dots, n_r'') \in \mathbb{Z}_r, \qquad q_{\vec{n}'} = (q_{\vec{n}'}^{(1)}, \dots, q_{\vec{n}'}^{(r)}) \in \mathbb{R}_r,$$
$$q_{\vec{n}''} = (q_{\vec{n}''}^{(1)}, \dots, q_{\vec{n}''}^{(r)}) \in \mathbb{R}_r, \qquad \vec{a} = (a^{(1)}, \dots, a^{(r)}) \in \mathbb{R}_r,$$
$$|q_{\vec{n}'} - q_{\vec{n}''} - \vec{a}|^2 = \sum_{S=1}^{r} (q_{\vec{n}'}^{(S)} - q_{\vec{n}''}^{(S)} - a^{(S)})^2.$$

1991 *Mathematics Subject Classification*. Primary 34A35; Secondary 70G25.

Furthermore, Γ is the set of nonordered pairs of r-dimensional integer vectors $\vec{n}'$ and $\vec{n}''$ such that $\sum_{S=1}^{r} |n'_S - n''_S| = 1$. We require that Γ contains only one of the two pairs $(\vec{n}', \vec{n}'')$, $(\vec{n}'', \vec{n}')$ and the second sum in the first term for $\mathcal{U}$ is taken over all pairs $(\vec{n}', \vec{n}'') \in \Gamma$.

We also assume that $\gamma > 0$ and $\chi > 0$ are parameters; the function $f^{(S)}(x)$ $(S = 1, \dots, r)$ is infinitely differentiable, periodic in x with period $a^{(S)} > 0$, and achieves the minimum at the points $x = na^{(S)}$ and maximum at the points $x = (n + \frac{1}{2})a^{(S)}$ (n runs over the set of integers). From the physical point of view the system represents a net of springs that are joined at the nodes of the lattice $\mathbb{Z}_r$ and are under an external field with the potential $\gamma \sum_{S=1}^{r} f^{(S)}(q_{\vec{n}}^{(S)})$.

If we denote the mass of the particle $P_{\vec{n}}$ by $m_{\vec{n}}$, its velocity vector by $v_{\vec{n}}$, and its momentum vector by $p_{\vec{n}} = m_{\vec{n}} v_{\vec{n}}$, then the formal Hamiltonian H of the system is

$$H = \sum_{\vec{n} \in \mathbb{Z}_r} \frac{|p_{\vec{n}}|^2}{2m_{\vec{n}}} + \mathcal{U}.$$

Therefore, the motion of particles is described by the infinite system of ordinary differential equations

$$m_{\vec{n}} \frac{d^2 q_{\vec{n}}^{(S)}}{dt^2} = -\gamma \varphi^{(S)}(q_{\vec{n}}^{(S)}) + 2\chi \left(-2r q_{\vec{n}}^{(S)} + \sum_{\vec{n}' \in \Gamma_{\vec{n}}} q_{\vec{n}'}^{(S)} \right), \tag{1}$$

where $S = 1, \dots, r$; $\vec{n} = (n_1 \dots, n_r) \in \mathbb{Z}_r$; $\varphi^{(S)}(q_{\vec{n}}^{(S)}) = \frac{df^{(S)}}{dx}(q_{\vec{n}}^{(S)})$; $\Gamma_{\vec{n}}$ is the set of all vectors $\vec{n}' = (n'_1 \dots, n'_r) \in \mathbb{Z}_r$, such that $\sum_{S=1}^{r} |n_S - n'_S| = 1$; t is time.

In the special case $r = 1$, the masses of all particles are equal to the same number $m > 0$. Moreover, we have $\vec{a} = a > 0$ and $f^{(S)}(x) = f(x) = 1 - \cos \frac{2\pi x}{a}$. This case has been studied in **[FK]** where the authors consider the behavior of the particles in the neighborhood of the stationary state $q_n(t) = na$, with $n \in \mathbb{Z}_1$. They introduced the variables ξ_n by the equations $q_n = na + \xi_n$ and rewrote the system of equations (1) as follows:

$$m \frac{d^2 \xi_n}{dt^2} = -\frac{2\pi\gamma}{a} \sin \frac{2\pi \xi_n}{a} + 2\chi(\xi_{n+1} - 2\xi_n + \xi_{n-1}). \tag{2_1}$$

In **[FK]**, the authors posed the problem of finding a solution $\xi_n \not\equiv 0$ of the system $(2)_1$ that satisfies, for all t and n, the equations

$$\xi_n(t) = \xi_{n+1}(t + \tau), \tag{3_1}$$

with a constant τ independent of t and n. In order to find such a solution they constructed a continuous approximation of equation (2_1) by a partial differential equation. In **[P1]**, the author found two different families of solutions of (2_1), which are periodic in time and satisfy (3_1). For these solutions the set of values of τ is shown to contain an interval on the line. Moreover, the amplitude of oscillations of solutions of the first family is bounded, while the amplitude of oscillations of solutions of the second family can be arbitrarily large. In the case $r = 1$ the main result in **[P1]** is the following: if the masses m_n decrease sufficiently fast as $|n| \to \infty$, then the stationary solutions (i.e., solutions for which all particles are located at maximums of $f(x)$) are stable in the sense of Lyapunov in the infinite-dimensional phase space; in any neighborhood of such a stationary solution there are infinite sets of almost periodic solutions with continual, countable, and an arbitrary finite number of rationally independent frequencies. In other words, in the case under consideration the infinite-dimensional analog of KAM-theory ([**K, A, M, SM**] holds.

It is also shown in [**P1**] that this statement is not valid in the case of nonvanishing but equal masses m_n. We note that in the case when $r = 1$ and all particles are located at minimums of the potential $\mathcal{U}$, the instability occurs even if all the masses m_n are equal to zero (see [**S**]).

The goal of the this paper is to generalize the above results to the case of any natural r. From the physical point of view the cases $r = 2$ and $r = 3$ correspond to the dislocation of atoms in two and three-dimensional rigid body. The dot above a function denotes differentiation.

§1. Construction of families of periodic solutions

The system (1) has infinitely many stationary solutions independent of time and masses of particles. The simplest of them are the solutions for which all particles are located at the points $\vec{q} = (q^{(1)}, \dots q^{(r)}) \in \mathbb{R}_r$ with coordinate $q^{(S)}$ $(S = 1, \dots, r)$ to be the minimum or the maximum of function $f^{(S)}(x)$.

We study the behavior of particles in the neighborhood of the stationary state $q_{\vec{n}} = (q_{\vec{n}}^{(1)}, \dots, q_{\vec{n}}^{(r)})$ for which the coordinate $q_{\vec{n}}^{(S)}$ satisfies the equality $q_{\vec{n}}^{(S)} = a^{(S)} n_S$ for $\vec{n} = (n_1, \dots, n_r)$, $S = 1, \dots, r$. We consider the special case when the masses of all particles are equal to a number $m > 0$, and $f^{(S)}(x) = 1 - \cos \frac{2\pi x}{a^{(S)}}$.

We introduce the variables $\xi_{\vec{n}} = (\xi_{\vec{n}}^{(1)}, \dots, \xi_{\vec{n}}^{(r)})$ by the equations $q_{\vec{n}}^{(S)} = a^{(S)} n_S + \xi_{\vec{n}}^{(S)}$ $(S = 1, \dots, r)$ and rewrite the system (1) as follows:

$$(2_S) \qquad m \frac{d^2 \xi_{\vec{n}}^{(S)}}{dt^2} = -\frac{2\pi\gamma}{a^{(S)}} \sin \frac{2\pi \xi_{\vec{n}}^{(S)}}{a^{(S)}} + 2\chi \Big(\sum_{\vec{n}' \in \Gamma_{\vec{n}}} q_{\vec{n}'}^{(S)} - 2r q_{\vec{n}}^{(S)} \Big).$$

DEFINITION. A solution of the system (1) is said to be space-homogeneous if for any $\vec{n} = (n_1, \dots, n_r) \in \mathbb{Z}_r$ the motion of the particle $P_{\vec{n}}$ determined by this solution with respect to the position $q_{\vec{n}}^* = (a^{(1)} n_1, \dots, a^{(r)} n_r)$ is independent of $\vec{n}$.

THEOREM 1. *The system* (2_S) *has two families of solutions* Π_1 *and* Π_2 *such that* $\Pi_1 \supset \Pi_2$ *and the following assertions hold*:

(1) *if* $\xi_{\vec{n}}(t) = (\xi_{\vec{n}}^{(1)}(t), \dots \xi_{\vec{n}}^{(r)}(t)) \in \Pi_1$, *then*

$$(3) \qquad \xi_{\vec{n}}^{(S)}(t) \not\equiv \text{const}, \quad \xi_{\vec{n}}^{(S)}(t) = \xi_{\vec{n}^{(S)}}^{(S)}(t + \tau_S), \quad S = 1, \dots, r,$$

where vectors $\vec{n} = (n_1, \dots, n_r)$ *and* $\vec{n}^{(S)} = (n_1^{(S)}, \dots, n_r^{(S)})$ *satisfy* $n_S^{(S)} = n_s + 1$, *the constant* τ_S *does not depend on* t *and* $\vec{n}$, *the set of values* τ_S *contains the interval* $\frac{m(a^{(S)})^2}{2\pi\gamma} < \tau_S$ *of the line, and the amplitude of the vibrations does not exceed* $\frac{a^{(S)}}{2}$;

(2) *if* $\xi_{\vec{n}}(t) \in \Pi_1$, *then the solution*

$$q_{\vec{n}}(t) = (a^{(1)} n_1 + \xi_{\vec{n}}^{(1)}(t), \dots, a^{(r)} n_r + \xi_{\vec{n}}^{(r)}(t))$$

of the system (1) *is space-homogeneous, and if* $\xi_{\vec{n}} \in \Pi_2$, *then the solution* $q_{\vec{n}}(t)$ *is periodic.*

In order to prove Theorem 1 we will find a solution $\xi_{\vec{n}}^{(S)}(t) = \xi^{(S)}$ of the system (2_S) that does not depend on $\vec{n}$. By (2_S) this solution $\xi^{(S)}$ satisfies the equation

$$(4) \qquad m\frac{d^2\xi^{(S)}}{dt^2} = -\frac{2\pi\gamma}{a^s}\sin\frac{2\pi\xi^{(S)}}{a^{(S)}}.$$

Theorem 1 will then follow from the properties of equation (4). The family of solutions Π_1 corresponds to bounded periodic solutions of equation (4) with τ_S to be the period of these periodic solutions, and the set of values τ_S contains the interval $I^{(S)} = \{\tau_s : \frac{m(a^{(S)})^2}{2\pi\gamma} < \tau_S\}$. The family Π_2 is the subset of Π_1 consisting of solutions $\xi_{\vec{n}}(t) = (\xi_{\vec{n}}^{(1)}(t), \dots, \xi_{\vec{n}}^{(r)}(t))$ for which the coordinate $\xi^{(S)} = \xi_{\vec{n}}^{(S)}(t)$, $S = 1, \dots, r$, is the periodic solution of the system (4) with period $\tau \in \bigcap_{S=1}^{r} I^{(S)}$, independent of S.

We now state our next result.

THEOREM 2. *The system* (2_S) *has a family of solution* Π *such that*:

(1) *if* $\xi_{\vec{n}}(t) = (\xi_{\vec{n}}^{(1)}(t), \dots, \xi_{\vec{n}}^{(r)}(t)) \in \Pi$, *then for* $S = 1, \dots, r$ *the function* $\xi_{\vec{n}}(t)$ *satisfies formula* (3) (*see Theorem* 1) *in which* $\tau_S > 0$ *and vectors* $\vec{n} = \{n_1 \dots, n_r)$ *and* $\vec{n}^{(S)} = (n_1^{(S)} \dots, n_r^{(S)})$ *are such that* $n_s^{(S)} = n_S + 1$;

(2) *the function* $\xi_{\vec{n}}^{(S)}(t)$ *has period* τ_S *and satisfies the equality* $\xi_{\vec{n}^{(S)}}^{(S)}(t) = -\xi_{\vec{n}}^{(S)}(t)$, *and the amplitude of the vibrations* $\xi_{\vec{n}}^{(S)}(t)$ *can be arbitrarily large.*

In order to prove Theorem 2 we find a solution of the system (2_S) such that for all t, $S = 1, \dots r$, and any vectors $\vec{n}$ and $\vec{n}^{(S)}$ the equality

$$\xi_{\vec{n}}^{(S)}(t) = -\xi_{\vec{n}^{(S)}}^{(S)}(t)$$

holds. By (2_S) for any $\vec{n}$ the function $\xi_{\vec{n}}^{(S)}(t)$ satisfies the equation

$$(5) \qquad m\frac{d^2\xi_{\vec{n}}^{(S)}}{dt^2} = -\frac{2\pi\gamma}{a^{(S)}}\sin\frac{2\pi\xi_{\vec{n}}^{(S)}}{a^{(S)}} - 8r\chi\xi_{\vec{n}}^{(S)}(t).$$

Theorem 2 follows from the properties of equation (5). The solutions of the system (2_S) that belong to Π are periodic solutions of equation (5) and the value τ_S is half of the period of these periodic solutions.

REMARK. Assertion 2 of Theorem 1 is valid for an arbitrary periodic function $f^{(S)}(x)$ and Theorem 2 holds for an arbitrary periodic function $f^{(S)}(x)$ that satisfies the condition

$$\frac{df^{(S)}}{dx}(x) = -\frac{df^{(S)}}{dx}(-x).$$

§2. Construction of almost periodic motions

The phase space Ω of the dynamical system under consideration is infinite-dimensional and consists of functions $q_{\vec{n}}^{(S)}(t)$ that are differentiable infinitely many times for all $S = 1, \dots, r$ and vectors $\vec{n} = (n_1, \dots, n_r) \in \mathbb{Z}_r$ whose first coordinate n_1 is equal to only 0 or 1, i.e., $n_1 = 0, 1$. Indeed, it follows from (1) that we can uniquely determine $q_{\vec{n}}^{(S)}(t)$ for any $\vec{n} \in \mathbb{Z}_r$, if the functions $q_{\vec{n}}^{(S)}(t)$ with $n_1 = 0, 1$ are known.

In this section we show that under some assumptions about the functions $f^{(S)}(x)$ $(S = 1, \dots, r)$ and the masses $m_{\vec{n}}$ there exist infinite sets of almost periodic solutions

$q_{\vec{n}}(t)$ of system (1) with continual, countable, and an arbitrary finite number of rationally independent frequencies. These solutions belong to an arbitrary neighborhood of a stationary solution $\widehat{q}_{\vec{n}}(t) = (\widehat{q}_{\vec{n}}^{(1)}(t), \dots, \widehat{q}_{\vec{n}}^{(r)}(t))$ whose coordinate $\widehat{q}_{\vec{n}}^{(S)}$ ($S = 1, \dots r$; $\vec{n} = (n_1, \dots, n_r)$) has the form

$$\widehat{q}_{\vec{n}}^{(S)}(t) \equiv \left(n_S - \frac{1}{2}\right) a^{(S)} \tag{6}$$

(and also in the neighbrhood of any other stationary solution $q_{\vec{n}}(t)$ for which the function $q_{\vec{n}}^{(S)}(t)$ takes on only values that are half-integral with respect to $a^{(S)}$, $S = 1, \dots r$). We construct these solutions explicitly.

We write the space Ω as the union $\Omega = \Omega^{(0)} \cup \Omega^{(1)}$ of two disjoint subsets of functions $q_{\vec{n}}^{(S)}(t)$, where

$$\Omega^{(0)} = \left\{ q_{\vec{n}^{(0)}}^{(S)}(t) : n_1^{(0)} = 0; \;\; S = 1, \dots, r; \;\; \vec{n}^{(0)} = (n_1^{(0)}, \dots, n_r^{(0)}) \right\},$$
$$\Omega^{(1)} = \left\{ q_{\vec{n}^{(1)}}^{(S)}(t) : n_1^{(1)} = 1; \;\; S = 1, \dots, r; \;\; \vec{n}^{(1)} = (n_1^{(1)}, \dots, n_r^{(1)}) \right\}.$$

Further, we consider the functions from $\Omega^{(0)}$ and $\Omega^{(1)}$ given in the form:

$$\begin{aligned} q_{\vec{n}^{(0)}}^{(S)}(t) &= \widehat{q}_{\vec{n}^{(0)}}^{(S)}(t) + \Delta_0^{(S)}(t) \in \Omega^{(0)}, \\ q_{\vec{n}^{(1)}}^{(S)}(t) &= \widehat{q}_{\vec{n}^{(1)}}^{(S)}(t) + \Delta_1^{(S)}(t) \in \Omega^{(1)}, \end{aligned} \tag{7}$$

where the functions $\widehat{q}_{\vec{n}^{(0)}}^{(S)}(t)$ and $\widehat{q}_{\vec{n}^{(1)}}^{(S)}(t)$ are defined in (6) for $\vec{n} = \vec{n}^{(0)}, \vec{n}^{(1)}$ and the functions $\Delta_0^{(S)}(t)$ and $\Delta_1^{(S)}(t)$ satisfy

$$|\Delta_v^{(S)}(t)| \le \kappa_0, \qquad \left| \frac{d^k \Delta_v^{(S)}}{dt^k}(t) \right| \le \kappa_k \tag{8}$$

for $v = 0, 1$; $S = 1, 2, \dots r$; $k = 1, 2, \dots$; κ_μ ($\mu = 0, 1, 2, \dots$) are fixed numbers independent of v and t.

THEOREM 3. *Suppose there exists $\delta_0 > 0$ such that the function $\varphi^{(S)}(x)$ is analytic in the complex region $|x - \frac{a^{(S)}}{2}| \le \delta_0$ for $S = 1, \dots, r$ and the following conditions are satisfied*:

1) $$-4 < \frac{\gamma}{2\chi} \dot{\varphi}^{(S)}\left(\frac{a^{(S)}}{2}\right) = \beta_S < 0;$$

2) $$\beta_s \ne -2 + 2\cos\left(\frac{2\pi l}{m}\right), \quad (l = 0, \pm 1, \dots, \pm m; \quad m = 1, \dots, 262);$$

3) $$\frac{\gamma}{2\chi} b_0(\lambda_S) \frac{d^3 \varphi^{(S)}}{dx^3}\left(\frac{a^{(S)}}{2}\right) + \frac{\gamma}{4\chi^2} b_1(\lambda_S) \left(\frac{d^2 \varphi^{(S)}}{dx^2}\left(\frac{a^{(S)}}{2}\right)\right)^2 \ne 0,$$

where

$$b_0(\lambda_S) = \frac{\lambda_S}{2(\lambda_S - \overline{\lambda}_S)},$$

$$b_1(\lambda_S) = \frac{\lambda_S^2}{(\lambda_S - \overline{\lambda}_S)^2(1-\lambda_S)} - \frac{1}{(\lambda_S - \overline{\lambda}_S)^2(1-\lambda_S)} + \frac{\lambda^2}{2(\lambda_S - \overline{\lambda}_S)^2(\lambda_S^2 - \lambda_S)} - \frac{1}{2(\lambda_S - \overline{\lambda}_S)^2(\lambda_S^2 - \overline{\lambda}_S)},$$

λ_S *is an eigenvalue of the matrix*

$$\begin{pmatrix} 1 + \frac{\gamma}{2\chi}\dot{\varphi}^{(S)}\left(\frac{a^{(S)}}{2}\right) & 1 \\ \frac{\gamma}{2\chi}\dot{\varphi}^{(S)}\left(\frac{a^{(S)}}{2}\right) & 1 \end{pmatrix}$$

with $\operatorname{Im}\lambda_S > 0$*, and* $\overline{\lambda}_S$ *is the complex conjugate to* λ_S*. Then there exist constants* $m^0_{\vec{n}} > 0$*,* $n = 0, \pm 1, \pm 2, \dots$ *that depend only on* κ_μ*,* $\mu = 0, 1, \dots,$ *and functions* $f^{(S)}(x)$*,* $S = 1, \dots, r$*, such that the following is true. If for some* n_* *and any vector* $\vec{n} = (n_1, \dots, n_r) \in \mathbb{Z}_r$*, with* $|n_1| \geq n_*$*, the masses* $m_{\vec{n}}$ *satisfy the condition* $0 \leq m_{\vec{n}} \leq m^0_{n_1}$*, then the stationary solution* $\widehat{q}_{\vec{n}}(t) = (\widehat{q}^{(1)}_{\vec{n}}(t), \dots \widehat{q}^{(r)}_{\vec{n}}(t))$ *defined by* (6) *is stable in the following sence: for any* $\varepsilon > 0$ *there exists* $\delta > 0$ *and a natural number* k_0 *such that if the functions* $q^{(S)}_{\vec{n}^{(0)}}(t)$*,* $q^{(S)}_{\vec{n}^{(1)}}(t)$ *are given by* (7)*, the functions* $\Delta^{(S)}_0(t)$ *and* $\Delta^{(S)}_1(t)$ *satisfy* (8)*,* $|\Delta^{(S)}_\nu(t)| \leq \delta$*, and* $\left|\frac{d^k \Delta^{(S)}_\nu}{dt^k}\right| \leq \delta$ *for* $k = 1, \dots, k_0$*,* $S = 1, \dots, r$*,* $\nu = 0, 1$*, and all* t*, then*

$$\left|q^{(S)}_{\vec{n}}(t) - \widehat{q}^{(S)}_{\vec{n}}(t)\right| \leq \varepsilon.$$

COROLLARY 1. *Assume that for* $S = 1, \dots, r$*,*

$$f^{(S)}(q) = 1 - \cos\frac{2\pi x}{a^{(S)}}, \quad \frac{\gamma}{\chi} < 2\frac{\left(a^{(s)}\right)^2}{\pi^2},$$

$$\frac{\gamma}{\chi} \neq \left(\frac{a^{(S)}}{\pi}\right)^2 - \left(\frac{a^{(S)}}{\pi}\right)^2 \cos\frac{2\pi l}{m}; \quad l = 0, \pm 1, \dots, \pm m; \quad m = 1, \dots, 262,$$

then Theorem 3 *holds.*

Corollary 1 follows directly from Theorem 3, since the conditions $\frac{d^2\varphi^{(S)}}{dx^2}\left(\frac{a}{2}\right) = 0$, $\frac{d^3\varphi^{(S)}}{dx^3}\left(\frac{a}{2}\right) = 0$, $\lambda_S = \overline{\lambda}_S$ are satisfied; this leads to the validity of Condition 3 of Theorem 3.

COROLLARY 2. *Theorem* 3 *holds when the perturbations* $\Delta^{(S)}_\nu(t)$ $(\nu = 0, 1;\ S = 1, \dots r)$ *in* (7) *have the form*

$$\Delta^{(S)}_\nu(t) = \int\limits_{|\tau| \leq N} h^{(S)}_\nu(\tau) e^{i\tau t}\, d\tau \quad \text{or} \quad \Delta^{(S)}_\nu(t) = \sum_k h^{(S)}_{\nu,k} e^{i\tau_k t},$$

where $h^{(S)}_\nu(\tau)$ *are absolutely integrable functions,* $h_\nu(\tau) \leq \delta$*,* $\sum\limits_k |h^{(k)}_\nu| \leq \delta$*,* $|\tau_k| \leq N$*,* k *ranges over a finite or countable set, and* $i = \sqrt{-1}$*.*

THEOREM 4. *Suppose that the conditions of Theorem 3 are satisfied. There exist constants* $m^0_{\vec{n}_1} > 0$ *such that if for some* n_* *and any vector* $\vec{n} = (n_1, \dots, n_r) \in \mathbb{Z}_r$, *with* $|n_1| \geq n_*$, *the masses* $m_{\vec{n}}$ *satisfy the condition* $0 \leq m_{\vec{n}} \leq m^0_{\vec{n}_1}$, *then for any* $\varepsilon > 0$ *there exist infinite sets of almost periodic solutions* $q_{\vec{n}}(t) = (q^{(1)}_{\vec{n}}(t), \dots, q^{(r)}_{\vec{n}}(t))$ *of the system* (1) *with continual, countable, or an arbitrary finite number of rationally independent frequencies for which, for* $S = 1, \dots, r$, *all* t *and* $\vec{n} \in \mathbb{Z}_r$,

$$|q^{(S)}_{\vec{n}}(t) - \widehat{q}^{(S)}_{\vec{n}}(t)| \leq \varepsilon.$$

One can derive Theorem 4 from Corollary 2. First note that the functions $q^{(S)}_{\vec{n}^{(0)}}(t)$ and $q^{(S)}_{\vec{n}^{(1)}}(t)$ in (7) are determined by the functions $\Delta^{(S)}_\nu(t)$ in Corollary 2 and thus are almost periodic with the number of frequencies specified by Theorem 4. Now it follows from (1) and the analyticity of $\varphi^{(S)}(x)$ that for all $S = 1, \dots, r$ and $\vec{n} \in \mathbb{Z}_r$, the functions $q^{(S)}_{\vec{n}}(t)$ are almost periodic with frequencies that are all possible integral combinations of the frequencies of the functions $\Delta^{(S)}_0(t)$ and $\Delta^{(S)}_1(t)$.

To prove Theorem 3, we need some lemmas. We introduce the following notation:

$$M_\Theta = \sup_{\substack{-\infty < x < \infty \\ k \leq \Theta \\ S=1,\dots,r}} \left| \frac{d^k \varphi^{(S)}}{dx^k}(x) \right|.$$

LEMMA 1. *Suppose that a function* $q(t)$ *is infinitely differentiable for all* t *and satisfies the condition* $\left|\frac{d^\Theta q}{dt^\Theta}\right| < c^{(\Theta)}$, $1 \leq c^{(\Theta)} \leq c^{(\Theta+1)}$, $G_S(t) = \frac{\gamma}{2\chi}\varphi^{(S)}(q(t))$ *for a natural* Θ. *Then*

$$\left|\frac{d^\Theta G_S}{dt^\Theta}\right| \leq M_\Theta \Theta! \left(c^{(\Theta)}\right)^\Theta \frac{\gamma}{2\chi}.$$

PROOF. Lemma 1 is a consequence of the following proposition.

PROPOSITION. *The expression for the derivative* $\frac{d^\Theta G_S}{dt^\Theta}$ *is the sum of at most* $\Theta!$ *monomials, each of which is the product of some derivative* $d^k\left(\frac{\gamma}{2\chi}\varphi^{(S)}(x)\right)/dx^k$ $(k \leq \Theta)$ *and derivatives of* $q(t)$ *of order not exceeding* Θ, *with the total number of factors not exceeding* Θ.

PROOF OF THE PROPOSITIONS. We use induction by Θ. For $\Theta = 1$, the result is obvious. We assume that it is true for some $\Theta \geq 1$ and calculate the derivative of the expression $\frac{d^\Theta G_S}{dt^\Theta}$. The derivative of each monomials consists of not more than $S + 1$ monomials. Each of them is the product of a derivative $d^k\left(\frac{\gamma}{2\chi}\varphi^{(S)}(x)\right)/dx^k$ $(k \leq \Theta + 1)$ and derivatives of $q(t)$ of order not exceeding $\Theta + 1$, with the total number of factors not exceeding $\Theta + 1$. This proves the statement for $\Theta + 1$ and completes the proof of the proposition.

LEMMA 2. *Suppose that for all natural* Θ; $S = 1, \dots r$; $\nu = n_1, n_1 - 1$ *and all vectors* $\vec{n}^0 = (n^0_1, \dots, n^0_r) \in \mathbb{Z}_r$ *with* $n^0_1 = \nu$ *the function* $q^{(S)}_{\vec{n}^0}(t)$ *is infinitely differentiable. Assume also that* $\left|\frac{d^\Theta q^{(S)}_{\vec{n}^0}}{dt^\Theta}\right| \leq c^{(\Theta)}_{n_1}$, $m_{\vec{n}^0} \leq \widehat{c}$, *where* $1 \leq c^{(\Theta)}_{n_1} \leq c^{(\Theta+1)}_{n_1}$, *and the functions*

$q^{(S)}_{\vec n^+}(t)$, $q^{(S)}_{\vec n^-}(t)$ *satisfy* (1) *with given functions* $q^{(S)}_{\vec n^0}(t)$ *and* $\vec n^+ = (n_1^+,\dots,n_r^+)\,\vec n^- = (n_1^-,\dots,n_r^-)$ *with* $n_1^+ = n_1+1, n_1^- = n_1-2$. *Then*

$$\left|\frac{d^\Theta q^{(S)}_{\vec n^+}}{dt^\Theta}(t), \frac{d^\Theta q^{(S)}_{\vec n^-}}{dt^\Theta}(t)\right| \le c^{(\Theta)}_{n_1+1},$$

where

$$c^{(\Theta)}_{n_1+1} = (4r-1)c^{(\Theta)}_{n_1} + M_\Theta \Theta!\left(c^{(\Theta)}_{n_1}\right)^\Theta \frac{\gamma}{2\chi} + \frac{\widehat{c}}{2\chi} c^{(\Theta+2)}_{n_1}.$$

PROOF. We prove Lemma 2 only for the function $\frac{d^\Theta q^{(S)}_{\vec n^+}}{dt^\Theta}(t)$. The proof for the function $\frac{d^\Theta q^{(S)}_{\vec n^-}}{dt^\Theta}(t)$ is similar. Let $\vec n = (n_1,\dots n_r) \in \mathbb{Z}_r$ be the vector such that $n_1 = n_1^+ - 1$ and $n_S = n_S^+$ for $S = 1,2,\dots,r$. Then by (1) we have the equation

$$\frac{d^\Theta q^{(S)}_{\vec n^+}}{dt^\Theta}(t) = 2r\frac{d^\Theta q^{(S)}_{\vec n}}{dt^\Theta}(t) - \sum_{\substack{\vec n' \in \Gamma_{\vec n}\\ \vec n' \ne \vec n'^+}} \frac{d^\Theta q^{(S)}_{\vec n'}}{dt^\Theta}(t) + \frac{\gamma}{2\chi}\frac{q^\Theta \varphi^{(S)}(q^{(S)}_{\vec n}(t))}{dt^\Theta}$$
$$+\frac{m_{\vec n}}{2\chi}\frac{d^{\Theta+2} q^{(S)}_{\vec n}}{dt}(t).$$

It follows that

$$\left|\frac{d^\Theta q^{(S)}_{\vec n^+}}{dt^\Theta}(t)\right| \le (4r-1)c^{(\Theta)}_{n_1} + \frac{\widehat{c}}{2\chi}c^{\Theta+2}_{n_1} + \left|\frac{\gamma}{2\chi}\frac{d^\Theta \varphi^{(S)}(q^{(S)}_{\vec n}(t))}{dt^\Theta}\right|.$$

Now applying Lemma 1, we obtain the desired statement.

LEMMA 3. *Suppose that for* $S = 1,\dots,r$ *the functions* $q^{(S)}_{\vec n^{(0)}}(t)$ *and* $q^{(S)}_{\vec n^{(1)}}(t)$ *are of the form* (7) *and satisfy conditions* (8). *Suppose also that* $q^{(S)}_{\vec n}(t)$ *is the solution of system* (1) *for all vectors* $\vec n = (n_1,\dots,n_r) \in \mathbb{Z}_r$ *and given functions* $q^{(S)}_{\vec n^{(0)}}(t)$ *and* $q^{(S)}_{\vec n^{(1)}}(t)$. *Then there exist constants* $m_{n_1^0} > 0$ *such that if for some* n_* *and any vector* $\vec n = (n_1,\dots n_r) \in \mathbb{Z}_r$, *with* $|n_1| \ge n_*$, *the masses* $m_{\vec n}$ *satisfy the condition* $0 \le m_{\vec n} \le m^0_{n_1}$, *then for all* $S = 1,\dots,r$ *and any fixed integer* $n_2,\dots,n_r$,

$$\sum_{n_1=-\infty}^{\infty} m_{n_1} \sup_{-\infty<t<\infty}\left|\frac{d^2 q^{(S)}_{n_1}(n_2,\dots,n_r,t)}{dt^2}\right| \le c^*,$$

where

$$q^{(S)}_{n_1}(n_2,\dots,n_r,t) = q^{(S)}_{\vec n}(t), \quad \vec n = (n_1,\dots,n_r),$$

c^* *is a constant independent of the functions* $q^{(S)}_{\vec n^{(0)}}(t)$, $q^{(S)}_{\vec n^{(1)}}(t)$ *and numbers* $n_2,\dots,n_r$.

Lemma 3 obviously follows from (8) and Lemma 2.

PROOF OF THEOREM 3. Suppose that the set of functions $q^{(S)}_{\vec n}(t)$ is a solution of (1). We fix a number S and coordinates $n_2,\dots,n_r$ of vector $\vec n = (n_1,\dots n_r)$ and use the notations $m_{n_1} = m_{\vec n}$, $q_{n_1}(t) = q^{(S)}_{\vec n}(t)$. We introduce a transformation A depending on S, and two sequences of families of transformations $B'_k(t) = B'_k(S,n_2,\dots,n_r,t)$, $B''_k(t) = B''_k(S,n_2,\dots,n_r,\,t)$ $(k = 1,2,\dots)$ of the (u,z)-plane depending on $S, n_2, \dots, n_r, t$. The transformation A is a mapping determined by the equality

$$(9) \qquad A : (u, z) \to (u', z') = \left(u + z', z + \frac{\gamma}{2\chi} \varphi^{(S)}(u) \right),$$

and $B'_k(t)$ and $B''_k(t)$ are mappings determined by the solution $q_{n_1}(t)$ for $n_1 = k$ and $n_1 = -k + 1$:

$$(10) \qquad B'_k(t) : (u, z) \to \left(u, z + \frac{m_k}{2\chi} \frac{d^2 q_k}{dt^2}(t) \right),$$

$$(11) \qquad B''_k(t) : (u, z) \to \left(u, z + \frac{m_{-k+1}}{2\chi} \frac{d^2 q_{-k+1}}{dt^2}(t) \right).$$

Let us introduce the functions

$$z'_{n_1+1}(t) = q_{n_1+1}(t) - q_{n_1}(t), \quad n_1 = 0, 1, 2, \dots,$$
$$z''_{n_1-1}(t) = q_{n_1-1}(t) - q_{n_1}(t), \quad n_1 = 1, 0, -1, -2, \dots.$$

By (1) we obtain the equalities

$$(q_{n_1+1}(t), z'_{n_1+1}(t)) = B'_{n_1}(t) A(q_{n_1}(t), z'_{n_1}(t)), \quad n_1 = 1, 2, \dots,$$
$$(q_{n_1-1}(t), z''_{n_1-1}(t)) = B_{1-n_1}(t) A(q_{n_1}(t), z''_{n_1}(t)), \quad n_1 = 0, -1, -2, \dots.$$

This implies

$$(12) \qquad (q_k(t), z'_k(t)) = B'_{k-1}(t) \circ A \circ B'_{k-2} \circ A \circ \dots \circ B'_1(t) \circ A(q_1(t), z'_1(t)) \qquad (k = 2, 3, \dots),$$

$$(13) \qquad (q_\mu(t), z''_\mu(t)) = B''_{-\mu}(t) \circ A \circ B''_{-\mu-1} \circ A \circ \dots \circ B''_1(t) \circ A(q_0(t), z''_0(t)) \qquad (\mu = -1, -2, \dots).$$

We introduce two sequences of transformations A'_k, A''_k $k = 1, 2, \dots$ of the (u, z)-plane as follows:

$$(14) \qquad A'_k : (u, z) \to (\widetilde{u}'_k, \widetilde{z}'_k) = A\left(ka^{(S)} - \frac{a^{(S)}}{2} + u, z + a^{(S)} \right) - \left(ka^{(S)} + \frac{a^{(S)}}{2}, a^{(S)} \right),$$

$$(15) \qquad A''_k : (u, z) \to (\widetilde{u}''_k, \widetilde{z}''_k) = A\left(-ka^{(S)} + \frac{a^{(S)}}{2} + u, z - a^{(S)} \right) - \left(-ka^{(S)} - \frac{a^{(S)}}{2}, -a^{(S)} \right).$$

By (9), (14), and (15) the transformations A'_k and A''_k do not depend on k and are identical to each other and to the following transformation $\widetilde{A}$:

$$(16) \qquad \widetilde{A} : (u, z) \to (\widetilde{u}, \widetilde{z}) = \left(u + \widetilde{z}, z + \frac{\gamma}{2\chi} \varphi^{(S)}\left(u - \frac{a^{(S)}}{2} \right) \right).$$

Substituting (14), (15), and (16) into (12) and (13) we reduce the proof of Theorem (3) to the following lemma.

LEMMA 4. *There exist constant* $m^0_{n_1} > 0$ $(n_1 = 0, \pm 1, \pm 2, \dots)$ *depending only on* $\kappa_i, i = 0, 1, \dots,$ *and functions* $\varphi^{(S)}(x)$, $S = 1, \dots, r$, *such that if for some* n_* *and any vector* $\vec{n} = (n_1, \dots, n_r) \in \mathbb{Z}_r$, *with* $|n_1| \geq n_*$, *the masses* m_{n_1} *satisfy the condition* $0 \leq m_{n_1} \leq m_{n_1^0}$, *then for any* $\varepsilon > 0$ *there exist* $\delta > 0$ *and a natural number* k_0 *with the property: if for* $v = 0$ *and* 1 *the functions* $\Delta_v^{(S)}(t)$ *satisfy* (8), $|\Delta_v^{(S)}(t)| \leq \delta$, *and* $\left|\frac{d^k \Delta_v^{(S)}}{dt^k}(t)\right| \leq \delta$ $(k = 1, \dots, k_0)$, *then for all* t *and* $k \geq 2$ *the functions* $\widehat{q}'_k(t), \widehat{z}'_k(t), \widehat{q}''_k(t), \widehat{z}''_k(t)$ *determined by*

$$(\widehat{q}'_k(t), \widehat{z}'_k(t)) = B'_{k-1}(t) \circ \widetilde{A} \circ \dots \circ B'_1(t) \circ \widetilde{A}(\Delta_1^{(S)}(t), \Delta_1^{(S)}(t) - \Delta_0^{(S)}(t)),$$
$$(\widehat{q}''_k(t), \widehat{z}''_k(t)) = B''_{k-1}(t) \circ \widetilde{A} \circ \dots \circ B''_1(t) \circ \widetilde{A}(\Delta_0^{(S)}(t), \Delta_0^{(S)}(t) - \Delta_1^{(S)}(t))$$

satisfy the conditions

$$|\widehat{q}'_k(t)| + |\widehat{z}'_k(t)| \leq \varepsilon, \quad |\widehat{q}''_k(t)| + |\widehat{z}''_k(t)| \leq \varepsilon.$$

PROOF OF LEMMA 4. We take $m^0_{n_1}$ as in Lemma 3 and show that for every t the two pairs of sequences of transformations $A'_k = \widetilde{A}$, $B'_k(t)$ and $A''_k = \widetilde{A}$, $B''_k(t)$ satisfy the conditions of Theorem 1 in **[P2]** in a neighborhood of the point $(0, 0)$. In our case this Theorem gives the following statements:

1) the transformation $\widetilde{A}$ is well defined in a complex neighborhood $V_\varepsilon = \{u, z : \max(|u|, |z|) \leq \varepsilon\}$ $(\varepsilon_0 > 0)$, and has $(0, 0)$ as a fixed point;

2) the transformation $\widetilde{A}$ is area-preserving, of general elliptic type at the point $(0, 0)$, and its Birkhoff normal form (see **[SM]**)

$$\alpha_1 = \alpha + \omega_0 + \omega_1 r + \dots + \omega_k r^k + \dots, \quad r_1 = r \tag{17}$$

(α is the angle and r is the radius) has the coefficient $\omega_1 \neq 0$;

3) the coefficient ω_0 in (17) satisfies the inequalities $\omega_0 \neq \frac{2\pi l}{m}$, $l = 0, \pm 1, \dots, \pm m$; $m = 1, \dots, 262$;

4) for all t and natural k, the mappings $B'_k(t)$ and $B''_k(t)$ are homeomorphisms of the region Re V_ε (real part of V_ε) given by

$$B'_k(t) : (u, z) \to (u'_k(t), z'_k(t)) = (u + b'_{1k}(u, z, t), z + b'_{2k}(u, z, t)),$$
$$B''_k(t) : (u, z) \to (u''_k(t), z''_k(t)) = (u + b''_{1k}(u, z, t), z + b''_{2k}(u, z, t)),$$

and

$$\sum_{k=1}^{\infty} \sup_{\substack{(u,z) \in \mathrm{Re}\, V_\varepsilon \\ -\infty < t < \infty}} (|b'_{1k}(u, z, t)| + |b'_{2k}(u, z, t)| + |b''_{1k}(u, z, t)| + |b''_{2k}(u, z, t)|) < \infty.$$

Proposition 1 obviously follows from (16) and Theorem 3 since the function $\varphi^{(S)}(x)$ is analytic in the complex region $\left|x - \frac{a^{(S)}}{2}\right| \leq \delta_0$. Proposition 4 follows from (10), (11), and Lemma 3.

By (16), the transformation $\widetilde{A}$ is area-preserving, and by Theorem 3 the trace Sp$d\widetilde{A}$ of the matrix $d\widetilde{A}(0, 0)$ satisfies $|\mathrm{Sp}\, d\widetilde{A}| < 2$. Therefore, $\widetilde{A}$ is of elliptic type at $(0, 0)$. Furthermore, Proposition 3 follows from Theorem 3. Proposition 2 also follows from Theorem 3 by direct reduction of transformation $\widetilde{A}$ to the Birkhoff normal form in the neighborhood of the $(0, 0)$. Thus the above statements 1–4 are valid and hence, for all t two pairs of sequences of transformations $A'_k, B'_k(t)$ and $A''_k, B''_k(t)$ $(k = 1, 2, \dots)$

satisfy the conditions of Theorem 1 in **[P2]**. Applying this theorem, we obtain that for any $\varepsilon > 0$ there exists $\widehat{\delta} > 0$ and natural numbers $\widehat{n}'$, $\widehat{n}''$ such that if $|u| \le \widehat{\delta}$, $|z| \le \widehat{\delta}$, then for all $k' \ge \widehat{n}'$, $k'' \ge \widehat{n}''$,

$$|B'_{k'}(t) \circ \widetilde{A} \circ \cdots \circ B'_{\widehat{n}'}(t)\widetilde{A}(u,z)| \le \varepsilon, |B''_{k''}(t) \circ \widetilde{A} \circ \cdots \circ B''_{\widehat{n}''}(t)\widetilde{A}(u,z)| \le \varepsilon$$

(here $|(u,z)| = |u| + |z|$). It follows also from Theorem 1 in **[P2]** and Proposition 4 that $\widehat{\delta}$, $\widehat{n}'$, and $\widehat{n}''$ do not depend on t. By the last two inequalities, to prove Lemma 4 it suffices to show that for $\widehat{\delta} > 0$, there exist $\delta > 0$ and a natural k_0 such that the following is true: if for $\nu = 0, 1$, $k = 1, \dots, k_0$, all t, and $S = 1, \dots, r$ the relations $|\Delta_\nu^{(S)}| \le \delta$, $\left|\frac{d^{(k)}\Delta_\nu^{(S)}(t)}{dt^k}\right| \le \delta$ hold, then for all t,

$$|B'_{\hat{n}'-1}(t) \circ \widetilde{A} \circ \cdots \circ B'_1(t) \circ \widetilde{A}(\Delta_1^{(S)}(t), \Delta_1^{(S)}(t) - \Delta_0^{(S)}(t)| \le \widetilde{\delta},$$
$$|B''_{\hat{n}''-1}(t) \circ \widetilde{A} \circ \cdots \circ B''_1(t) \circ \widetilde{A}(\Delta_0^{(S)}(t), \Delta_0^{(S)}(t) - \Delta_1^{(S)}(t)| \le \widetilde{\delta}.$$

These inequalities obviously follow from the definitions of transformations $\widetilde{A}$, $B'_k(t)$, $B''_k(t)$, and Lemma 2. Thus, Lemma 4 and Theorem 3 are proved.

References

[A] V. I. Arnol′d, *Proof of a theorem of A. N. Kolmogorov on the conservation of quasiperiodic motions under a small change of the Hamiltonian function*, Uspekhi Mat. Nauk **18** (1963), no. 5, 13–40; English transl., Russian Math. Surveys **18** (1963), 9–36.

[FK] Ya. I. Frenkel′ and T. A. Kontorova, Zh. Eksp. Teor. Fiz. **8** (1938), 89–97. (Russian)

[K] A. N. Kolmogorov, *On the conservation of conditionally periodic motions for a small change in Hamilton's function*, Dokl. Akad. Nauk SSSR **98** (1954), 525–530; English transl., Lecture Notes in Physics **93** (1979), 51–56.

[M] J. Moser,, *Convergent series expansions for quasiperiodic motions*, Math. Ann. **169** (1967), 136–176.

[P1] L. D. Pustyl′nikov, *A dynamical system with infinitely many degrees of freedom and solution of the Frenkel′–Kontorova problem*, Teor. Mat. Fiz. **68** (1986), no. 1, 58–68; English transl., Teor. Mat. Fiz. (1987), 673–681.

[P2] ______, *Unbounded growth of an action variable in some physical models*, Trudy Moskov. Mat. Obshch. **46** (1983), 187–200; English transl, Trans. Moscow Math. Soc. **1984**, no. 2(46).

[S] Ya. G. Sinai, *Commensurate–incommensurate phase transitions in one-dimensional chains*, J. Statist. Physics **29** (1982), no. 3, 401–425.

[SM] C. L. Siegel and J. Moser, *Lectures on celestial mechanics*, Springer-Verlag, Heidelberg, 1971.

Moscow Institute of Engeneering Construction, Moscow, Russia

Keldysh Institute of Applied Mathematics, Moscow, Russia

Amer. Math. Soc. Transl.
(2) Vol. **171**, 1996

Ground States of a Boson Quantum Lattice Model

Alexander E. Mazel and Yurii M. Suhov

ABSTRACT. A boson system on $\mathbb{Z}^d$, $d \geq 2$ is considered with the Hamiltonian

$$H = -1/2 \sum_y a_y^+ (\Delta a)_y + \sum_{\langle y, y' \rangle} U(N_y, N_{y'}).$$

Here a_y^+, a_y are the creation and annihilation operators, and $N_y = a_y^+ a_y$. The nearest-neighbor potential U is given by: $U(n, n') = 0$ if $n + n' \leq m$ and $U(n, n') = +\infty$ if $n + n' > m$ (m is a fixed positive integer). The Gibbs ensemble is determined by $\exp -\beta(H - \mu N)$, where $N = \sum N_y$. Using a polymer expansion technique, we prove that if $d \geq 2$ and $\mu > 0$, $\beta > 0$ are large enough, the system has (exactly) two translation-periodic pure phases when m is odd and a single pure phase when m is even. We prove that the same is true of the ground states, and establish that the ground states are singular (degenerate) and correspond to 'dim' chess board configurations.

§1. Introduction. The model and the main result

In contrast with the theory of phase transitions in classical lattice systems, its quantum counterpart is much less developed. Few models have so far been studied at a rigorous level, and the results are not as complete as in classical cases. One of the major difficulties is that the analysis of the structure of the ground states — the starting point of any modern "classical" work — is more difficult for a quantum model even when it is defined as an "immediate" analogue of its classical prototype (which itself is often highly nontrivial). A striking example is the standard Heisenberg model (see **[FSiSp]** and **[DyLSi]**). Similar problems arise in dealing with the Hubbard models (see **[KL]**, **[MeMir-So]**, and the references therein).

1991 *Mathematics Subject Classification*. Primary 82B20; Secondary 81V25.

Key words and phrases. quantum lattice boson models, hard-core interaction, Gibbs states and measures, ground states, uniqueness and nonuniqueness, Feynman-Kac formula, classical and quantum contours, polymer expansion.

This research was supported in part by the Institute for Mathematics and its Applications, University of Minnesota, Minneapolis, MN 5545, USA and by grant NSF-DMR 92-13424.

A.E.M. thanks St John's College, Cambridge, for its warm hospitality and support. Both authors thank the Isaac Newton Institute, University of Cambridge, for its warm hospitality during the Random Spatial Processes program. Y.M.S. thanks the Institute for Mathematics and its Applications, University of Minnesota, for support towards this work.

This paper deals with a class of boson quantum lattice models with a (generalized) hard-core interaction. Some of these models were treated two decades ago by Ginibre **[G1]** (a different method was used in **[R]**; Ginibre's method was improved, in some situations, in **[M2]**). We use a technically different approach that allows us to extend Ginibre's result to a wider class of systems.

As usual, $\mathbb{Z}^d$, $d \geq 2$, denotes the cubic lattice in $\mathbb{R}^d$. The notation V, V', etc., is used below for a finite "volume" (subset) of $\mathbb{Z}^d$. Given V, we denote by $\mathcal{H}_V$ the boson Fock–Hilbert space in V and by $\mathcal{B}_V$ the C^*-algebra of bounded operators in $\mathcal{H}_V$. For $V' \subset V, \mathcal{H}_V$ is identified with $\mathcal{H}_{V'} \times \mathcal{H}_{V \setminus V'}$. We use a standard orthonormal basis in $\mathcal{H}_V$ labeled by the (classical) configurations X_V in V. That is, X_V is a function $\mathbb{Z}^d \to \mathbb{Z}^1_+$, with $X(y) = 0$ for y outside V. The notation X, Y, etc., is used for configurations in $\mathbb{Z}^d$ (i.e., functions $\mathbb{Z}^d \to \mathbb{Z}^1_+$). Given a configuration X, we denote by X_V the restriction of X to V. By $|X_V|$ we denote the number of particles in X_V: $|X_V| = \sum_{y \in V} X_V(y)$. By a_y^+ and a_y we denote, respectively, the boson creation and annihilation operators and by N_y the particle-number operator $a_y^+ a_y$, at site $y \in \mathbb{Z}^d$. $N_V = \sum_{y \in V} N_y$ stands for the particle number operator in V. By $\mathcal{B}$ we denote the inductive limit $\lim_{V \nearrow \mathbb{Z}^d} \operatorname{ind} \mathcal{B}_V$ (the C^*-algebra of the quasilocal observables). The action of the space-translation group (e.g., on $\mathcal{B}$) is defined in a standard way. We refer to the property of translation-periodicity when we have translation-invariance for a subgroup of finite index. In particular, we talk about period two when the subgroup is of index two. The ergodicity (of a state of $\mathcal{B}$) is considered below with respect to the space-translation group; in a translation-periodic case it is with respect to the corresponding subgroup.

The *Hamiltonian* H_V of the system in a finite volume V reads as

$$H_V = H_V^0 + H_V^{\text{int}}. \tag{1.1}$$

Here H_V^0 is the kinetic part

$$H_V^0 = -\frac{1}{d} \sum_{\langle y, y' \rangle \in V} a_y^+ a_y + \frac{1}{2d} N_V; \tag{1.2}$$

the sum $\sum_{\langle y, y' \rangle \in V}$ is over the (unordered) nearest-neighbor pairs $y, y' \in V$. Observe that H_V^0 is nothing but the (second-quantized) discrete Laplacian in V, with the Dirichlet's boundary conditions.

Furthermore, H_V^{int} is the potential part:

$$H_V^{\text{int}} = \sum_{\langle y, y' \rangle \subset V} U(N_y, N_{y'}). \tag{1.3}$$

Here, the nearest-neighbor potential U prevents the accumulation of the particles:

$$U(n, n') = \begin{cases} 0, & \text{if } n + n' \leq m, \\ +\infty, & \text{otherwise}, \end{cases} \tag{1.4}$$

parameter m is a fixed positive integer. In the physical language, U describes a kind of generalized hard-core repulsion (the case $m = 1$ corresponds to a 'true' hard core and was considered by Ginibre **[G1]**).

The models under consideration are natural quantum analogues of the classical models considered in **[MaSu]**. The classical model is obtained by omitting the kinetic part H_V^0 from the Hamiltonian H_V.

Formally, the operator H_V acts in the subspace of $\mathcal{H}_V$ spanned by the basis vectors corresponding to admissible configurations X_V. A configuration X_V (or X) is called admissible if $X_V(y) + X_V(y') \le m$ $(X(y) + X(y') \le m)$ for any nearest-neighbor pair $\langle y, y'\rangle \subset V$ $(\langle y, y'\rangle \subset \mathbb{Z}^d)$. In fact, we deal with the operators acting in a smaller space, $\mathcal{H}_V^{(Y)}$ generated by those configurations X_V for which $X_V + Y_{V^c}$ is admissible. Here Y is a fixed configuration (in $\mathbb{Z}^d$). This gives a kind of "potential" boundary condition: the particles in V are subject to the influence of a "frozen" configuration Y_{V^c}. Of particular interest is the case where Y coincides with one of the "close-packed" admissible configurations $Y^{(l)}, l = 0, \dots, m$:

$$Y^{(l)}(y) = \begin{cases} l, & \text{if } y \in \mathbb{Z}^d \text{ is even,} \\ m - l, & \text{if } y \in \mathbb{Z}^d \text{ is odd.} \end{cases} \tag{1.5}$$

Here and below we call $y = (y_1, \dots, y_d)$ even (odd) if the sum $\sum_i y_i$ is even (odd).

The Gibbs ensemble in V is determined by a standard expression

$$\exp(-\beta(H_V - \mu N_V)), \tag{1.6}$$

where $\beta \ge 0$ is the inverse temperature and $\mu \in \mathbb{R}^1$ the chemical potential of the system. More precisely, we will deal with the *Gibbs state* $\varphi_V^{(Y)}$ in V, with a boundary condition Y, which is given by

$$\varphi_V^{(Y)}(A) = \operatorname{tr}_V A\Pi_V^{(Y)}, \quad A \in \mathcal{B}_V, \tag{1.7}$$

where $\Pi_V^{(Y)}$ is the *density matrix*:

$$\Pi_V^{(Y)} = (\Xi^{(Y)}(V))^{-1}\Theta_V^{(Y)} \exp\big(-\beta(H_V^{(Y)} - \mu N_V)\big)\Theta_V^{(Y)}, \tag{1.8}$$

$\Theta_V^{(Y)}$ is the orthogonal projection $\mathcal{H}_V \to \mathcal{H}_V^{(Y)}$ and tr_V stands for the trace in $\mathcal{H}_V$. The "reduced" Hamiltonian $H_V^{(Y)}$ is defined by

$$H_V^{(Y)} = \Theta_V^{(Y)} H_V \Theta_V^{(Y)}, \tag{1.9}$$

and $\Xi^{(Y)}(V)$ is the corresponding partition function:

$$\Xi^{(Y)}(V) = \operatorname{tr}_V \Theta_V^{(Y)} \exp\big(-\beta(H_V^{(Y)} - \mu N_V)\big)\Theta_V^{(Y)}. \tag{1.10}$$

In the case $Y = Y^{(l)}$ we will use the simplified notations $\varphi_V^{(l)}, \mathcal{H}_V^{(l)}, \Theta_V^{(l)}, \Xi^{(l)}(V)$, etc.

Our goal is to study the w^*-limit

$$\lim_{V \nearrow \mathbb{Z}^d} \varphi_V^{(Y)} \tag{1.11}$$

of the Gibbs states $\varphi_V^{(Y)}$. If the limit (1.11) exists, it gives a state of the C^*-algebra $\mathcal{B}$. A standard low-density (or high-temperature) expansion method (see, e.g., [**BR,** Vol. 2, 3.2.5]) makes it possible to prove that there exists $\mu_0 = \mu_0(d, m) \in \mathbb{R}^1$ such that for $\mu \le \mu_0$ and any $\beta > 0$ the limit (1.11) exists and the limiting state φ does not depend on the boundary condition. In fact, one can prove a somewhat stronger assertion that the KMS-state for the corresponding derivation of $\mathcal{B}$ (see (1.13)) is unique provided that the condition $\mu \le \mu_0$ is fulfilled. In this paper we consider an opposite situation where μ is large. The result is the following theorem.

THEOREM 1. *Let $d \geq 2$ and a positive integer m be fixed. There exists $\bar{\mu} = \bar{\mu}(d) > 0$ and $\bar{\beta} = \bar{\beta}(d) > 0$, such that, for any pair (β, μ) with $\mu \geq \mu^{(0)} = \bar{\mu}\, m$ and $\beta \geq \beta^{(0)} = \bar{\beta} m$, the following assertions hold.*

(i) *If m is even, the limit* (1.11) *exists for any periodic boundary condition Y. The limiting state φ^0 does not depend on Y and is translation-invariant.*

(ii) *If m is odd, there exist exactly two different translation-periodic, ergodic limiting states* (1.11), *φ^+ and φ^-. More precisely,*

$$\varphi^+ = \lim_{V \nearrow \mathbb{Z}^d} \varphi_V^{(l)} \quad \text{for} \quad l = \frac{1}{2}(m+1), \dots, m, \tag{1.12a}$$

and

$$\varphi^- = \lim_{V \nearrow \mathbb{Z}^d} \varphi_V^{(l)} \quad \text{for} \quad l = 0, \dots, \frac{1}{2}(m-1), \tag{1.12b}$$

and the limit (1.11) *coincides, for any translation-periodic Y, with a mixture of φ^+ and φ^-. States $\varphi^\pm$ are transformed into each other by the unit space-shifts; they are translation-periodic with period two and ergodic.*

To explain the result contained in Theorem 1, observe that for $\mu > 0$ the classical system with potential U has the translation-periodic ground states that are precisely the close-packed configurations $Y^{(l)}$. The classical analogue of Theorem 1 (see **[MaSu]**) states that among the classical ground states those that generating translation-periodic thermodynamical phases are precisely $Y^{(m/2)}$ for m even and $Y^{(m/2 \pm 1/2)}$ for m odd. The point is that these ground states are 'dominant', in the sense that they have greater entropy of low-energy excitations. The authors showed in **[MaSu]** that there exists a bound $(\beta\mu)^0$, uniform in m, such that for $\beta\mu \geq (\beta\mu)^0$, only the dominant ground states generate the translation-periodic Gibbs measures associated with H_V^{int}.

In the quantum case, the nature of the dominance is slightly changed: the main factor extracting the dominant states is the entropy of "jumps" related to the kinetic part of the Hamiltonian. Because of that difference between the classical and quantum cases, we fail to obtain a bound for β^0 and μ^0 which would be uniform in m.

It is possible to prove (we do not do so in this paper) that the limiting states mentioned in Theorem 1 satisfy the KMS boundary condition, with respect to the derivation

$$\delta(A) = \lim_{V \nearrow \mathbb{Z}^d} i[H_V - \mu N_V, A]. \tag{1.13}$$

[More precisely, we have to modify slightly the standard definition of a KMS-state (see **[BR,** Vol. 2, Chap. 5.3]), because of the hard-core condition.] However, it is still an open question that the states indicated in Theorem 1 are the unique translation-periodic KMS-states of the system.

The proof of Theorem 1 is carried in §§2–4. In §2 we introduce, using the Feynman–Kac (FK) formula, a geometrical representation of the models under consideration, in terms of quantum contours (which are quantum analogues of the classical contours (or cylinders, in the terminology used in **[MaSu]**)). In §3 we develop a specific scheme of polymer expansion, and in §4 complete the proof.

We should note that the FK-formula (in various versions) was used, to study quantum models, for a long time by many authors. We refer the reader to papers **[G1]**, **[G2]**, **[M2]**, **[P]**, **[Su]**, **[OPiSu]**, and book **[BR]** (see also the references therein).

Our second result concerns the quantum *ground states* defined here as the limits of the Gibbs states as $\beta \to \infty$. We say that a state ψ is *classical* if there exists a probability measure Ψ on the space of the admissible configurations X in $\mathbb{Z}^d$ such that, for any finite $V \subset \mathbb{Z}^d$ and any $A \in \mathcal{B}_V$,

$$\psi(A) = \int \Psi(dX) f_A(X), \tag{1.14}$$

where the function f is given by the diagonal matrix elements of A in our standard basis in $\mathcal{H}_V$: $f_A(X) = A(X_V, X_V)$. We call a classical state ψ *singular* if measure Ψ is concentrated at a single configuration.

THEOREM 2. *For any $\mu > \mu^{(0)} = \bar{\mu}\, m$ (with constant $\bar{\mu}$ from Theorem 1) there exist the w^*-limits of the states φ^0 and $\varphi^{\pm}$ constructed in Theorem 1:*

$$\bar{\varphi}^0 = \lim_{\beta \to \infty} \varphi^0, \quad \bar{\varphi}^{\pm} = \lim_{\beta \to \infty} \varphi^{\pm} \tag{1.15}$$

which are singular classical states corresponding to the configurations $Y^{(l)}$ with $l = m/2$ and $l = (m \pm 1)/2$, respectively.

A well-known definition of a ground state refers to a positive spectrum condition for the generator of the unitary group induced, in the GNS-representation, by derivation (1.13). See, e.g., [**BR,** Vol. 2, §5.3.3]. The classical singular states corresponding to the close-packed configurations $Y^{(l)}, l = 0, \dots, m$, and in particular, the states indicated in Theorem 2, obviously satisfy this condition (provided that μ is large enough). Again, it is an open question whether or not these states are the only pure translation-periodic states with this property. Furthermore, it is interesting to clarify what additional conditions are needed to extract, from the whole family of classical singular close-packed states, those mentioned in Theorem 2.

The proof of Theorem 2 is obtained as a by-product in the course of proving Theorem 1.

We conclude this section by noting that the convergence in (1.12 a,b) and (1.15) is equivalent to the convergence of the matrix elements of the *reduced* density matrices. In the case of (1.12 a,b) we have to check that for any fixed finite $V^0 \subset \mathbb{Z}^d$ and any pair of admissible configurations X_{V^0}, X'_{V^0} in V^0, the limits

$$\lim_{V \nearrow \mathbb{Z}^d} \pi^{(l)}_{V, V^0}(X_{V^0}, X'_{V^0}) = \pi^{(l)}_{V^0}(X_{V^0}, X'_{V^0}) \tag{1.16}$$

exist and give the corresponding states. Here $\pi^{(l)}_{V,V^0}(X_{V^0}, X'_{V^0})$ stands for the matrix element of the reduced density matrix $\pi^{(l)}_{V,V^0}$ that is related to the density matrix $\Pi^{(l)}_V$ (see (1.8)) by

$$\pi^{(l)}_{V,V^0} = \mathrm{tr}_{V \setminus V^0}\, \Pi^{(l)}_V. \tag{1.17}$$

The translation-invariance and translation-periodicity properties of a limiting state are equivalent to the corresponding properties of the family $\{\pi^{\cdot}_{V^0}\}$ of the limiting reduced matrices. [Here $\pi^{\cdot}_{V^0}$ denotes one of $\pi^0_{V^0} = \pi^{(m/2)}_{V^0}$ and $\pi^{\pm}_{V^0} = \pi^{(m/2 \pm 1/2)}_{V^0}$.] The ergodicity follows from a correlation-decay property for the matrix elements $\{\pi^{\cdot}_{V^0}(X_{V^0}, X'_{V^0})\}$. See again [**BR,** Vol. 1].

In the case of (1.15) we have to check that the limiting reduced density matrices $\pi^{\cdot}$ obey

$$\lim_{\beta\to\infty} \pi^0_{V^0}(X_{V^0}, X'_{V^0}) = \begin{cases} 1, & \text{if } X_{V^0} = X'_{V^0} = Y^{(m/2)}_{V^0}, \\ 0, & \text{otherwise,} \end{cases} \tag{1.18}$$

and

$$\lim_{\beta\to\infty} \pi^{\pm}_{V^0}(X_{V^0}, X'_{V^0}) = \begin{cases} 1, & \text{if } X_{V^0} = X'_{V^0} = Y^{(m/2\pm 1/2)}_{V^0}, \\ 0, & \text{otherwise.} \end{cases} \tag{1.19}$$

§2. Some technical tools

The Feynman–Kac formula.. As mentioned in §1, one of the main tools used for proving Theorems 1 and 2 is the FK-formula. In a standard form, it may be found, e.g., in **[G2]** or [**BR**, Vol. 2, §§3.3.2, 3.3.3]. We use a version of the FK-formula based on the concept of a quantum configuration. More precisely, we consider sequences $\mathbb{X}_V = (X^0_V, X^1_V, \dots, X^n_V)$ of admissible classical configurations in V, with X^s_V and X^{s+1}_V differing only at two nearest-neighbor lattice sites, $y^{\pm}_s$, with $X^{s+1}_V(y^{\pm}_s) - X^s_V(y^{\pm}_s) = \pm 1, s = 0, \dots, n-1$. Such families are called *quantum configurations* in V.

LEMMA 2.1. *For any β, μ, and V, any pair of configurations X_V, X'_V in V, and any configuration Y, the matrix element $E^{(Y)}_V(X_V, X'_V)$ of the corresponding operator $\Theta^{(Y)}_V \exp\left(-\beta(H^{(Y)}_V - \mu N_V)\right)\Theta^{(Y)}_V$ is given by*

(2.1)

$$E^{(Y)}_V(X_V, X'_V) = \begin{cases} \exp(\beta\mu|\mathbb{X}_V| - \beta|\mathbb{X}_V|) \sum\limits_{n=0}^{\infty} \frac{1}{n!}\left(\frac{\beta}{2d}\right)^n \\ \quad \times \sum\limits_{\substack{\mathbb{X}_V=(X^0_V, X^1_V,\dots,X^n_V):\\ X^0_V = X_V,\ X^n_V = X'_V \\ X^s_V + Y_{V^c} \text{ is admissible}}} \prod\limits_{i=0}^{n-1} X^i_V(y^-_i), & \text{if } |X_V| = |X'_V|, \\ 0, & \text{otherwise.} \end{cases}$$

Here $|\mathbb{X}_V|$ is the number of particles in (any of) classical configurations X^s_V representing $\mathbb{X}_V$.

We omit the proof of Lemma 2.1: it is of standard probabilistic character and may be immediately deduced from the "usual" FK-formula (see the above references).

By using Lemma 2.1, we obtain the following formulas for the partition function $\Xi^{(Y)}(V)$ and the matrix elements of the reduced density matrix $\pi^{(Y)}_{V,V^0}$:

$$\Xi^{(Y)}(V) = \sum_{X_V} E^{(Y)}_V(X_V, X_V) \tag{2.2}$$

and for any admissible configurations X_{V^0}, X'_{V^0} in V^0,

$$\pi^{(Y)}_{V,V^0}(X_{V^0}, X'_{V^0}) = \frac{\Xi^{(Y)}(V, V^0; X_{V^0}, X'_{V^0})}{\Xi^{(Y)}(V)}. \tag{2.3}$$

Here,

$$(2.4) \qquad \Xi^{(Y)}(V, V^0; X_{V^0}, X'_{V^0}) = \sum_{X_{V\setminus V^0}} E_V^{(Y)}(X_{V^0} + X_{V\setminus V^0}, X'_{V^0} + X_{V\setminus V^0}).$$

The sum in (2.2) is over the configurations X_V such that $X_V + Y_{V^c}$ is admissible, and in (2.4) over the configurations $X_{V\setminus V^0}$ such that $X_{V^0} + X_{V\setminus V^0} + Y_{V^c}$ and $X'_{V^0} + X_{V\setminus V^0} + Y_{V^c}$ are admissible.

Lemma 2.1 and formulas (2.2)–(2.4) allow us to treat the quantum model under consideration as a system of "classical" statistical mechanics, where one deals with quantum configurations (and later with quantum contours). The *statistical weight* $w(\mathbb{X}_V)$ of a quantum configuration $\mathbb{X}_V = (X_V^0, \dots, X_V^n)$ (where each $X_V^s + Y_{V^c}$ is admissible) is given by

$$(2.5) \qquad w(\mathbb{X}_V) = \exp(\beta(\mu - 1)|\mathbb{X}_V| - \beta(\mu - 1)|Y_V|) \left(\frac{\beta}{2d}\right)^n \frac{1}{n!} \prod_{i=0}^{n-1} X_V^i(y_i^-),$$

and the partition function (2.2) may be rewritten as

$$(2.6) \qquad \Xi^{(Y)}(V) = \exp(\beta(\mu - 1)|Y_V|) \sum_{\substack{\mathbb{X}_V:\ X_V^0 = X_V^n, \\ X_V^s + Y_{V^c} \text{ is admissible}}} w(\mathbb{X}_V).$$

Theorems 1 and 2 may therefore be reformulated as statements about the structure of the set of the *Gibbs* (or *DLR*) *measures* for an ensemble of quantum configurations. Furthermore, the properties of the ensemble of quantum configurations may be expressed in terms of an appropriate ensemble of quantum contours.

Henceforth, we depart from the original quantum model and concentrate on the analysis of the corresponding ensembles of quantum configurations, with statistical weight (2.5). We focus attention on the studying of *cyclic* quantum configurations $\mathbb{X}_V$ (with $X_V^0 = X_V^n$) figuring in (2.6). The extension of the argument to the general case needed for (2.4) leads to a "local" perturbation of the space of quantum configurations, which does not change its phase diagram; the corresponding technical details are omitted from the paper.

Geometry of the classical configurations. Our main goal in §2 is to introduce the concept of a quantum contour which will then form a base for an appropriate polymer expansion. The definition and properties of quantum contours are related to the definition and properties of classical contours (or briefly, contours). A *contour* γ is a triple $(\widetilde{\gamma}, E, I)$, where

(i) $\widetilde{\gamma} = \widetilde{\gamma}(\gamma)$ is a finite connected set of the plaquettes of the dual lattice $\widetilde{\mathbb{Z}}^d = \{y + (\frac{1}{2}, \dots, \frac{1}{2}) :\ y \in \mathbb{Z}^d\}$ such that
  - (i.1) either 0 or 2 or 4 plaquettes from $\widetilde{\gamma}$ meet at any $(d-2)$-dimensional face of $\widetilde{\mathbb{Z}}^d$,
  - (i.2) if precisely two plaquettes from $\widetilde{\gamma}$ meet at a given $(d-2)$-dimensional face of $\widetilde{\mathbb{Z}}^d$, then they are not parallel;

(ii) $E = E(\gamma)$ and $I = I(\gamma)$ are two integers from $\{0, 1, \dots, m\}$ such that
  - (ii.1) $I > E$ for $\widetilde{\gamma}$ even,
  - (ii.2) $I < E$ for $\widetilde{\gamma}$ odd.

The set $\widetilde{\gamma} = \widetilde{\gamma}(\gamma)$ in (i) is called a (classical) *interface*. We call an interface $\widetilde{\gamma}$ even (odd) when there exists an odd (even) $x \in \mathbb{Z}^d$ such that

(a) x is adjacent to some of the plaquettes from $\widetilde{\gamma}$;

(b) for any natural n there exists a self-avoiding path of length n, along the bonds of $\mathbb{Z}^d$, such that it does not intersect the plaquettes from $\widetilde{\gamma}$.

It is not hard to see that every interface $\widetilde{\gamma}$ separates a finite connected and simple-connected subset of $\mathbb{Z}^d$, called the interior of $\widetilde{\gamma}$ and denoted by $\operatorname{Int}\widetilde{\gamma}$, from its complement called the exterior of $\widetilde{\gamma}$ and denoted by $\operatorname{Ext}\widetilde{\gamma}$. In the case where $\widetilde{\gamma}$ is considered as the component of a contour γ, we speak about the interior and exterior of a contour and use the notations $\operatorname{Int}(\gamma)$ and $\operatorname{Ext}(\gamma)$.

Two contours, $\gamma' = (\widetilde{\gamma}', E', I')$ and $\gamma'' = (\widetilde{\gamma}'', E'', I'')$, are called *compatible* if one of the following six conditions holds

(i) $\widetilde{\gamma}' \cap \widetilde{\gamma}'' = \varnothing$, $\operatorname{Int}\widetilde{\gamma}' \subset \operatorname{Int}\widetilde{\gamma}''$ and $I'' = E'$;

(ii) $\widetilde{\gamma}' \cap \widetilde{\gamma}'' = \varnothing$, $\operatorname{Int}\widetilde{\gamma}'' \subset \operatorname{Int}\widetilde{\gamma}'$ and $I' = E''$;

(iii) $\widetilde{\gamma}' \cap \widetilde{\gamma}'' = \varnothing$, $\operatorname{Int}\widetilde{\gamma}' \cap \operatorname{Int}\widetilde{\gamma}'' = \varnothing$ and $E'' = E'$;

(iv) $\widetilde{\gamma}' \cap \widetilde{\gamma}'' \neq \varnothing$, $\operatorname{Int}\widetilde{\gamma}' \subset \operatorname{Int}\widetilde{\gamma}''$ and either $E'' > I'' = E' > I'$ or $E'' < I'' = E' < I'$;

(v) $\widetilde{\gamma}' \cap \widetilde{\gamma}'' \neq \varnothing$, $\operatorname{Int}\widetilde{\gamma}'' \subset \operatorname{Int}\widetilde{\gamma}'$ and either $E' > I' = E'' > I''$ or $E' < I' = E'' < I''$;

(vi) $\widetilde{\gamma}' \cap \widetilde{\gamma}'' \neq \varnothing$, $\operatorname{Int}\widetilde{\gamma}' \cap \operatorname{Int}\widetilde{\gamma}'' = \varnothing$ and either $I'' > E'' = E' > I'$ or $I'' < E'' = E' < I'$.

A finite collection $\{\gamma_i\}$ of contours is *compatible* if any two contours from the collection not separated by any third contour are compatible. [We say that a contour γ separates contours γ' and γ'' if the corresponding interfaces $\widetilde{\gamma}$, $\widetilde{\gamma}'$ and $\widetilde{\gamma}''$ obey $\operatorname{Int}\widetilde{\gamma}' \subset \operatorname{Int}\widetilde{\gamma}$ and $\operatorname{Int}\widetilde{\gamma}'' \subset \operatorname{Ext}\widetilde{\gamma}$ or $\operatorname{Int}\widetilde{\gamma}'' \subset \operatorname{Int}\widetilde{\gamma}$ and $\operatorname{Int}\widetilde{\gamma}' \subset \operatorname{Ext}\widetilde{\gamma}$.] Note that for any finite compatible collection $\{\gamma_i\}$ all external contours have the same value E which we denote by $E(\{\gamma_i\})$. Here and below, we call a contour γ external (in a given collection) if $\operatorname{Int}\widetilde{\gamma}$ is not contained in the interior of any other contour from the collection.

An important fact is that there exists a one-to-one correspondence between the admissible configurations X coinciding with some $Y^{(l)}$ outside a finite subset of $\mathbb{Z}^d$ and the finite compatible collections of contours $\{\gamma_i\}$ with $E(\{\gamma_i\}) = l$. To describe this correspondence, it is convenient to equip the set of the contours γ with a partial order determined by the set-theoretical inclusion of the interiors $\operatorname{Int}(\gamma)$. Given $y \in \mathbb{Z}^d$ and a compatible collection of contours $\{\gamma_i\}$, consider the contours from $\{\gamma_i\}$ with the interors containing y and select among them a minimal one in the sense of the above order. Denote by $\gamma(y) = (\widetilde{\gamma}(y), E(y), I(y))$ the minimal contour, if it exists, and set $\gamma(y) = \varnothing$, otherwise. If $\gamma(y) \neq \varnothing$, then the admissible configuration X corresponding to $\{\gamma_i\}$ has the value $X(y) = I(y)$. On the other hand, if $\gamma(y) = \varnothing$ then $X(y) = E(\{\gamma_i\})$.

The correspondence just introduced is especially useful because of the following identity:

$$|Y_V^{(l)}| - |X_V| = \frac{1}{2d} \sum_i |\widetilde{\gamma}_i| \cdot |E(\gamma_i) - I(\gamma_i)|, \tag{2.7}$$

where V is an (arbitrary) finite volume such that $X_{V^c} = Y_{V^c}^{(l)}$. Here and below, $|\widetilde{\gamma}_i|$ denotes the number of plaquettes in $\widetilde{\gamma}_i$.

Geometry of quantum configurations. In this subsection we construct, for quantum configurations, an analogue of contour representation. From now on we are working with cyclic quantum configurations. Recall that in a cyclic quantum configuration $\mathbb{X}_V = (X_V^0, \dots, X_V^n)$, $n = n(\mathbb{X}_V)$ is always even, and $X_V^0 = X_V^n$. In the special case $n(\mathbb{X}_V) = 0$, the cyclic quantum configuration consists of a single classical configuration X_V. We assume that in a given quantum configuration $\mathbb{X}_V = (X_V^0, \dots, X_V^n)$ in V, the classical configurations $X_V^i + Y_{V^c}^{(l)}$ are admissible for some $l = 0, \dots, m$, and write $X^i = X_V^i + Y_{V^c}^{(l)}$, $i = 1, \dots, n$. Furthermore, we write $\mathbb{X} = (X^0, \dots, X^n)$ and call $\mathbb{X}$ a quantum configuration in $\mathbb{Z}^d$.

As was pointed out before, each classical configuration X^i corresponds to a unique compatible collection of contours $\{\gamma_j^i\}$, with $E(\{\gamma_j^i\}) = l$. Consider the union $\bigcup_{i,j} \widetilde{\gamma}_j^i$ of the interfaces from all configurations X^i, $i = 1, \dots, n$ (briefly, the interfaces from $\mathbb{X}$). This is a finite set of plaquettes of $\widetilde{\mathbb{Z}}^d$, and it can be decomposed into the connected components. Let $K = K(\mathbb{X})$ be the number of these components. We index them by a superscript k taking values $1, \dots, K$ (in an arbitrary order). Next, we partition the collection of the contours γ_j^i, from all configurations X^i, $i = 1, \dots, n$ (briefly, the collection of the contours from $\mathbb{X}$) into subcollections, $\{\gamma_t^{(k)}\}$, $k = 1, \dots, K$, in a such a way that the interfaces of the contours $\gamma_t^{(k)}$, with a fixed k, are exactly the interfaces from the k-th connected component of $\bigcup_{i,j} \widetilde{\gamma}_j^i$. A single subcollection $\{\gamma_t^{(k)}\}$ is called a quantum *interface*. This procedure leads us further to the concept of a quantum contour.

A *quantum contour* ζ is a finite family of classical contours $\{\gamma_j^i,\ i = 0, \dots, n(\zeta),\ j = 1, \dots, n_i(\zeta)\}$ such that

(i) for any fixed $i = 0, 1, \dots, n(\zeta)$, the subcollection $\{\gamma_j^i : j = 1, \dots, n_i(\zeta)\}$ is a compatible collection of contours with $E(\{\gamma_j^i\}) = l$, which defines an admissible classical configuration $X^i = X^i(\zeta)$ coinciding with $Y^{(l)}$ outside some finite $V \subset \mathbb{Z}^d$;

(ii) the collection $\mathbb{X}_V(\zeta) = (X_V^0, \dots, X_V^{n(\zeta)})$ is a cyclic quantum configuration in V (or equivalently, $\mathbb{X}(\zeta) = (X^0, \dots, X^{n(\zeta)})$ is a cyclic quantum configuration in $\mathbb{Z}^d$);

(iii) the union $\widetilde{\zeta} = \bigcup_{i,j} \widetilde{\gamma}_j^i(\zeta)$ of the interfaces $\widetilde{\gamma}_j^i(\zeta)$ of all contours from ζ is a connected set of plaquettes of zd.

Observe that $\widetilde{\zeta} = \widetilde{\zeta}(\zeta)$ is a unique quantum interface in quantum configuration $\mathbb{X}_V(\zeta)$ (or $\mathbb{X}(\zeta)$). Also note that the quantum configuration $\mathbb{X}(\zeta)$ is uniquely determined by ζ.

A collection $\{\zeta_r\}$ of quantum contours is *compatible* if the quantum interfaces $\widetilde{\zeta}(\zeta_r)$ do not intersect each other for different r's and $\{\gamma_j^0(\zeta_r) : j = 1, \dots, n_0(\zeta_r);\ r = 0, 1, \dots\}$ is a compatible collection of classical contours.

The construction described before allows us to assign to any quantum configuration $\mathbb{X}$ in $\mathbb{Z}^d$ a unique compatible collection of quantum contours (by decomposing the union $\bigcup_{i,j} \widetilde{\gamma}_j(X^i(\mathbb{X}))$ into the connected components). However, unlike the compatible collections of classical contours, a finite compatible collection of quantum contours defines, in general, not a single quantum configuration in $\mathbb{Z}^d$ but a family of configurations. To explain this phenomenon, note that any quantum configuration describes nothing more than the sequence of particle jumps. A quantum contour ζ (that has a unique corresponding quantum configuration $\mathbb{X}(\zeta)$) is a sequence of $n(\zeta)$

jumps. Now observe that for any compatible collection of quantum contours, the jumps corresponding to distinct quantum contours are independent. Hence, with a given compatible collection of quantum contours $\{\zeta_r\}$ one can precisely associate

$$\frac{(\sum_r n(\zeta_r))!}{\prod_r n(\zeta_r)!} \tag{2.8}$$

quantum configurations that correspond to various possibilities to intermit the jumps from different $\mathbb{X}(\zeta_r)$'s with each other.

Formally, the procedure is as follows. Fix a compatible collection $\{\zeta_r\}$ of quantum contours. For each i and r we have a family of classical contours $\{\gamma_j^i(\zeta_r) : j = 1, \dots, n_i(\zeta_r)\}$, which determines a (unique) admissible configuration $X^i(\mathbb{X}(\zeta_r))$. Reindex the pairs (i, r) by a single label s, taking values $0, \dots, n(\{\zeta_r\})$, $n(\{\zeta_r\}) = \sum_r n(\zeta_r)$, so that for any (i_1, r), (i_2, r) with $i_1 < i_2$ the corresponding labels s_1, s_2 obey $s_1 < s_2$. It is not hard to check that the number of different labelings with this property is given by (2.8). Given such a labeling, we construct the (unique) corresponding quantum configuration $\mathbb{X}(\{\zeta_r\}) = (X^0(\{\zeta_r\}), \dots, X^{n(\{\zeta_r\})}(\{\zeta_r\}))$ inductively, as follows.

(a) Configuration $X^0(\{\zeta_r\})$ is determined by the collection of classical contours $\{\gamma_j^0(\zeta_r)\colon\ j = 1, \dots, n_0(\zeta_r);\ r = 0, 1, \dots\}$.

(b) Suppose that configuration $X^s(\{\zeta_r\})$ is already determined. Take the pair (i_0, r_0) labeled by value $s+1$. We construct the configuration $X^{s+1}(\{\zeta_r\})$ by replacing, in the compatible collection of contours corresponding to $X^s(\{\zeta_r\})$, the subcollection $\{\gamma_j^{i_0-1}(\zeta_{r_0}) : j = 1, \dots, n_{i_0-1}(\zeta_{r_0})\}$ with $\{\gamma_j^{i_0}(\zeta_{r_0})\colon j = 1, \dots, n_{i_0}(\zeta_{r_0})\}$.

Now assign to a compatible collection of quantum contours $\{\zeta_r\}$ the statistical weight

$$w(\{\zeta_r\}) = \sum_{\mathbb{X}(\{\zeta_r\})} w(\mathbb{X}(\{\zeta_r\})), \tag{2.9}$$

where the sum is extended to all quantum configurations $\mathbb{X}(\{\zeta_r\})$ associated with $\{\zeta_r\}$. [Formally, statistical weight (2.5) is assigned only to the quantum configurations in a finite volume. But it is easy to check that there exists a unique l such that for any s the configuration $X^s(\mathbb{X}(\{\zeta_r\}))$ coincides with $Y^{(l)}$ outside some finite V. Substituting this V and $Y = Y^{(l)}$ into (2.5), we get the correct expression for $w(\mathbb{X}(\{\zeta_r\}))$.] Obviously, the statistical weight is the same for each quantum configuration $\mathbb{X}(\{\zeta_r\})$ associated with $\{\zeta_r\}$. Moreover,

$$\frac{(\sum_r n(\zeta_r))!}{\prod_r n(\zeta_r)!} w(\mathbb{X}(\{\zeta_r\})) = \prod_r w(\zeta_r), \tag{2.10}$$

where, for an arbitrary quantum contour ζ,

$$w(\zeta) = w(\mathbb{X}(\zeta)), \tag{2.11}$$

and $w(\mathbb{X}(\zeta))$ is calculated by means of (2.5) with the corresponding V and $Y = Y^{(l)}$. Now, for any l, expression (2.6) can be rewritten as

$$\Xi^{(l)}(V) = \exp\left(\beta(\mu - 1)|Y_V^{(l)}|\right) \sum_{\substack{\{\zeta_r\}:\ \{\widetilde{\zeta}_r\} \subset V, \\ X_V^s(\mathbb{X}(\{\zeta_r\})) + Y_{V^c}^{(l)} \\ \text{is admissible}}} \prod_r w(\zeta_r). \tag{2.12}$$

The external factor $\exp(\beta(\mu-1)|Y_V|)$ in (2.6) (and $\exp(\beta(\mu-1)|Y_V^{(l)}|)$ in (2.12)) may be omitted, and we use the previous notation for the partition functions with this factor canceled.

Denoting for a classical contour $\gamma = (\widetilde{\gamma}, E, I)$,

$$\|\gamma\| = |\widetilde{\gamma}| \cdot |E - I|, \tag{2.13}$$

and for a compatible collection of contours $\{\gamma_j\}$,

$$\|\{\gamma_j\}\| = \sum_j \|\gamma_j\|, \tag{2.14}$$

we obtain, by virtue of (2.7), the following lemma.

LEMMA 2.2. *The statistical weight of a quantum contour ζ equals*

$$w(\zeta) = \exp\left(-\frac{1}{2d}\beta(\mu-1)\|\{\gamma_j^0(\zeta)\}\|\right)\left(\frac{\beta}{2d}\right)^{n(\zeta)}\frac{1}{n(\zeta)!}\prod_{i=0}^{n(\zeta)-1} X^i(y_i^-), \tag{2.15}$$

where $X^i = X^i(\mathbb{X}(\zeta))$. The partition function $\Xi^{(l)}(V)$, for any finite V and $l = 0, \dots, m$, equals

$$\Xi^{(l)}(V) = \sum_{\substack{\{\zeta_r\}:\ \{\widetilde{\zeta}_r\}\subset V,\\ X_V^s(\mathbb{X}(\{\zeta_r\}))+Y_{V^c}^{(l)}\\ \text{is admissible}}} \prod_r w(\zeta_r). \tag{2.16}$$

Formula (2.16) provides the so-called contour representation of the partition function: it allows us to introduce the ensemble of quantum contours in V. This is the starting point of a polymer expansion used for proving Theorems 1 and 2. In the next section we derive an appropriate scheme leading to this expansion, and in §4 complete the proof.

§3. A polymer expansion for abstract contour models and its application to contour models on $\mathbb{Z}^d$

Formally, the content of this section does not depend on that of §§1 and 2. However, we use a notation and terminology that provide a transparent connection with the preceding section and serve as a prototype for §4.

Abstract contours. Consider a countable set Θ, whose elements of which are called (abstract) *contours* and denoted θ, θ', etc.. A finite subset of Θ is denoted by Λ. We fix some reflexive and symmetric relation $\not\sim$ on $\Theta \times \theta$. A pair $\theta, \theta' \in \Theta$ is called incompatible if $\theta \not\sim \theta'$ and compatible otherwise (i.e., $\theta \sim \theta'$). A collection $\{\theta_i\}$ of contours is called compatible if any pair of its contours is compatible. We assign to every contour θ a (generally speaking) complex-valued *statistical weight* $w(\theta)$, and for every Λ define an (abstract) *partition function*

$$Z(\Lambda) = \sum_{\{\theta_i\}\subseteq\Lambda} \prod_i w(\theta_i), \tag{3.1}$$

where the sum is extended to all compatible collections of contours $\theta_i \in \Lambda$. The empty collection is compatible by definition, and it is included in $Z(\Lambda)$ with the statistical weight 1. A *polymer* π is an (unordered) finite collection, $\pi = [\theta_i^{\alpha_i}]$, of different contours $\theta_i \in \Theta$ taken with positive integer multiplicities α_i, such that for every pair θ', $\theta'' \in \pi$ there exists a sequence $\theta' = \theta_{i_1}, \theta_{i_2}, \dots, \theta_{i_k} = \theta'' \in \pi$ with $\theta_{i_j} \not\sim \theta_{i_{j+1}}$, $1 \le j < k$. The notation $\pi \subseteq \Lambda$ used below is self-explanatory. Our result can be formulated as follows.

THEOREM 3.1. *Suppose that there exists a function $a : \Theta \mapsto \mathbb{R}^+$,*

$$\sum_{\theta' :\, \theta' \not\sim \theta} |w(\theta')| \exp(a(\theta')) \le a(\theta) \tag{3.2}$$

for each contour θ. Then, for any finite Λ,

$$\log Z(\Lambda) = \sum_{\pi \subseteq \Lambda} r(\pi) w(\pi), \tag{3.3}$$

where

(i) *the statistical weight of a polymer $\pi = [\theta_i^{\alpha_i}]$ equals*

$$w(\pi) = \prod_i w(\theta_i)^{\alpha_i}; \tag{3.4}$$

(ii) *the combinatorial factor $r(\pi)$ is calculated as*

$$r(\pi) = \prod_i (\alpha_i!)^{-1} \sum_{G(\pi)} (-1)^{|G(\pi)|}. \tag{3.5}$$

Here the sum $\sum_{G(\pi)}$ is taken over all connected graphs $G(\pi)$ with $\sum_i \alpha_i$ vertices

$$v_{i,j}(G(\pi)), \quad 1 \le j \le \alpha_i, \tag{3.6}$$

which are labeled by the contours θ_i from π counted with their multiplicities, such that an edge can link two vertices $v(G(\pi))$ and $v'(G(\pi))$ only if the corresponding contours are incompatible, and $|G(\pi)|$ denotes the number of the edges in $G(\pi)$.

Moreover, the series (3.3) for $\log Z(\Lambda)$ is absolutely convergent, and for any contour θ

$$\sum_{\pi :\, \pi \ni \theta} \alpha(\theta, \pi) |r(\pi) w(\pi)| \le |w(\theta)| \exp(a(\theta)), \tag{3.7}$$

where $\alpha(\theta, \pi)$ is a multiplicity of the contour θ in the polymer π.

REMARK. The meaning of Theorem 3.1 that distinguishes it from other results on general polymer expansions (see, e.g., **[M1]**, **[KoPr]**, **[S]**, **[MMi]**) is that it combines a convenient condition (3.2) of the absolute convergence of the expansion (3.3) with a standard form of the polymers and their statistical weights. [In the references above, one or another of these tasks was not fully accomplished. For example, in **[KoPr]** (where condition (3.2) was introduced for the first time), the polymer expansion is established in terms of partition functions in subsets $\Lambda' \subseteq \Lambda$; this is not convenient when one is interested in the detailed properties of the model.]

PROOF. Formula (3.3) for $\log Z(\Lambda)$ can be easily obtained as a result of calculations with the formal series (see, e.g., **[S]**), and the only question is the convergence of the series at the right-hand side of (3.3). It can be verified that for every $\pi = [\theta_i^{\alpha_i}]$,

$$\left| \sum_{G(\pi)} (-1)^{|G(\pi)|} \right| \le \sum_{T(\pi)} 1, \tag{3.8}$$

where $T(\pi)$ is defined similarly to $G(\pi)$ with an additional requirement that it is a tree (see [S]). From (3.8), we have the following bound

$$\sum_{\pi:\ \pi\ni\theta} \alpha(\theta,\pi)|r(\pi)w(\pi)| \le \sum_{\pi:\ \pi\ni\theta} \alpha(\theta,\pi) \sum_{T(\pi)} \prod_i (\alpha_i!)^{-1}|w(\theta_i)|^{\alpha_i}. \tag{3.9}$$

For any tree $T(\pi)$ choose one of the vertices labeled by θ as the root $r(T)$. There exists exactly $\alpha(\theta,\pi)$ different possibilities to select such a root. Hence,

$$\sum_{\pi:\ \pi\ni\theta} \alpha(\theta,\pi)|r(\pi)w(\pi)| \le \sum_{\pi:\ \pi\ni\theta} \sum_{T_r(\pi)} \prod_i (\alpha_i!)^{-1}|w(\theta_i)|^{\alpha_i}, \tag{3.10}$$

where the second sum is taken over the trees with the specified root labeled by one of the copies of θ. Observe that (3.10) is less than or equal to

$$\sum_{T:\ l(r(T))=\theta} \prod_{v\in T} |w(l(v))|, \tag{3.11}$$

where the sum is over finite *θ-rooted labeled trees* T with the following properties:

(i) for every vertex $v \in T$, there is a contour θ' associated with v (we write this as $\theta' = l(v)$ and call θ' a label of vertex v);

(ii) for any edge of T the contours labeling the vertices adjacent to this edge are incompatible;

(iii) one of the vertices of T is fixed as the root of T, this vertex is denoted by $r(T)$ and it is labeled by contour θ (i.e., $l(r(T)) = \theta$).

Note that two labeled trees are considered isomorphic (and not distinguished) if and only if there is a one-to-one map from the vertex set of one labeled tree onto that of another one, which takes the root of one tree to that of another and preserves both the adjacency and the labeling. Obviously, (3.11) is equal to

$$|w(\theta)| \sum_{T:l(r(T))=\theta} \prod_{v\in T,\ v\neq r(T)} |w(l(v))| = |w(\theta)| \sum_{T:l(r(T))=\theta} w(T), \tag{3.12}$$

where

$$w(T) = \prod_{v\in T, v\neq r(T)} |w(l(v))|. \tag{3.13}$$

We define the level of a vertex v of a rooted tree T as the length of the path from $r(T)$ to v; the highest level $n(T)$ in T is the maximal length of the paths from $r(T)$ to the leaves of T. In the sequel, when speaking about an edge (v,v') of a rooted tree T, we assume that the level of vertex v (the beginning of the edge) is one less than the level of vertex v' (the end of the edge).

The edges of θ-rooted tree T can be uniquely decomposed into bunches. A *bunch* is defined as a set of edges having a common beginning at some vertex v, and a single contour θ' labeling their ends (we say that a bunch grows from v and is labeled by θ'). To make the partition unique, every bunch should include the maximal possible number of edges. We denote a bunch by $b = \langle v, \theta', \alpha'\rangle$, where $\alpha' = \alpha'(b)$ is the number of edges in the bunch. We also use a short-hand notation $b = \langle\theta', \alpha'\rangle$, when it is clear or not important what the initial vertex v is. Our definition of the bunch implies that, for given contour θ' and vertex v, at most one bunch labeled by θ' can grow from v.

Every θ-rooted labeled tree can be obtained as a result of the following procedure of growing bunches. To form the level-one part of a θ-rooted labeled tree, we fix a finite collection of contours $\{\theta_i'\}$, $\theta_i' \not\sim \theta$, and corresponding values $\alpha_i' \geq 1$ and join the set of bunches $b_i = \langle r, \theta_i', \alpha_i' \rangle$, $i = 1, 2, \ldots$, with the root r labeled by θ. To form the level-two part, we repeat this procedure with the ends of every bunch constructed at the previous step. That is, for a given end v labeled by θ_i', we choose a finite collection of contours $\{\theta_j''\}$, $\theta_j'' \not\sim \theta_i'$, and corresponding values $\alpha_j'' \geq 1$ and join the set of bunches $b_j = \langle v, \theta_j'', \alpha_j'' \rangle$, $j = 1, 2, \ldots$, with vertex v. Repeating the procedure several times, we can construct any finite θ-rooted labeled tree. Observe that for any θ-rooted labeled tree T with a collection of bunches $b_k(T) = \langle \alpha_k(T), \theta_k(T) \rangle$, $k = 1, 2, \ldots$, our procedure produces $\prod_k \alpha_k(T)!$ isomorphic copies of T because it treats the vertices of a given bunch as distinct. In fact, any permutation of the edges of a bunch gives us an isomorphic copy of the θ-rooted labeled tree

We take the statistical weight of bunch $b = \langle \theta', \alpha' \rangle$ to be equal to

$$w(b) = \frac{|w(\theta'(b))|^{\alpha'(b)}}{\alpha'(b)!} \tag{3.14}$$

and denote

$$f(\theta, k) = \sum_{T:\, l(r(T))=\theta,\, n(T)\leq k} w(T), \tag{3.15}$$

where $n(T)$ is the number of levels in T. By induction in m we prove below that

$$f(\theta, k) \leq \exp(a(\theta)). \tag{3.16}$$

First we estimate the quantity $f(\theta, 1)$. According to the initial step of the bunch-growing procedure,

$$\begin{aligned} f(\theta, 1) &= \sum_{T:\ l(r(T))=\theta,\, n(T)\leq 1} w(T) = \prod_{\theta' \not\sim \theta} \sum_{\alpha'=0}^{\infty} w(\langle \theta', \alpha' \rangle) \\ &= \prod_{\theta' \not\sim \theta} \exp|w(\theta')| = \exp \sum_{\theta' \not\sim \theta} |w(\theta')|. \end{aligned} \tag{3.17}$$

By (3.2), $\sum_{\theta' \not\sim \theta} |w(\theta')| \leq a(\theta)$, and hence

$$f(\theta, 1) \leq \exp(\alpha(\theta)). \tag{3.18}$$

Assume now that

$$f(\theta, k-1) \leq \exp(a(\theta)). \tag{3.19}$$

Again using the bunch-growing procedure, one can write a recursive representation for the right-hand side of (3.15):

$$f(\theta, k) = \prod_{\theta' \not\sim \theta} \sum_{\alpha'=0}^{\infty} w(\langle \theta', \alpha' \rangle) f(\theta', k-1)^{\alpha'}. \tag{3.20}$$

Using the induction hypothesis for $f(\theta, k-1)$, we obtain
(3.21)

$$
\begin{aligned}
f(\theta, k) &= \prod_{\theta' \not\sim \theta} \sum_{\alpha'=0}^{\infty} w(\langle \theta', \alpha' \rangle) f(\theta', k-1)^{\alpha'} \le \prod_{\theta' \not\sim \theta} \sum_{\alpha'=0}^{\infty} w(\langle \theta', \alpha' \rangle) \exp(\alpha' a(\theta')) \\
&= \prod_{\theta' \not\sim \theta} \exp(|w(\theta')| \exp(a(\theta'))) = \exp\Big(\sum_{\theta' \not\sim \theta} |w(\theta')| \exp(a(\theta')) \Big) \le \exp(a(\theta)),
\end{aligned}
$$

which proves (3.6).

We will apply general Theorem 3.1 to various contour models on lattice $\mathbb{Z}^d$. Our aim here is to derive some general assertions, which are then applied to the specific situation of the quantum system under consideration. The corresponding contour model deals with complicated geometrical objects, and to make the whole exposition easier, we begin with somewhat simpler models. More precisely, we derive a series of statements for contour models of increasing complexity, with each model being a natural generalization of the previous one.

Contour Model 1. We start with the following model. Let a *bulky* contour $\widetilde{\Omega}$ be defined as a finite connected subset of lattice $\mathbb{Z}^d$, $d \ge 2$. Define the boundary $\partial\widetilde{\Omega}$ as the set of unit plaquettes of dual lattice $\widetilde{\mathbb{Z}}^d$ separating $\widetilde{\Omega}$ and its complement $\widetilde{\Omega}^c = \mathbb{Z}^d \setminus \widetilde{\Omega}$. Then $\partial\widetilde{\Omega}$ can be uniquely decomposed in connected components $\partial_i\widetilde{\Omega}$, $i = 0, 1, \dots, n(\widetilde{\Omega})$, alternatively denoted by $\widetilde{\omega}_i(\widetilde{\Omega})$. We always assume that $\widetilde{\omega}_0(\widetilde{\Omega})$ stands for the (unique) external component of $\partial\widetilde{\Omega}$.

We say that $\widetilde{\Omega}$ and $\widetilde{\Omega}'$ are compatible if $\widetilde{\Omega} \cap \widetilde{\Omega}' = \varnothing$. A finite collection $\{\widetilde{\Omega}_i\}$ is called compatible if any pair $\widetilde{\Omega}_{i_1}, \widetilde{\Omega}_{i_2}$ with $i_1 \neq i_2$ is compatible.

Assign to $\widetilde{\Omega}$ a statistical weight $w(\widetilde{\Omega})$ and suppose that it satisfies

$$
|w(\widetilde{\Omega})| \le \exp\Big(-p \sum_{i=0}^{n(\widetilde{\Omega})} |\widetilde{\omega}_i(\widetilde{\Omega})| - q|\widetilde{\Omega}| \Big), \tag{3.22}
$$

where $p, q > 0$ are constants. Here and below $|\widetilde{\omega}_i|$ is the number of plaquettes in $\widetilde{\omega}_i$ and $|\widetilde{\Omega}|$ the number of lattice sites in $\widetilde{\Omega}$.

For a finite $V \subset \mathbb{Z}^d$, define the partition function in V by

$$
Z(V) = \sum_{\{\widetilde{\Omega}_j\} \subseteq V} \prod_j w(\widetilde{\Omega}_j). \tag{3.23}
$$

Here, the sum $\sum_{\{\widetilde{\Omega}_j\} \subseteq V}$ is over the compatible collections $\{\widetilde{\Omega}\}$, where each $\widetilde{\Omega}_j \subseteq V$.

Contour models of such type usually serve to describe two-phase physical models, with one phase dominating another. The simplest example is the Ising model with the external magnetic field. To see the correspondence, suppose that the field is positive and fix the $+$ boundary condition. Then bulky contours $\widetilde{\Omega}$ are connected clusters of the $-$'s in a sea of the $+$'s. Parameters p, q in this case are the coupling constant and the magnetic field, respectively.

The following statement is a corollary of Theorem 3.1.

PROPOSITION 3.2. *Under the condition* (3.22) *with*

$$
p \ge 2 + \log(2d), \; q \ge 8d \exp(-p/2), \tag{3.24}
$$

the logarithm of the partition function (3.23) *admits a polymer expansion of the form* (3.3), *where the property* (3.6) *is valid, with*

$$a(\widetilde{\Omega}) = h|\widetilde{\Omega}|, \; h = 2d\exp(-p/2). \tag{3.25}$$

PROOF. According to Theorem 3.1, it is sufficient to verify that $a(\widetilde{\Omega})$ given by (3.25) satisfies condition (3.2).

To carry out the summation in (3.2), we provide every $\widetilde{\Omega}$ with a tree-like structure. For this purpose we consider $\widetilde{\Omega}$ as a closed domain in $\mathbb{R}^d$, i.e., as the union of the unit cubes of $\mathbb{Z}^d$ centered at the points of $\widetilde{\Omega}$. Next we mark, in each connected component $\widetilde{\omega}_i$ of boundary $\partial\widetilde{\Omega}$, the center of an arbitrarily selected plaquette and through each marked point draw a line parallel to the first coordinate axis L. Together with $\partial\widetilde{\Omega}$, the union of these lines represents a connected subset of $\mathbb{R}^d$. Every line is partitioned by $\partial\widetilde{\Omega}$ into intervals. As the final step of the construction, we choose, in an arbitrary way, a minimal number of the intervals such that, being united with $\partial\widetilde{\Omega}$, these intervals still form a connected subset of $\mathbb{R}^d$. Obviously, all selected intervals belong to $\widetilde{\Omega}$ treated as a subset of $\mathbb{R}^d$.

The object just constructed has a tree-like structure in the following sense. The component $\widetilde{\omega}_0(\widetilde{\Omega})$ is by definition the root of the tree. The components $\widetilde{\omega}_{1,k}(\widetilde{\Omega})$, which are linked by selected intervals directly with $\widetilde{\omega}_0(\widetilde{\Omega})$, are the vertices of the first level (from now on we use the double-indexed numeration for the components of $\partial\widetilde{\Omega}$). The components $\widetilde{\omega}_{j,k}(\widetilde{\Omega})$ which are linked by intervals directly with $\bigcup_k \widetilde{\omega}_{j-1,k}(\widetilde{\Omega})$ are the vertices of the j-th level. Denote by $L_{j,k}$ the interval linking $\widetilde{\omega}_{j,k}(\widetilde{\Omega})$ with $\bigcup_k \widetilde{\omega}_{j-1,k}(\widetilde{\Omega})$ and denote by $|L_{j,k}|$ the number of points of $\mathbb{Z}^d$ belonging to $L_{j,k}$. We treat the intervals as the edges of our tree. The bound

$$|w(\widetilde{\Omega})| \le \exp\Big(-(p+q/(4d))|\widetilde{\omega}_0(\widetilde{\Omega})| - (p+q/(4d))\sum_{j,k}|\widetilde{\omega}_{j,k}(\widetilde{\Omega})| - q/2\sum_{j,k}|L_{j,k}|\Big) \tag{3.26}$$

follows from (3.22), because, by construction,

$$\sum_{j,k}|L_{j,k}| \le |\widetilde{\Omega}| \;\text{ and }\; 2d|\widetilde{\Omega}| \ge \sum_{j,k}|\widetilde{\omega}_{j,k}(\widetilde{\Omega})\,| + |\,\widetilde{\omega}_0(\widetilde{\Omega})\,|.$$

Now consider $\widetilde{\Omega}$ as a tree-like object, $T = (\widetilde{\omega}_0, \{\widetilde{\omega}_{j,k}, L_{j,k}\})$, where $\widetilde{\omega}_0$ and $\widetilde{\omega}_{j,k}$ are finite connected sets of plaquettes of $\widetilde{\mathbb{Z}}^d$ and $L_{j,k}$ are the intervals parallel to L (see above); $L_{j,k}$ links the center of a plaquette from $\widetilde{\omega}_{j,k}$ with the center of a plaquette from $\bigcup_k \widetilde{\omega}_{j-1,k}$. Assign to every T the statistical weight

$$w(T) = \exp\Big(-(p+q/(4d))|\widetilde{\omega}_0| - (p+q/(4d))\sum_{j,k}|\widetilde{\omega}_{j,k}| - q/2\sum_{j,k}|L_{j,k}|\Big) \tag{3.27}$$

and set

$$a(T) = h\sum_{j,k}|L_{j,k}(T)|. \tag{3.28}$$

Then, for every fixed $\widetilde{\omega}$,

$$\sum_{\widetilde{\Omega}:\,\widetilde{\omega}_0(\widetilde{\Omega})=\widetilde{\omega}} |w(\widetilde{\Omega})|\exp\Big(a(\widetilde{\Omega})\Big) \le \sum_{T:\,\widetilde{\omega}_0(T)=\widetilde{\omega}} w(T)\exp(a(T)) = S(\widetilde{\omega}), \tag{3.29}$$

because $h \le q/4$.

LEMMA 3.3. *The following bound holds true*:

$$\sum_{T:\,\widetilde{\omega}_0(T)=\widetilde{\omega}} w(T)\,\exp(a(T)) \leq \exp(-p|\widetilde{\omega}|).$$

PROOF OF LEMMA 3.3. We proceed by induction on the number of levels $n(T)$ in T. Suppose that, for some $J \geq 1$, the bound

$$\sum_{T:\widetilde{\omega}_0(T)=\widetilde{\omega},\ n(T)\leq J-1} w(T)\exp(a(T)) \leq \exp(-p|\widetilde{\omega}|) \tag{3.30}$$

has been established, and consider an arbitrary T with $n(T) \leq J$. For a given vertex $\widetilde{\omega}_{1,k}$ of the first level in T, denote by $T_{1,k}$ the maximal subtree of T that has its root at $\widetilde{\omega}_{1,k}$. Then

$$\begin{aligned} S(\widetilde{\omega}) &= \sum_{T:\widetilde{\omega}_0(T)=\widetilde{\omega},\ n(T)\leq J} w(T)\exp(a(T)) \\ &= \exp\Big[-(p+q/(4d))|\widetilde{\omega}|\Big]\sum_{s=0}^{|\widetilde{\omega}|}\binom{|\widetilde{\omega}|}{s} \\ &\quad\times\prod_{k=1}^{s}\Big(\sum_{|L_{1,k}|=1}^{\infty}\exp\Big[-(q/2-h)|L_{1,k}|\Big] \\ &\quad\times\sum_{\widetilde{\omega}_{1,k}}\ \sum_{\substack{T_{1,k}:\,\widetilde{\omega}_0(T_{1,k})=\widetilde{\omega}_{1,k},\\ n(T_{1,k})\leq J-1}} w(T_{1,k})\exp(a(T_{1,k}))\Big), \end{aligned} \tag{3.31}$$

where s is the number of the vertices of the first level in T and $\binom{|\widetilde{\omega}|}{s}$ describes the ways to choose the centers of plaquettes that serve as the starting points of intervals $L_{1,k}$, $k = 1,\ldots,s$. Applying the induction assumption to the last sum in (3.31), we obtain

$$\begin{aligned} S(\widetilde{\omega}) &\leq \exp\Big[-(p+q/(4d))|\widetilde{\omega}|\Big]\sum_{s=0}^{|\widetilde{\omega}|}\binom{|\widetilde{\omega}|}{s} \\ &\quad\times\prod_{k=1}^{s}\Big(\sum_{|L_{1,k}|=1}^{\infty}\exp\Big[-(q/2-h)|L_{1,k}|\Big]\sum_{\widetilde{\omega}_{1,k}}\exp(-p|\widetilde{\omega}_{1,k}|)\Big) \\ &\leq \exp\Big[-(p+q/(4d))|\widetilde{\omega}|\Big]\sum_{s=0}^{|\widetilde{\omega}|}\binom{|\widetilde{\omega}|}{s} \\ &\quad\times\prod_{k=1}^{s}\Big(\frac{2d\exp(-p)}{1-2d\exp(-p)}\sum_{|L_{1,k}|=1}^{\infty}\exp\Big[-(q/2-h)|L_{1,k}|\Big]\Big) \\ &\leq \exp\Big[-(p+q/(4d))|\widetilde{\omega}|\Big]\sum_{s=0}^{|\widetilde{\omega}|}\binom{|\widetilde{\omega}|}{s} \end{aligned} \tag{3.32}$$

$$\times \prod_{k=1}^{s} \left(\frac{2d\exp(-p)}{1-2d\exp(-p)} \cdot \frac{\exp(-(q/2-h))}{1-\exp(-(q/2-h))} \right)$$

$$= \exp\Big[-(p+q/(4d))|\widetilde{\omega}|\Big] \sum_{s=0}^{|\widetilde{\omega}|} \binom{|\widetilde{\omega}|}{s}$$

$$\times \left(\frac{2d\exp(-p)}{1-2d\exp(-p)} \cdot \frac{\exp(-(q/2-h))}{1-\exp(-(q/2-h))} \right)^{s}$$

$$= \exp\Big[-(p+q/(4d))|\widetilde{\omega}|\Big]$$

$$\times \left(1+ \frac{2d\exp(-p)}{1-2d\exp(-p)} \cdot \frac{\exp(-(q/2-h))}{1-\exp(-(q/2-h))} \right)^{|\widetilde{\omega}|}$$

$$\leq \exp(-p|\widetilde{\omega}|).$$

The last bound is valid because for $p \geq 2+\log(2d)$ and $h \leq \frac{q}{4}$ we have

$$1-2d\exp(-p) \geq \frac{1}{2} \tag{3.33}$$

and hence

$$\begin{aligned} 1+\frac{2d\exp(-p)}{1-2d\exp(-p)} \cdot \frac{\exp(-(q/2-a))}{1-\exp(-(q/2-a))} &\leq 1+\frac{4d\exp(-p)}{q/2-a} \\ \leq 1+\frac{16d\exp(-p)}{q} &\leq 1+q/(4d) \leq \exp(q/(4d)). \end{aligned} \tag{3.34}$$

To establish the initial step of the induction, observe that for T with $n(T)=0$, we obviously have $w(T)=\exp(-p|\widetilde{\omega}|)$. □

Now we can check condition (3.2):

$$\begin{aligned} \sum_{\widetilde{\Omega}:\ \widetilde{\Omega}'\not\sim\widetilde{\Omega}} |w(\widetilde{\Omega}')| \exp\Big(a(\widetilde{\Omega}')\Big) &= \sum_{\widetilde{\omega}':\widetilde{\omega}'\cap\widetilde{\Omega}\neq\varnothing}\ \sum_{\widetilde{\Omega}':\ \widetilde{\omega}_0(\widetilde{\Omega}')=\widetilde{\omega}'} |w(\widetilde{\Omega}')| \exp\Big(a(\widetilde{\Omega}')\Big) \\ &\leq \sum_{\widetilde{\omega}':\ \widetilde{\omega}'\cap\widetilde{\Omega}\neq\varnothing} \exp(-p|\widetilde{\omega}'|) \leq |\widetilde{\Omega}| \frac{2d\exp(-p)}{1-2d\exp(-p)} \\ &\leq 4d\exp(-p)|\widetilde{\Omega}| \leq 2d\exp(-p/2)|\widetilde{\Omega}| = h|\widetilde{\Omega}|. \end{aligned} \tag{3.35}$$

[Symbol $\not\sim$, here and below, is used to indicate incompatibility in the sense of a current contour model.] This completes the proof of Proposition 3.2. □

Contour Model 2. The idea behind our second model is that the difference in the statistical weight of lowest exitations can lead to an asymmetry between two phases, in the same way as the external field does. [An example of a model of this kind is the so-called ANNNI model (see, e.g., [**DMaSin**])]. We now define a bulky contour $\widetilde{\Omega}$ as a finite connected subset of $\mathbb{Z}^d$, with the additional condition

$$\operatorname{diam} |\widetilde{\omega}_i(\widetilde{\Omega})| \geq u > 1, \tag{3.36}$$

and consider a statistical weight $W(\widetilde{\Omega})$ satisfying

$$(3.37)\qquad |W(\widetilde{\Omega})| \leq \exp\Big(-p\sum_{i=0}^{n(\widetilde{\omega})}|\widetilde{\omega}_i(\widetilde{\Omega})|\Big).$$

Moreover, we suppose that, in addition, the model contains other objects denoted $\widetilde{\varepsilon}^+$ and $\widetilde{\varepsilon}^-$ and called *short* contours. Both $\widetilde{\varepsilon}^+$ and $\widetilde{\varepsilon}^-$ are finite connected sets of plaquettes of dual lattice $\widetilde{\mathbb{Z}}^d$ such that they are the boundaries of connected and simple-connected subsets of lattice $\mathbb{Z}^d$, with

$$(3.38)\qquad \operatorname{diam}\widetilde{\varepsilon}^{\pm} < u.$$

The difference between $\widetilde{\varepsilon}^+$ and $\widetilde{\varepsilon}^-$ is twofold. First, it is manifested in the modified compatibility rule for the possible pairs $(\widetilde{\Omega},\widetilde{\Omega}')$, $(\widetilde{\Omega},\widetilde{\varepsilon}^{\pm})$, and $(\widetilde{\varepsilon}'^{\pm},\widetilde{\varepsilon}''^{\pm})$, which now requires

(i) $\widetilde{\Omega}\cap\widetilde{\Omega}' = \varnothing$;
(ii.1) $\widetilde{\varepsilon}^{\pm}\cap\widetilde{\omega}_i(\widetilde{\Omega}) = \varnothing$,
(ii.2) $\widetilde{\varepsilon}^{-}\cap\widetilde{\Omega}^c = \varnothing$,
(ii.3) $\widetilde{\varepsilon}^{+}\cap\widetilde{\Omega} = \varnothing$;
(iii) $\widetilde{\varepsilon}'^{\pm}\cap\widetilde{\varepsilon}''^{\pm} = \varnothing$.

The definition of a compatible collection $\{\widetilde{\Omega}_i,\widetilde{\varepsilon}_j^+,\widetilde{\varepsilon}_k^-\}$ is based, as before, on pairwise compatibility.

Secondly, the statistical weights of $\widetilde{\varepsilon}^{\pm}$ are small:

$$(3.39)\qquad |w(\widetilde{\varepsilon}^{\pm})| \leq \exp\left(-p|\widetilde{\varepsilon}^{\pm}|\right)$$

and are different for $\widetilde{\varepsilon}^+$ and $\widetilde{\varepsilon}^-$. More precisely, we assume that for any finite V,

$$(3.40)\qquad \frac{Z^-(V)}{Z^+(V)} \leq \exp\Big[-|V|8d\exp(-(p-2)(du-u+1))\Big],$$

where the *restricted* (or *metastable*) partition functions $Z^{\pm}(V)$ are given by

$$(3.41)\qquad Z^{\pm}(V) = \sum_{\{\widetilde{\varepsilon}_i^{\pm}\}\subseteq V}\prod_i w(\widetilde{\varepsilon}_i^{\pm}).$$

Here the sum $\sum_{\{\widetilde{\varepsilon}_i^{\pm}\}\subseteq V}$ is over the compatible collections of the $\widetilde{\varepsilon}^+$'s or $\widetilde{\varepsilon}^-$'s (all of the same sign), where each $\widetilde{\varepsilon}_j$ is contained in set V treated as a closed domain in $\mathbb{R}^d$ (in the same way as before). We will briefly say that $\widetilde{\varepsilon}_j$ is contained in V in the $\mathbb{R}^d$-sense. Although condition (3.40) looks difficult to check, it may be verified in many situations where bound (3.39) is valid. One method of proving the validity of (3.40) is to use Theorem 3.1 (more precisely, formula (3.3) for $\log Z^{\pm}(V)$), with a direct bound for the difference of statistical weights of corresponding polymers.

The full partition function of the model is defined as

$$(3.42)\qquad Z(V) = \sum_{\{\widetilde{\Omega}_i,\widetilde{\varepsilon}_j^+,\widetilde{\varepsilon}_k^-\}\subseteq V}\prod_i W(\widetilde{\Omega}_i)\prod_j w(\widetilde{\varepsilon}_j^+)\prod_k w(\widetilde{\varepsilon}_k^-).$$

The sum $\sum_{\{\widetilde{\Omega}_i,\widetilde{\varepsilon}_j^+,\widetilde{\varepsilon}_k^-\}\subseteq V}$ is over the compatible collections of the $\widetilde{\Omega}$'s, $\widetilde{\varepsilon}^+$'s, and $\widetilde{\varepsilon}^-$'s, where each $\widetilde{\Omega}_i \subseteq V$ in a usual sense and each $\widetilde{\varepsilon}_j^+$ and $\widetilde{\varepsilon}_k^-$ is contained in V in the $\mathbb{R}^d$-sense. [The same meaning is used below, unless otherwise specified.]

As it was already mentioned, a two-phase ("+"-phase and "−"-phase) physical model lies behind our contour model. The correspondence between physical and contour models is the following. In the physical model the short contours $\widetilde{\varepsilon}^+$ ($\widetilde{\varepsilon}^-$) are the boundaries of small bubbles of "−"-phase ("+"-phase) inside the sea of "+"-phase ("−"-phase). Generally speaking, a small bubble of "+"-phase can be placed inside another small bubble of "−"-phase and so on. Hence the natural compatibility rule for the short contours requires the alteration of the signs for embedded short contours. Note that in (3.41) and (3.42) we use the compatibility rule stating that the sign of any short contour embedded into given short contour $\widetilde{\varepsilon}^+$ ($\widetilde{\varepsilon}^-$) should also be "+" ("−"). In fact, such a transformation of the compatibility rule becomes an identical transformation of the physical model (i.e., it keeps all partition function unchanged) if the initial statistical weight of every short contour $\widetilde{\varepsilon}^+$ ($\widetilde{\varepsilon}^-$) is modified by the multiplication by the ratio of the partition functions calculated for the volume enclosed by $\widetilde{\varepsilon}^+$ ($\widetilde{\varepsilon}^-$) with $-$ $(+)$ and $+$ $(-)$ boundary conditions, respectively. Usually the property which selects the short contours from the family of all boundaries separating "+"-phase and "−"-phase is that the aforementioned ratio is of order 1. For the physical model with the stable "+"-phase and the metastable "−"-phase this ratio can not be of order 1 for the volumes large enough. So we describe the large regions of the metastable "−"-phase by means of bulky contours $\widetilde{\omega}$. This allows us to represent the physical partition function with the $+$ boundary condition in form (3.42), i.e., as the partition function of the contour model.

PROPOSITION 3.4. *Under conditions* (3.37)–(3.40), *with*

$$p \geq 4 + \log(2d), \tag{3.43}$$

the logarithm of partition function (3.42) *admits a polymer expansion of form* (3.3), *where property* (3.6) *is valid with*

$$a(\widetilde{\varepsilon}) = |\widetilde{\varepsilon}|, \quad a(\widetilde{\Omega}) = \sum_{i=0}^{n(\widetilde{\Omega})} |\widetilde{\omega}_i(\widetilde{\Omega})| + h|\widetilde{\Omega}|, \tag{3.44}$$

and $h = 2d\exp(-(p-2)(du-u+1))$.

PROOF. We will show that the current lattice contour model is similar to Model 1. Indeed,

$$\begin{aligned} Z(V) &= \sum_{\{\widetilde{\Omega}_i\}\subseteq V} Z^+\left(\left[V\setminus\left(\bigcup_i \widetilde{\Omega}_i\right)\right]\right)\prod_i W(\widetilde{\Omega}_i)Z^-([\widetilde{\Omega}_i]) \\ &= \sum_{\{\widetilde{\Omega}_i\}\subseteq V} Z^+\left(\left[V\setminus\left(\bigcup_i \widetilde{\Omega}_i\right)\right]\right)\prod_i W(\widetilde{\Omega}_i)\frac{Z^-([\widetilde{\Omega}_i])}{Z^+([\widetilde{\Omega}_i])}Z^+([\widetilde{\Omega}_i]). \end{aligned} \tag{3.45}$$

Here and below given $D \subset \mathbb{Z}^d$, $[D]$ stands for the set $\{x \in D : \operatorname{dist}(x, D^c) \geq 1\}$. Denoting

$$w(\widetilde{\Omega}) = W(\widetilde{\Omega}) \frac{Z^-([\widetilde{\Omega}])}{Z^+([\widetilde{\Omega}])}, \tag{3.46}$$

we obtain

$$\begin{aligned} Z(V) &= \sum_{\{\widetilde{\Omega}_i\} \subseteq V} Z^+\left(\left[V \setminus \left(\bigcup_i \widetilde{\Omega}_i\right)\right]\right) \prod_i w(\widetilde{\Omega}_i) Z^+([\widetilde{\Omega}_i]) \\ &= \sum_{\{\widetilde{\Omega}_i, \widetilde{\varepsilon}_j^+\} \subseteq V} \prod_i w(\widetilde{\Omega}_i) \prod_j w(\widetilde{\varepsilon}_j^+). \end{aligned} \tag{3.47}$$

At the right-hand side of (3.47) the sum $\sum_{\{\widetilde{\Omega}_i, \widetilde{\varepsilon}_j^+\} \subseteq V}$ is taken over compatible collections of bulky contours $\widetilde{\Omega}_i$ and short contours $\widetilde{\varepsilon}_j^+$, with a simplified binary compatibility rule:

(i) $\widetilde{\Omega} \cap \widetilde{\Omega}' = \varnothing$;

(ii) $\widetilde{\varepsilon}^+ \cap \widetilde{\omega}_i(\widetilde{\Omega}) = \varnothing$;

(iii) $\widetilde{\varepsilon}'^+ \cap \widetilde{\varepsilon}''^+ = \varnothing$.

To stress this difference between the compatibility rules, and to simplify the notation, we omit the superscript $+$. The final form of the partition function is

$$Z(V) = \sum_{\{\widetilde{\Omega}_i, \widetilde{\varepsilon}_j\} \subseteq V} \prod_i w(\widetilde{\Omega}_i) \prod_j w(\widetilde{\varepsilon}_j). \tag{3.48}$$

Now, according to (3.40),

(3.49)

$$|w(\widetilde{\Omega})| \leq \exp\left(- p \sum_{i=0}^{n(\widetilde{\Omega})} |\widetilde{\omega}_i(\widetilde{\Omega})| - q |\, [\widetilde{\Omega}] \,| \right) \leq \exp\left(- (p-1) \sum_{i=0}^{n(\widetilde{\Omega})} |\widetilde{\omega}_i(\widetilde{\Omega})| - q |\widetilde{\Omega}| \right)$$

for $q = 8d \exp(-(p-2)(du - u + 1)) < 1$. The difference between (3.48) and (3.23) is due only to the presence of short contours, which is easy to deal with. We will check condition (3.2), with $a(\widetilde{\varepsilon})$ and $a(\widetilde{\Omega})$ given by (3.44). Namely, the following bound holds

$$\begin{aligned} \sum_{\widetilde{\varepsilon}' :\, \widetilde{\varepsilon}' \not\sim \widetilde{\varepsilon}} w(\widetilde{\varepsilon}') \exp\left(a(\widetilde{\varepsilon}')\right) &\leq \sum_{\widetilde{\varepsilon}' :\, \widetilde{\varepsilon}' \not\sim \widetilde{\varepsilon}} \exp\left(-(p-2)|\widetilde{\varepsilon}'|\right) \\ &\leq |\widetilde{\varepsilon}| \sum_{n=1}^{\infty} \left(2d \exp(-(p-2))\right)^n \leq 1/2 |\widetilde{\varepsilon}|. \end{aligned} \tag{3.50}$$

On the other hand,

$$\begin{aligned} &\sum_{\widetilde{\Omega}' :\, \widetilde{\Omega}' \not\sim \widetilde{\varepsilon}} w(\widetilde{\Omega}') \exp(a(\widetilde{\Omega}')) \\ &\leq \sum_{\widetilde{\Omega}' :\, \widetilde{\Omega}' \not\sim \widetilde{\varepsilon}} \exp\left(- (p-2) \sum_{i=0}^{n(\widetilde{\Omega}')} |\widetilde{\omega}_i(\widetilde{\Omega}')| - (q-h)|\widetilde{\Omega}'| \right) \end{aligned}$$

$$\leq \sum_{\widetilde{\omega}':\, \widetilde{\omega}'\cap\widetilde{\varepsilon}\neq\varnothing} \; \sum_{\widetilde{\Omega}':\, \widetilde{\omega}_0(\widetilde{\Omega}')=\widetilde{\omega}'} \exp\Big(-(p-2+q/(4d)) \times \sum_{i=0}^{n(\widetilde{\Omega}')} |\widetilde{\omega}_i(\widetilde{\Omega}')| - (q/2-h)|\widetilde{\Omega}'|\Big) \tag{3.51}$$

$$\leq \sum_{\widetilde{\omega}':\, \widetilde{\omega}'\cap\widetilde{\varepsilon}\neq\varnothing} \exp(-(p-2)|\widetilde{\omega}'|)$$

$$\leq |\widetilde{\varepsilon}| \sum_{n=2du-2u+2}^{\infty} \big(2d\exp(-(p-2))\big)^n \leq \frac{1}{2}|\widetilde{\varepsilon}|.$$

Here we used an analogue of Lemma 3.3 with $(p-2)$ instead of p. The proof of this statement is similar to that of Lemma 3.3, with the only exception that (3.32) is replaced by

$$S(\widetilde{\omega}) \leq \exp\Big[-(p-2+q/(4d))|\widetilde{\omega}|\Big] \sum_{s=0}^{|\widetilde{\omega}|} \binom{|\widetilde{\omega}|}{s} \times \prod_{k=1}^{s} \Big(\sum_{|L_{1,k}|=1}^{\infty} \exp\Big[-(q/2-h)|L_{1,k}|\Big] \sum_{\widetilde{\omega}_{1,k}} \exp\Big[-(p-2)|\widetilde{\omega}_{1,k}|\Big]\Big)$$

$$\leq \exp\Big[-(p-2+q/(4d))|\widetilde{\omega}|\Big] \sum_{s=0}^{|\widetilde{\omega}|} \binom{|\widetilde{\omega}|}{s} \times \prod_{k=1}^{s} \Big(\frac{2d\exp\big[-(p-2)(2du-2u+2)\big]}{1-2d\exp(-p)} \sum_{|L_{1,k}|=1}^{\infty} \exp\Big[-(q/2-h)|L_{1,k}|\Big]\Big)$$

$$\leq \exp\Big[-(p-2+q/(4d))|\widetilde{\omega}|\Big] \sum_{s=0}^{|\widetilde{\omega}|} \binom{|\widetilde{\omega}|}{s} \times \prod_{k=1}^{s} \Big(\frac{2d\exp\big[-(p-2)(2du-2u+2)\big]}{1-2d\exp -p} \cdot \frac{\exp(-(q/2-h))}{1-\exp(-(q/2-h))}\Big) \tag{3.52}$$

$$= \exp\Big[-(p-2+q/(4d))|\widetilde{\omega}|\Big] \sum_{s=0}^{|\widetilde{\omega}|} \binom{|\widetilde{\omega}|}{s} \times \Big(\frac{2d\exp\Big[-(p-2)(2du-2u+2)\Big]}{1-2d\exp(-p)} \cdot \frac{\exp(-(q/2-h))}{1-\exp(-(q/2-h))}\Big)^s$$

$$= \exp\Big[-(p-2+q/(4d))|\widetilde{\omega}|\Big] \times \Big(1 + \frac{2d\exp\big[-(p-2)(2du-2u+2)\big]}{1-2d\exp(-p)} \cdot \frac{\exp(-(q/2-h))}{1-\exp(-(q/2-h))}\Big)^{|\widetilde{\omega}|}$$

$$\leq \exp(-(p-2)|\widetilde{\omega}|),$$

and (3.34) by

$$\begin{aligned}
&1+\frac{2d\exp\Big[-(p-2)(2du-2u+2)\Big]}{1-2d\exp(-p)}\cdot\frac{\exp(-(q/2-h))}{1-\exp(-(q/2-h))}\\
(3.53)\qquad &\le 1+\frac{4d\exp\Big[-(p-2)(2du-2u+2)\Big]}{q/2-h}\\
&\le 1+\frac{16d\exp\Big[-(p-2)(2du-2u+2)\Big]}{q}\\
&\le 1+q/(4d)\le\exp(q/(4d)).
\end{aligned}$$

In (3.51)–(3.53) we use the fact that, for any $\widetilde{\omega}$ with $\operatorname{diam}\widetilde{\omega}\ge u$, the number of plaquettes in $\widetilde{\omega}$ is greater than or equal to $2du-2u+2$. From (3.50) and (3.51) we get

$$(3.54)\qquad \sum_{\widetilde{\varepsilon}':\,\widetilde{\varepsilon}'\not\sim\widetilde{\varepsilon}} w(\widetilde{\varepsilon}')\exp(a(\widetilde{\varepsilon}'))+\sum_{\widetilde{\Omega}':\,\widetilde{\Omega}'\not\sim\widetilde{\varepsilon}} w(\widetilde{\Omega}')\exp(a(\widetilde{\Omega}'))\le\frac{1}{2}|\widetilde{\varepsilon}|+\frac{1}{2}|\widetilde{\varepsilon}|=|\widetilde{\varepsilon}|.$$

In a similar way

$$\begin{aligned}
&\sum_{\widetilde{\varepsilon}':\,\widetilde{\varepsilon}'\not\sim\widetilde{\Omega}} w(\widetilde{\varepsilon}')\exp(a(\widetilde{\varepsilon}'))+\sum_{\widetilde{\Omega}':\,\widetilde{\Omega}'\not\sim\widetilde{\Omega}} w(\widetilde{\Omega}')\exp\Big(a(\widetilde{\Omega}')\Big)\\
&\quad\le\sum_{\widetilde{\varepsilon}':\,\widetilde{\varepsilon}'\cap\partial\widetilde{\Omega}\ne\varnothing} w(\widetilde{\varepsilon}')\exp(a(\widetilde{\varepsilon}'))\\
&\qquad+\sum_{\widetilde{\omega}':\,\widetilde{\omega}'\cap\widetilde{\Omega}\ne\varnothing}\;\sum_{\widetilde{\Omega}':\,\widetilde{\omega}_0(\widetilde{\Omega}')=\widetilde{\omega}'}|w(\widetilde{\Omega}')|\exp\Big(a(\widetilde{\Omega}')\Big)\\
(3.55)\qquad&\quad\le|\partial\widetilde{\Omega}|+\sum_{\widetilde{\omega}':\,\widetilde{\omega}'\cap\widetilde{\Omega}\ne\varnothing}\exp\big(-(p-2)|\widetilde{\omega}'|\big)\\
&\quad\le|\partial\widetilde{\Omega}|+|\widetilde{\Omega}|\frac{2d\exp\Big[-(p-2)(2du-2u+2)\Big]}{1-2d\exp(-p)}\\
&\quad\le|\partial\widetilde{\Omega}|+4d\exp\Big[-(p-2)(2du-2u+2)\Big]|\widetilde{\Omega}|\\
&\quad\le|\partial\widetilde{\Omega}|+2d\exp\Big[-(p-2)(du-u+1)\Big]|\widetilde{\Omega}|=|\partial\widetilde{\Omega}|+h|\widetilde{\Omega}|,
\end{aligned}$$

where

$$(3.56)\qquad |\partial\widetilde{\Omega}|=\sum_{i=0}^{n(\widetilde{\Omega})}|\widetilde{\omega}_i(\widetilde{\Omega})|.\quad\square$$

Contour Model 3. Our next contour model describes a situation where one has a finite number of metastable phases, with a single stable phase that is dominating, owing to its more substantial low-energy excitations. The model contains contours of two types, long contours ω and short contours ε. Both long and short contours are triples, $\omega = (\widetilde{\omega}, E(\Omega), I(\Omega))$ and $\varepsilon = (\widetilde{\varepsilon}, E(\varepsilon), I(\varepsilon))$, where

(i) $\widetilde{\omega}$ and $\widetilde{\varepsilon}$ are sets of the plaquettes of the dual lattice $\widetilde{\mathbb{Z}}^d$ which are the boundaries of finite connected and simple-connected sets on $\mathbb{Z}^d$;

(ii) $(E(\Omega), I(\Omega))$ and $(E(\varepsilon), I(\varepsilon))$ are two pairs of distinct numbers from the set $\{0, \dots, m\}$.

In fact, the set $\{0, \dots, m\}$ may be replaced by an arbitrary finite set. Physically, sets $\widetilde{\omega}$ and $\widetilde{\varepsilon}$ (called, as before, interfaces) separate different "phases" (outside and inside $\widetilde{\omega}$ and $\widetilde{\varepsilon}$, respectively), and numbers $E(\cdot)$ and $I(\cdot)$ label these phases. As in Model 2, we suppose that for any long contour ω,

$$\text{diam } \widetilde{\omega} \geq u > 1, \tag{3.57}$$

while for short contours ε,

$$\text{diam } \widetilde{\varepsilon} < u. \tag{3.58}$$

As in §2, we denote by $\text{Int}\,\widetilde{\omega}$ and $\text{Int}\,\widetilde{\varepsilon}$ the subsets of $\mathbb{Z}^d$ enclosed by $\widetilde{\omega}$ and $\widetilde{\varepsilon}$, respectively, and by $\text{Ext}\,\widetilde{\omega}$ and $\text{Ext}\,\widetilde{\varepsilon}$ the complements $\mathbb{Z}^d \setminus \text{Int}\,\widetilde{\omega}$ and $\mathbb{Z}^d \setminus \text{Int}\,\widetilde{\varepsilon}$, respectively.

The compatibility rules for collections of ω's and ε's in Model 3 are more complicated (and not even reduced to binary ones). We state these rules for any pair of long or short contours from a given collection, which are *not separated* by any third long contour from the same collection. Here, $\omega' = (\widetilde{\omega}', E', I')$ and $\omega'' = (\widetilde{\omega}'', E'', I'')$ are said to be separated by ω if either $\text{Int}\,\widetilde{\omega}' \subset \text{Int}\,\widetilde{\omega}, \text{Int}\,\widetilde{\omega}'' \subset \text{Ext}\,\widetilde{\omega}$ or $\text{Int}\,\widetilde{\omega}'' \subset \text{Int}\,\widetilde{\omega}, \text{Int}\,\widetilde{\omega}' \subset \text{Ext}\,\widetilde{\omega}$, and similarly for pairs ε', ω'' and $\varepsilon', \varepsilon''$. Thus, two long contours, $\omega' = (\widetilde{\omega}', E', I')$ and $\omega'' = (\widetilde{\omega}'', E'', I'')$, not separated by a third long contour, are compatible if

(i.1) $\widetilde{\omega}' \cap \widetilde{\omega}'' = \varnothing, \text{Int}\,\widetilde{\omega}' \subset \text{Int}\,\widetilde{\omega}''$ and $I'' = E'$;

(i.2) $\widetilde{\omega}' \cap \widetilde{\omega}'' = \varnothing, \text{Int}\,\widetilde{\omega}'' \subset \text{Int}\,\widetilde{\omega}'$ and $I' = E''$;

(i.3) $\widetilde{\omega}' \cap \widetilde{\omega}'' = \varnothing, \text{Int}\,\widetilde{\omega}' \cap \text{Int}\,\widetilde{\omega}'' = \varnothing$ and $E'' = E'$;

(i.4) $\widetilde{\omega}' \cap \widetilde{\omega}'' \neq \varnothing, \text{Int}\,\widetilde{\omega}' \subset \text{Int}\,\widetilde{\omega}''$ and either $E'' > I'' = E' > I'$ or $E'' < I'' = E' < I'$;

(i.5) $\widetilde{\omega}' \cap \widetilde{\omega}'' \neq \varnothing, \text{Int}\,\widetilde{\omega}'' \subset \text{Int}\,\widetilde{\omega}'$ and either $E' > I' = E'' > I''$ or $E' < I' = E'' < I''$;

(i.6) $\widetilde{\omega}' \cap \widetilde{\omega}'' \neq \varnothing, \text{Int}\,\widetilde{\omega}' \cap \text{Int}\,\widetilde{\omega}'' = \varnothing$ and either $I'' > E'' = E' > I'$ or $I'' < E'' = E' < I'$.

Two short contours, $\varepsilon' = (\widetilde{\varepsilon}', E', I')$ and $\varepsilon'' = (\widetilde{\varepsilon}'', E'', I'')$, not separated by a long contour, are compatible if

(ii.1) $\widetilde{\varepsilon}' \cap \widetilde{\varepsilon}'' = \varnothing$ and $E' = E''$;

(ii.2) $\widetilde{\varepsilon}' \cap \widetilde{\varepsilon}'' \neq \varnothing, \text{Int}\,\widetilde{\varepsilon}' \subset \text{Int}\,\widetilde{\varepsilon}''$ or $\text{Int}\,\widetilde{\varepsilon}'' \subset \text{Int}\,\widetilde{\varepsilon}', E' = E''$ and either $E' > I', E'' > I''$ or $E' < I'$, $E'' < I''$;

(ii.3) $\widetilde{\varepsilon}' \cap \widetilde{\varepsilon}'' \neq \varnothing$, $\text{Int}\,\widetilde{\varepsilon}'' \cap \text{Int}\,\widetilde{\varepsilon}' = \varnothing$, $E' = E''$ and either $E' > I'$, $E'' < I''$ or $E' < I'$, $E'' > I''$.

Finally, long contour $\omega' = (\widetilde{\omega}', E', I')$ and short contour $\varepsilon'' = (\widetilde{\varepsilon}'', E'', I'')$ not separated by another long contour, are compatible if

(iii.1) $\widetilde{\omega}' \cap \widetilde{\varepsilon}'' = \varnothing$, $\text{Int}\,\widetilde{\varepsilon}'' \subset \text{Int}\,\widetilde{\omega}'$ and $I' = E''$;

(iii.2) $\widetilde{\omega}' \cap \widetilde{\varepsilon}'' = \varnothing$, $\text{Int}\,\widetilde{\omega}' \cap \text{Int}\,\widetilde{\varepsilon}'' = \varnothing$ and $E'' = E'$;

(iii.3) $\widetilde{\omega}' \cap \widetilde{\varepsilon}'' \neq \varnothing$, $\text{Int}\,\widetilde{\varepsilon}'' \subset \text{Int}\,\widetilde{\omega}'$ and either $E' > I' = E'' > I''$ or $E' < I' = E'' < I''$;

(iii.4) $\widetilde{\omega}' \cap \widetilde{\varepsilon}'' \neq \varnothing$, $\operatorname{Int}\widetilde{\omega}' \cap \operatorname{Int}\widetilde{\varepsilon}'' = \varnothing$ and either $I'' > E'' = E' > I'$ or $I'' < E'' = E' < I'$.

Note that the situation $\operatorname{Int}\widetilde{\omega} \subset \operatorname{Int}\widetilde{\varepsilon}$ is impossible, and the inclusion $\subset$ is always understood in the strict sense.

We say that a collection $\{\varepsilon_i, \omega_j\}$ of short and long contours is compatible if any two of the contours not separated by a third long contour from the collection are compatible. The motivation of such a definition of compatibility is as follows. We give the definition in such a way that any compatible collection of long contours $\{\omega_j\}$ uniquely defines a *configuration* $X = X(\{\omega_j\}) : x \in \mathbb{Z}^d \mapsto X_x \in \{0, 1, \dots, m\}$, with every $\widetilde{\omega}_j$ separating the region with $X_x = E(\omega_j)$ from the region with $X_x = I(\omega_j)$. The short contours ε_i, with $E(\varepsilon_i) = k$, are situated inside the set $\{x \in \mathbb{Z}^d \mid X_x = k\}$; they do not change the configuration defined by $\{\omega_j\}$.

Conditions (i.1)–(i.3), (ii.1). and (iii.1)–(iii.2) seem to be the most natural part of the compatibility rule above. Conditions (i.4)–(i.6), (ii.2)–(ii.3) and (iii.3)–(iii.4) cover the case when the interfaces of short or long contours have some plaquettes in common. We add these items to the compatibility rule to have a wider class of the corresponding configurations $X(\{\omega_j\})$.

In the current model, we suppose that the statistical weights of long and short contours satisfy the bounds

$$|w(\omega)| \leq \exp\big[-p|\widetilde{\omega}| \cdot |E(\omega) - I(\omega)|\big] \tag{3.59}$$

and

$$|w(\varepsilon)| \leq \exp\big[-p|\widetilde{\varepsilon}| \cdot |E(\varepsilon) - I(\varepsilon)|\big]. \tag{3.60}$$

This means that the statistical weights of the contours depend not only on the geometry of their interfaces but also on their labels $E(\cdot)$ and $I(\cdot)$. For every value $k = 0, \dots, m$, the metastable partition function in V, with the external phase k, is defined as

$$Z^{(k)}(V) = \sum_{\substack{\{\varepsilon_i\}:\ \{\widetilde{\varepsilon}_i\} \subseteq V,\\ E(\varepsilon_i) = k}} \prod_i w(\widetilde{\varepsilon}_i). \tag{3.61}$$

Here, the summation is extended to the compatible collections $\{\varepsilon_i\}$ where each $\widetilde{\varepsilon}_j$ is contained in V in the $\mathbb{R}^d$-sense, and $E(\varepsilon_j) \equiv k$. Further, we suppose that m is even and phase $m/2$ is the dominant one, in the sense that

$$\frac{Z^{(k)}(V)}{Z^{(m/2)}(V)} \leq \exp\Big[-|V| 16d \exp\Big(-(p-2)(du - u + 1)\Big)\Big]. \tag{3.62}$$

The full partition function of the contour Model 3 is

$$\Xi^{(k)}(V) = \sum_{\substack{\{\varepsilon_i, \omega_j\}:\ \{\widetilde{\varepsilon}_i, \widetilde{\omega}_j\} \subseteq V,\\ E(\varepsilon_i^{\text{ext}}) = E(\omega_j^{\text{ext}}) = k}} \prod_i w(\varepsilon_i) \prod_j w(\omega_j). \tag{3.63}$$

Here, the summation is extended to the compatible collections $\{\varepsilon_i, \omega_j\}$ such that each $\widetilde{\varepsilon}_i$ and $\widetilde{\omega}_j$ is contained in V in the $\mathbb{R}^d$-sense and $E(\varepsilon_i^{\text{ext}}) \equiv E(\omega_j^{\text{ext}}) \equiv k$. Furthermore, $\varepsilon_i^{\text{ext}}$ and ω_j^{ext} are the external (i.e., not enclosed by another one) long and short contours, respectively. We are interested in constructing a polymer expansion for the logarithm of the partition function $\Xi^{(m/2)}(V)$ corresponding to the dominant phase. To simplify the notation, we set below $Z(V) = \Xi^{(m/2)}(V)$.

It is convenient to construct, in the current model, an analogue of the bulky contour from Models 1 and 2 . This is a compatible collection $\Omega = \{\omega_j\}$, of long contours with $E(\{\omega_j\}) = \frac{m}{2}$, such that the set $\widetilde{\Omega} = \widetilde{\Omega}(\{\omega_j\})$ given by

$$(3.64) \qquad \widetilde{\Omega} = \{x \in \mathbb{Z}^d | \ X_x(\{\omega_j\}) \neq m/2\}$$

is connected. Denote by $\partial_r\Omega$ the long contour from collection $\{\omega_j(\Omega)\}$ whose interface is $\partial_r\widetilde{\Omega}$. As before, we denote by $\partial_0\Omega$ the unique external long contour in collection $\{\partial_r\Omega\}$. By construction, $E(\partial_0\Omega) = I(\partial_r\Omega) = \frac{m}{2}$, $r \neq 0$ while $E(\omega_j(\Omega)), I(\omega_j(\Omega)) \neq \frac{m}{2}$ for $\omega_j(\Omega)$ from $\{\omega_j(\Omega)\} \setminus \{\partial_r\Omega\}$. The collection $\{\omega_j(\Omega)\} \setminus \{\partial_r\Omega\}$ partitions $\widetilde{\Omega}$ into subsets $\widetilde{\Omega}^{(k)}$, $k \neq \frac{m}{2}$ with $X_x(\Omega) = k$ when $x \in \widetilde{\Omega}^{(k)}$.

Every compatible collection $\{\varepsilon_i, \omega_j\}$ can now be treated as a compatible collection $\{\varepsilon_i, \Omega_n\}$, and (3.63) can be rewritten as

$$\begin{aligned} Z(V) &= \sum_{\{\varepsilon_i,\Omega_n\}:\ \{\widetilde{\varepsilon}_i,\widetilde{\Omega}_n\}\subseteq V} \prod_i w(\varepsilon_i) \prod_n \prod_j w(\omega_j(\Omega_n)) \\ (3.65) \qquad &= \sum_{\{\Omega_n\}:\ \{\widetilde{\Omega}_n\}\subseteq V} Z^{(m/2)}\left(V \setminus \left(\bigcup_n \widetilde{\Omega}_n\right) | \{\partial_r\Omega_n\}\right) \\ &\quad \times \prod_n \prod_j w(\omega_j(\Omega_n)) \prod_k Z^{(k)}(\widetilde{\Omega}_n^{(k)} | \Omega_n). \end{aligned}$$

Here the conditioned metastable partition functions are defined as follows:

$$\begin{aligned} &Z^{(m/2)}\left(V \setminus \left(\bigcup_n \widetilde{\Omega}_n\right) | \{\partial_r\Omega_n\}\right) \\ (3.66) \qquad &= \sum_{\substack{\{\varepsilon_i\}:\ \{\widetilde{\varepsilon}_i\}\subseteq V\setminus(\bigcup_n \widetilde{\Omega}_n), E(\varepsilon_i)=m/2, \\ \text{any pair } \varepsilon_i, \partial_r\Omega_n \text{ is compatible}}} \prod_i w(\varepsilon_i) \end{aligned}$$

and

$$(3.67) \qquad Z^{(k)}(\widetilde{\Omega}_n^{(k)} | \Omega_n) = \sum_{\substack{\{\varepsilon_i\}:\ \{\widetilde{\varepsilon}_i\}\subseteq\widetilde{\Omega}_n^{(k)}, E(\varepsilon_i)=k, \\ \text{any pair } \varepsilon_i, \omega_j(\Omega_n) \text{ is compatible}}} \prod_i w(\varepsilon_i),$$

cf. (3.61). The meaning of the notation $\{\widetilde{\varepsilon}_i, \widetilde{\Omega}_n\} \subseteq V$, $\{\widetilde{\Omega}_n\} \subseteq V$, etc, is the same as in Model 2.

Given a bulky contour $\Omega \subset \mathbb{Z}^d$ we define

$$(3.68) \qquad w(\Omega) = \prod_j w(\omega_j(\Omega)) \prod_k \frac{Z^{(k)}(\widetilde{\Omega}^{(k)} | \Omega)}{Z^{(m/2)}(\widetilde{\Omega}^{(k)} | \Omega^*)}.$$

Here the collection $\Omega^* = \{\omega_j^*\}$ is obtained from $\Omega = \{\omega_j\}$ by setting $\widetilde{\omega}_j^* = \widetilde{\omega}_j$, $I(\omega_j^*) = \frac{m}{2}$, and $E(\omega_j^*) = \frac{m}{2} + \operatorname{sign}(E(\omega_j) - I(\omega_j))$. Then (3.65) can be rewritten as

$$\begin{aligned} Z(V) &= \sum_{\{\varepsilon_i,\Omega_n\}:\ \{\widetilde{\varepsilon}_i,\widetilde{\Omega}_n\}\subseteq V} \prod_i w(\varepsilon_i) \prod_n \prod_j w(\omega_j(\Omega_n)) \\ &= \sum_{\{\Omega_n\}:\ \{\widetilde{\Omega}_n\}\subseteq V} Z^{(m/2)}\Big(V \setminus \Big(\bigcup_n \widetilde{\Omega}_n\Big) | \{\partial_r \Omega_n\}\Big) \\ &\qquad \times \prod_n w(\Omega_n) \prod_k Z^{(m/2)}(\widetilde{\Omega}_n^{(k)} | \Omega_n^*), \end{aligned} \tag{3.69}$$

and expanding $Z^{(m/2)}(\cdot)$ in form (3.66) and (3.67) we get the representation

$$Z(V) = \sum_{\{\varepsilon_i,\Omega_n\}:\ \{\widetilde{\varepsilon}_i,\widetilde{\Omega}_n\}\subseteq V} \prod_i w(\varepsilon_i) \prod_n w(\Omega_n), \tag{3.70}$$

which is an analogue of (3.48). Here we suppose that $E(\varepsilon_i) = \frac{m}{2}$ for all i. The compatibility rule for the ε's and Ω's is an appropriate analogue of that in (3.48). More precisely, the rule is the binary one and it requires for any pair (Ω, Ω'), $(\varepsilon, \varepsilon')$, and (Ω, ε):

(i) $\widetilde{\Omega} \cap \widetilde{\Omega}' = \varnothing$;
(ii) $\widetilde{\varepsilon} \cap \widetilde{\varepsilon}' = \varnothing$;
(iii) either
(iii.1) $\operatorname{Int}\widetilde{\varepsilon} \subset \widetilde{\Omega}^c$ and, for any r, the long contour $\partial_r\Omega$ and short contour ε are compatible, or
(iii.2) $\operatorname{Int}\widetilde{\varepsilon} \subset \widetilde{\Omega}$ and, for any j, the long contour $(\widetilde{\omega}_j(\Omega), \frac{m}{2} + \operatorname{sign}(E(\omega_j) - I(\omega_j)), \frac{m}{2})$ constructed from $\omega_j(\Omega) = (\widetilde{\omega}_j(\Omega), E(\omega_j(\Omega)), I(\omega_j(\Omega)))$, and short contour ε are compatible.

PROPOSITION 3.5. *Under conditions* (3.59), (3.60), *and* (3.62), *with*

$$p \geq 4 + \log(2d), \tag{3.71}$$

the logarithm of partition function (3.63), *with* $k = m/2$, (*or equivalently, the logarithm of partition function* (3.70)) *admits a polymer expansion of form* (3.3), *where property* (3.6) *is valid with*

$$\begin{aligned} a(\varepsilon) &= |\widetilde{\varepsilon}|, \\ a(\Omega) &= \sum_j |\widetilde{\omega}_j(\Omega)| + h|\widetilde{\Omega}|, \end{aligned} \tag{3.72}$$

and $h = 2d \exp(-(p-2)(du - u + 1))$.

PROOF. We proceed in a way similar to Propositions 3.2 and 3.4. In analogy with (2.13) we define

$$\|\varepsilon\| = |\widetilde{\varepsilon}| \cdot |E(\varepsilon) - I(\varepsilon)|, \quad \|\Omega\| = |\widetilde{\omega}| \cdot |E(\Omega) - I(\Omega)|. \tag{3.73}$$

A minor difference with the proof of the above propositions is caused by the fact that, in all calculations, $|\cdot|$ should be replaced by $\|\cdot\|$, and an additional summation over $|E(\cdot) - I(\cdot)|$ is carried on. This does not change the final bounds.

Another difference is that, instead of (3.22), we now have, in view of (3.62), the bound

$$|w(\Omega)| \le \exp\Big(- p \sum_j \|\omega_j(\Omega)\| - q' \sum_k |\, [\widetilde{\Omega}^{(k)}]\, |\Big), \tag{3.74}$$

where $q' = 16d \exp(-(p-2)(du - u + 1))$.

LEMMA 3.6. *For* $q = 8d \exp(-(p-2)(du-u+1))$, *the following bound holds true*:

$$\sum_{\Omega:\, \{\partial_r\Omega\} \text{ is fixed}} |w(\Omega)| \le \exp\Big(-(p-1)\sum_r \|\partial_r\Omega\| - q|\widetilde{\Omega}|\Big). \tag{3.75}$$

PROOF. The sum in the left-hand side of (3.75) is, in fact, over all possibilities to complete $\{\partial_r\Omega\}$ to $\{\omega_j(\Omega)\}$. We have

$$\begin{aligned}
\sum_{\Omega:\, \{\partial_r\Omega\} \text{ is fixed}} |w(\Omega)| &\le \Big(1 + \sum_{\omega:\, \mathrm{Int}\,\widetilde{\omega}\ni 0} \exp(-p\|\Omega\|)\Big)|\widetilde{\Omega}| \\
&\le \Big(1 + \sum_{\widetilde{\omega}:\, \mathrm{Int}\,\widetilde{\omega}\ni 0} \sum_{I=0}^{m} \exp(-p|\omega| \cdot |E - I|)\Big)^{|\widetilde{\Omega}|} \\
&\le \Big(1 + 4 \sum_{\widetilde{\omega}:\, \mathrm{Int}\,\widetilde{\omega}\ni 0} \exp(-p|\omega|)\Big)^{|\widetilde{\Omega}|} \\
&\le \Big(1 + \frac{8d\exp(-p(2du - 2u + 2))}{1 - 2d\exp(-p)}\Big)^{|\widetilde{\Omega}|} \\
&\le \exp\big[|\widetilde{\Omega}| 8d \exp\big(- (p-2)(du - u + 1)\big)\big].
\end{aligned} \tag{3.76}$$

Combining (3.76) with (3.74), we obtain the lemma. □

Now (3.22) is valid for $\sum_{\Omega:\, \{\partial_r\Omega\} \text{ is fixed}} |w(\Omega)|$, and we can proceed further exactly as in Propositions 3.2 and 3.4. □

Usually, a "physical" model can be described directly in terms of a contour model only when it possesses a unique limiting Gibbs measure. In fact, in many cases the description by means of a contour model is a natural way to prove uniqueness. To analyse physical models with more than one stable phase, one usually needs the rather involved machinery of Pirogov–Sinai Theory (see **[PS]**, **[Si]**, **[Z]**). A situation where it can be avoided arises in models with two symmetric dominant phases. The simplest example is the symmetry between + and − phases of the Ising model with zero external magnetic field. Our next contour model provides a description of such situations.

Contour Model 4. The setting is similar to Model 3, with the following addenda reflecting the symmetry between two selected phases, say $\frac{m}{2} - \frac{1}{2}$ and $\frac{m}{2} + \frac{1}{2}$ (the value m is now supposed to be odd):

(i) there exists a permutation, ν, of the set $\{0, 1, \dots, m\}$, such that $\frac{m}{2} + \frac{1}{2} = \nu(\frac{m}{2} - \frac{1}{2})$ and every long contour ω with $E(\Omega) = k$ and $I(\Omega) = l$ has the same statistical weight as a similar long contour with $E(\Omega) = \nu(k)$, $I(\Omega) = \nu(l)$; every short contour ε with $E(\varepsilon) = k$ and $I(\varepsilon) = l$ has the same statistical weight as an analogous short contour with $E(\varepsilon) = \nu(k)$, $I(\varepsilon) = \nu(l)$;

(ii) phases $\frac{m}{2} - \frac{1}{2}$ and $\frac{m}{2} + \frac{1}{2}$ are the dominant ones, in the sense that

$$Z^{(m/2-1/2)}(V) = Z^{(m/2+1/2)}(V),$$

and for $k \neq \frac{m}{2} \pm \frac{1}{2}$,

$$\frac{Z^{(k)}(V)}{Z^{(m/2-1/2)}(V)} \leq \exp\left[-|V|32d\exp\left(-(p-2)(du-u+1)\right)\right]. \tag{3.77}$$

Similarly to (3.63), the full partition functions in V are defined as

$$\Xi^{(k)}(V) = \sum_{\substack{\{\varepsilon_i,\omega_j\}:\ \{\tilde{\varepsilon}_i,\tilde{\omega}_j\}\subseteq V,\\ E(\varepsilon_i^{\rm ext})=E(\omega_j^{\rm ext})=k}} \prod_i w(\varepsilon_i)\prod_j w(\omega_j), \tag{3.78}$$

where the range of summation is the same as in (3.63). The symmetry immediately gives that

$$Z^{(m/2-1/2)}(V) = Z^{(m/2+1/2)}(V), \quad \Xi^{(m/2-1/2)}(V) = \Xi^{(m/2+1/2)}(V). \tag{3.79}$$

PROPOSITION 3.7. *Under conditions* (3.59), (3.60), *and* (3.77), *with p satisfying condition* (3.71), *the logarithm of partition function* (3.78), *with $k = \frac{m}{2} \pm \frac{1}{2}$, admits a polymer expansion of form* (3.3), *with $a(\cdot)$ defined as in Proposition* 3.5.

PROOF. Due to symmetry between phases $\frac{m}{2} - \frac{1}{2}$ and $\frac{m}{2} + \frac{1}{2}$, one can consider long contours ω with $E(\Omega) = \frac{m}{2} - \frac{1}{2}$ and $I(\Omega) = \frac{m}{2} + \frac{1}{2}$ as not changing the phase (i.e., one can put $E(\Omega) = I(\Omega) = \frac{m}{2} - \frac{1}{2}$ keeping the partition functions $\Xi^{(m/2-1/2)}(V)$ unchanged). This allows us to include such contours in the class of short contours of phase $\frac{m}{2} - \frac{1}{2}$. We still have, for $k \neq \frac{m}{2} \pm \frac{1}{2}$,

$$\frac{Z^{(k)}(V)}{Z^{(m/2-1/2)}(V)} \leq \exp\left[-|V|16d\exp(-(p-2)(du-u+1))\right], \tag{3.80}$$

because the new short contours can add, to the exponent at the right-hand side of (3.79), the correction of the order $|V|d\exp(-p(2du-2u+2))$ at most. Thus, Proposition 3.7 is reduced to Proposition 3.5. □

Our final models (5 and 6) have a more complicated geometry of long and short contours. We do not have an obvious explanation of the origin of this geometry, apart from the fact that it covers the case of the quantum system under consideration.

Contour Model 5. We first consider the case of a unique dominant phase. Model 5, like Models 3 and 4, contains long and short contours. A long contour is now denoted by ζ and defined as a pair $(\{\omega_j\}, \sigma)$. Here,

(i) each $\omega_j = \omega_j(\zeta)$ is a contour (either long or short) from the previous model, and $\{\omega_j\}$ is a finite, nonempty compatible collection, again in the sense of the previous model;

(ii) $\sigma = \sigma(\zeta)$ is a finite set of plaquettes of dual lattice $\widetilde{\mathbb{Z}}^d$;

(iii) the union $(\bigcup_j \tilde{\omega}_j(\zeta)) \bigcup \sigma(\zeta)$ denoted below by $\tilde{\zeta} = \tilde{\zeta}(\zeta)$ is a connected set of plaquettes of $\widetilde{\mathbb{Z}}^d$;

(iv) for at least one $\omega_j(\zeta)$,

$$\text{diam } \widetilde{\omega}_j(\zeta) \geq u. \tag{3.81}$$

A short contour is denoted, as before, by ε and defined again as a pair $(\{\omega_j\}, \sigma)$ (or $(\{\omega_j(\varepsilon)\}, \sigma(\varepsilon))$) for which the above conditions (i) – (iii) are valid, but (iv) is not:

(iv$'$) for each $\omega_j(\varepsilon)$,

$$\text{diam } \widetilde{\omega}_j(\varepsilon) < u. \tag{3.82}$$

The union $(\bigcup_j \widetilde{\omega}_j(\varepsilon)) \bigcup \sigma(\varepsilon)$ is denoted by $\widetilde{\varepsilon} = \widetilde{\varepsilon}(\varepsilon)$.

A collection $\{\varepsilon_i, \zeta_j\}$ of short and long contours is now called compatible if

(i) the $\widetilde{\varepsilon}_i$'s and $\widetilde{\zeta}_j$'s form a collection of pairwise nonintersecting subsets of $\widetilde{\mathbb{Z}}^d$;

(ii) the whole collection of the $\omega_s(\varepsilon_i)$'s and $\omega_t(\zeta_j)$'s from contours ε_i and ζ_j is compatible, in the sense of Model 4.

In the current model we suppose that the statistical weights of long and short contours satisfy the bounds

$$|w(\zeta)| \leq \exp\Big(- p \sum_i |\widetilde{\omega}_i(\zeta)| \cdot |E(\omega_i(\zeta)) - I(\omega_i(\zeta))| - t|\sigma(\zeta)|\Big) \tag{3.83}$$

and

$$|w(\varepsilon)| \leq \exp\big[- p|\widetilde{\omega}_i(\varepsilon)| \cdot |E(\omega_i(\varepsilon)) - I(\omega_i(\varepsilon))| - t|\sigma(\varepsilon)|\big], \tag{3.84}$$

where $t > 0$.

As in the previous models, we introduce the metastable partition functions

$$Z^{(k)}(V) = \sum_{\substack{\{\varepsilon_j\}:\, \{\widetilde{\varepsilon}_j\} \subseteq V,\\ E(\omega_i(\varepsilon_j))=k}} \prod_j w(\varepsilon_j), \tag{3.85}$$

with the range of summation similar to (3.61). We also assume that $Z^{(m/2)}(V)$ is dominating in the sense that

(3.86)
$$\frac{Z^{(k)}(V)}{Z^{(m/2)}(V)} \leq \exp\big[-(|\,[V]_v\,| - v|\,\partial V\,|)\, 16d \exp\big(-(p - 2\log(2d) - 2)(du - u + 1)\big)\big]$$

for each $k = 0, \ldots, m$ with $k \neq m/2$ and $[V]_v = \{x \in V |\ \text{dist}\,(x, V^c) \geq v + 1\}$, where $v \geq 1$ is a constant.

The full partition functions are defined as

$$\Xi^{(k)}(V) = \sum_{\substack{\{\varepsilon_i, \zeta_j\}:\, \{\widetilde{\varepsilon}_i, \widetilde{\zeta}_j\} \subseteq V,\\ E(\varepsilon_i^{\text{ext}}) = E(\omega_n^{\text{ext}}(\zeta_j)) = k}} \prod_i w(\varepsilon_i) \prod_j w(\zeta_j). \tag{3.87}$$

Like (3.63), the summation in (3.87) is extended to the compatible collections $\{\varepsilon_i, \zeta_j\}$ such that each $\widetilde{\varepsilon}_i$ and $\widetilde{\zeta}_j$ is contained in V in the $\mathbb{R}^d$-sense and $E(\varepsilon_i^{\text{ext}}) \equiv E(\zeta_j^{\text{ext}}) \equiv k$.

As before, it is convenient to introduce an analogue of a bulky contour. It is again a compatible collection, $\Omega = \{\zeta_i\}$, of long contours, with $E(\{\zeta_i\}) = E(\{\omega_j(\zeta_i)\}) = m/2$, such that the set $\widetilde{\Omega} = \widetilde{\Omega}(\{\zeta_i\})$ given by

$$\widetilde{\Omega} = \{x \in \mathbb{Z}^d |\ X(\{\zeta_i\}) \neq m/2\} \tag{3.88}$$

has the following property. The union of the boundary of $\widetilde{\Omega}$ with $\bigcup_i \sigma(\zeta_i)$ forms a connected set of plaquettes of $\widetilde{\mathbb{Z}}^d$. As before, $X(\{\zeta_i\}) = X(\omega_j(\{\zeta_i\}))$ is a configuration uniquely determined by the compatible collection $\{\zeta_i\}$.

Repeating the construction from the previous models (see (3.65) – (3.70)), replacing the ω_j's (the long contours in the previous models) by the ζ_i's, one can write, for $Z(V) = \Xi^{(m/2)}(V)$, a formula

$$Z(V) = \sum_{\{\varepsilon_i,\Omega_n\}:\ \{\widetilde{\varepsilon}_i,\widetilde{\Omega}_n\}\subseteq V} \prod_i w(\varepsilon_i) \prod_n w(\Omega_n) \tag{3.89}$$

identical to (3.70).

PROPOSITION 3.8. *Under conditions* (3.83)–(3.86), *with*

$$p \geq 4 + 2\log(2d) + \frac{6\log 2 + \log v}{du - u + 1}, \ t \geq 2\log(2d) + 1, \tag{3.90}$$

the logarithm of partition function (3.87), *with* $k = m/2$ (*or equivalently, the logarithm of partition function* (3.89)), *admits a polymer expansion of form* (3.3), *with*

$$\begin{aligned} a(\varepsilon) &= |\widetilde{\varepsilon}| = |\sigma(\varepsilon)| + \sum_j |\widetilde{\omega}_j(\varepsilon)| \\ a(\Omega) &= \sum_i \left(\sum_j |\widetilde{\omega}_j(\zeta_i(\Omega))| + |\sigma(\zeta_i(\Omega))| \right) + h|\widetilde{\Omega}|, \end{aligned} \tag{3.91}$$

and $h = d\exp(-(p - 2\log(2d) - 2)(du - u + 1)$.

PROOF. The idea of the proof is to repeat the proof of Proposition 3.5, with corresponding modifications. We put

$$q = 8d\exp\big[-(p - 2\log(2d) - 2)(du - u + 1)\big] \tag{3.92}$$

and substitute $p - 2\log(2d) - 2$ in place of $(p - 2)$ in all corresponding calculations. To make it possible, one needs a summability of statistical weights $w(\zeta)$.

LEMMA 3.9. *For* $p \geq 4 + 2\log(2d)$ *and* $t \geq 2\log(2d)$

$$\sum_{\zeta:\ \bigcup_j \widetilde{\omega}_j(\zeta) \bigcup \sigma(\zeta) \ni 0} |w(\zeta)| \leq \exp\big[-(p - 2\log(2d))(2du - 2u + 2)\big]. \tag{3.93}$$

PROOF. As $\bigcup_j \widetilde{\omega}_j(\zeta) \bigcup \sigma(\zeta)$ is connected and, for at least one value of j, the diameter of $\widetilde{\omega}_j(\zeta)$ is greater than or equal u, we have

$$\begin{aligned} \sum_{\zeta:\ \bigcup_j \widetilde{\omega}_j(\zeta) \bigcup \sigma(\zeta) \ni 0} |w(\zeta)| &\leq \sum_{\zeta:\ \bigcup_j \widetilde{\omega}_j(\zeta) \bigcup \sigma(\zeta) \ni 0} \sum_{I=-m}^{m} \exp\Bigg(-p\sum_j |\widetilde{\omega}_j(\zeta)| \\ &\qquad \times |E(\omega_j(\zeta)) - (I + E(\omega_j(\zeta)))| - t|\sigma(\zeta)| \Bigg) \\ &\leq 4 \sum_{\zeta:\ \bigcup_j \widetilde{\omega}_j(\zeta) \bigcup \sigma(\zeta) \ni 0} \exp\Bigg(-p\sum_j |\widetilde{\omega}_j(\zeta)| - t|\sigma(\zeta)| \Bigg) \end{aligned}$$

$$(3.94)\qquad \begin{aligned} &\le 4\exp\Big[-(p-t)(2du-2u+2)\Big] \\ &\quad\times \sum_{\zeta:\,\bigcup_j \widetilde{\omega}_j(\zeta)\bigcup\sigma(\zeta)\ni 0} \exp\Big(-t\sum_j|\widetilde{\omega}_j(\zeta)|-t|\sigma(\zeta)|\Big) \\ &\le 4\exp\Big[-(p-t)(2du-2u+2)\Big]\frac{2d\exp(-t)}{1-2d\exp(-t)} \\ &\le \exp\Big[-(p-2\log(2d))(2du-2u+2)\Big]. \quad \square \end{aligned}$$

The proof of Proposition 3.8 can now proceed along the same line of arguments as that of Proposition 3.5. The only place where a modification is needed is in an analogue of bound (3.74). Because of the difference between (3.62) and (3.86), the analogue of (3.74) takes the form

(3.95)
$$|w(\Omega)|\le\exp\Big(-\sum_i\Big((p-vq')\sum_j\|\omega_j(\zeta_i(\Omega))\|+t|\sigma(\zeta_i(\Omega))|\Big)-q'\sum_k|\,[\widetilde{\Omega}^{(k)}]_v\,|\Big),$$

where $q'=16d\exp\Big[-(p-2\log(2d)-2)(du-u+1)\Big]$ and $w(\Omega)$ is defined similarly to (3.68).

Obviously $|\widetilde{\omega}\setminus(\bigcup_k[\widetilde{\Omega}^{(k)}]_v)|$ is less than $2v\sum_i\sum_j|\widetilde{\omega}_j(\zeta_i(\Omega))|$, and

$$(3.96)\qquad 1\ge 64vd\exp(-(p-2\log(2d)-2)(du-u+1))$$

for $p\ge 4+2\log(2d)+\dfrac{6\log 2+\log v}{du-d+1}$. Hence

(3.97)
$$|w(\Omega)|\le\exp\Big(-\sum_i\Big((p-1)\sum_j\|\omega_j(\zeta_i(\Omega))\|+t|\sigma(\zeta_i(\Omega))|\Big)-q'\sum_k|\widetilde{\Omega}^{(k)}|\Big).$$

With this remark, one can proceed further exactly as in the proof of Proposition 3.5. $\square$

Contour Model 6. The last contour model of this section describes a situation with two symmetric dominant phases in the setup of Model 5. Here, an analogue of Proposition 3.7 may be proved, which we refer to as Proposition 3.10, without stating it in an explicit form.

4. Proof of Theorems 1 and 2

In this section we use Propositions 3.8 and 3.10 in a special case of the ensemble of quantum contours (more precisely, of compatible collections of quantum contours) determined by (2.15), (2.16). As was noted before, our aim is to construct a polymer expansion for $\log\Xi^{(l)}(V)$ with $l=\frac{m}{2}$ for m even and $l=\frac{m}{2}\pm\frac{1}{2}$ for m odd, and obtain thereby the existence (and ergodicity) of states φ^0 and $\varphi^{\pm}$ figuring in Theorem 1. The remaining statements of Theorem 1 require an additional construction provided at the end of the section. The term probability is used below in the sense of the ensemble of quantum contours.

To match the notation from §3 we set for a quantum contour ζ:

$$(4.1)\qquad \widetilde{\omega}_j(\zeta)=\widetilde{\gamma}^0_j(\zeta),\;\; j=1,\dots,n_0(\zeta),$$

and

$$\sigma(\zeta) = \bigcup_{i=1}^{n(\zeta)-1} \bigcup_{j=1}^{n_i(\zeta)} \widetilde{\gamma}_j^i(\zeta). \tag{4.2}$$

LEMMA 4.1. *Bound* (3.83) *holds for a quantum contour* ζ, *with*

$$t = 2\log(2d) + 1,$$

$$p = \frac{\beta}{2d}\left((\mu - 1) - \frac{(m+1)}{4d} - (2\log(2d)+1)(2ed)^{4d}(m+1)\right).$$

PROOF. First, we estimate the factor $\prod_{i=0}^{n(\zeta)-1} X^i(y_i^-)$ in (2.15) from above. Given a quantum contour ζ, fix a sequence $0 \le i_1 < \cdots < i_k \le n(\zeta)$ of the indices of classical configurations $X^i = X^i(\mathbb{X}(\zeta))$ such that $X^{i_j}(y_{i_j}^-) \ge \frac{m}{2} + 1$. The sites $y_{i_j}^-$ may be partitioned into two groups. The first group consists of the sites with $X^{i_j}(y_{i_j}^-) = X^0(y_{i_j}^-)$. The second one consists of the sites with $X^{i_j}(y_{i_j}^-) \ne X^0(y_{i_j}^-)$. The cyclic property of ζ implies that for every site $y_{i_j}^-$ from the first group, there exists a conjugate site, $y_{i'_j}^-$, from $\{y_1^-, \dots, y_{n(\zeta)}^-\} \setminus \{y_{i_1}^-, \dots, y_{i_k}^-\}$, such that $i'_j > i_j$, $y_{i_j}^- = y_{i'_j}^+$ and

$$X^{i'_j}(y_{i_j}^-) = X^{i_j}(y_{i_j}^-) - 1. \tag{4.3}$$

Clearly,

$$X^{i'_j}(y_{i'_j}^+) + X^{i'_j}(y_{i'_j}^-) \le m \tag{4.4}$$

and

$$X^{i_j}(y_{i_j}^-) + X^{i'_j}(y_{i'_j}^-) \le m + 1. \tag{4.5}$$

Similarly, for every site $y_{i_j}^-$ from the second group, there exists a conjugate site, $y_{i'_j}^-$, from $\{y_1^-, \dots, y_{n(\zeta)}^-\} \setminus \{y_{i_1}^-, \dots, y_{i_k}^-\}$, such that $i'_j < i_j$, $y_{i_j}^- = y_{i'_j}^+$ and

$$X^{i'_j}(y_{i_j}^-) = X^{i_j}(y_{i_j}^-) - 1, \tag{4.6}$$

which yields

$$X^{i_j}(y_{i_j}^-) + X^{i'_j}(y_{i'_j}^-) \le m + 1. \tag{4.7}$$

By construction, we have $X^i(y_i^-) < \frac{m}{2}$ for any site y_i from $\{y_1^-, \dots, y_{n(\zeta)}^-\} \setminus \{y_{i_1}^-, \dots, y_{i_k}^-\}$. Obviously

$$\max_{\substack{g=[m/2]+1,\dots,m \\ h=1,\dots,m \\ g+h \le m+1}} (gh) = \frac{(m+1)(m+1)}{4}. \tag{4.8}$$

This implies the bound

$$\prod_{i=0}^{n(\zeta)-1} X^i(y_i^-) \leq \left(\frac{(m+1)(m+1)}{4}\right)^{n(\zeta)/2}. \tag{4.9}$$

Now assume that

$$n(\zeta) \geq \frac{1}{2d}(2ed)^{4d}\beta(m+1). \tag{4.10}$$

Then

$$\begin{aligned} \left(\frac{\beta}{2d}\right)^{n(\zeta)} \frac{1}{n(\zeta)!} \prod_{i=0}^{n(\zeta)-1} X^i(y_i^-) &\leq \left(\frac{\beta(m+1)}{4d}\right)^{n(\zeta)} \frac{1}{n(\zeta)!} \\ &\leq \left(\frac{\beta e(m+1)}{4dn(\zeta)}\right)^{n(\zeta)} \leq \left(\frac{e}{2(2ed)^{4d}}\right)^{n(\zeta)}. \end{aligned} \tag{4.11}$$

Observe that

$$|\sigma(\zeta)| \leq 2dn(\zeta), \tag{4.12}$$

because

$$\left|\bigcup_{j=1}^{n_i(\zeta)} \widetilde{\gamma}_j^i(\zeta)\right| + 2d \geq \left|\bigcup_{j=1}^{n_{i+1}(\zeta)} \widetilde{\gamma}_j^{i+1}(\zeta)\right|. \tag{4.13}$$

In turn, bound (4.13) means that, while passing from the classical configuration $X^i(\zeta)$ to $X^{i+1}(\zeta)$ (corresponds to a single jump) the number of plaquettes in the interface cannot increase by more than $2d$. Bounds (4.11) and (4.12) together give

$$\left(\frac{\beta}{2d}\right)^{n(\zeta)} \frac{1}{n(\zeta)!} \prod_{i=0}^{n(\zeta)-1} X^i(y_i^-) \leq \left(\frac{e}{2(2ed)^{4d}}\right)^{|\sigma(\zeta)|/(2d)} \leq \exp(-t|\sigma(\zeta)|). \tag{4.14}$$

Hence, under condition (4.10),

$$w(\zeta) \leq \exp\left(-\beta(\mu-1)\frac{1}{2d}\|\{\omega_j(\zeta)\}\| - t|\sigma(\zeta)|\right). \tag{4.15}$$

On the other hand, if

$$n(\zeta) < \frac{1}{2d}(2ed)^{4d}\beta(m+1), \tag{4.16}$$

then

$$|\sigma(\zeta)| < (2ed)^{4d}\beta(m+1). \tag{4.17}$$

Therefore,
(4.18)

$$w(\zeta) \le \exp\left(-\beta(\mu-1)\frac{1}{2d}\|\{\omega_j(\zeta)\}\| + \frac{\beta(m+1)}{4d}\right)$$
$$\le \exp\left[-\beta\Big((\mu-1)\frac{1}{2d}\|\{\omega_j(\zeta)\}\| - \frac{m+1}{4d} - t(2ed)^{4d}(m+1)\Big) - t|\sigma(\zeta)|\right]$$
$$\le \exp\left[-\beta\Big((\mu-1) - \frac{m+1}{4d} - t(2ed)^{4d}(m+1)\Big) \times \frac{1}{2d}\|\{\omega_j(\zeta)\}\| - t|\sigma(\zeta)|\right],$$

because $\|\{\omega_j(\zeta)\}\| \ge 2d$. Bounds (4.15) and (4.18) immediately give the lemma. □

We now have to choose values of u and v in Propositions 3.8 and 3.10. Namely, set $u = 4$ and $v = e\beta m$. Recall that in §3 the choice of u specifies the long and short contours. In this section we apply the general scheme from §3 to a particular situation of the quantum contours. Following §3, we use, for a quantum contour, the notation ε and ζ, depending on the category into which it falls. Namely, we call a quantum contour ε short if

$$\operatorname{diam} \widetilde{\omega}_i(\varepsilon) < u, \tag{4.19}$$

cf. (3.82). The quantum contours for which condition (4.19) does not hold are called long and denoted by ζ.

The restricted partition functions $Z^{(l)}(V)$ are defined as

$$Z^{(l)}(V) = \sum_{\{\varepsilon_i\}:\, \{\widetilde{\varepsilon}_i\}\subseteq V,\, E(\omega_j(\varepsilon_i))=l} \prod_i w(\varepsilon_i), \tag{4.21}$$

cf. (3.85).

LEMMA 4.2. *There exist positive constants, $\bar{\mu} = \bar{\mu}(d)$ and $\bar{\beta} = \bar{\beta}(d)$ such that, for $u = 4$, $v = e\beta m$, p, and t as in Lemma 4.1, and $\mu \ge \bar{\mu}\, m$, $\beta \ge \bar{\beta}\, m$, the following bounds hold*:

(i) *for m even,*

$$\frac{Z^{(l)}(V)}{Z^{(m/2)}(V)} \le \exp\Big[-(|[V]_v| - v|\partial V|) \times 16d\exp(-(p - 2\log(2d) - 2)(du - u + 1))\Big],$$
$$l = 0,\dots,m, \quad l \ne \frac{m}{2}, \tag{4.22}$$

(ii) *for m odd,*

$$\frac{Z^{(l)}(V)}{Z^{(m/2\pm1/2)}(V)} \le \exp\Big[-(|[V]_v| - v|\partial V|) \times 32d\exp(-(p - 2\log(2d) - 2)(du - u + 1))\Big],$$
$$l = 0,\dots,m, \quad l \ne \frac{m}{2} \pm \frac{1}{2}. \tag{4.23}$$

PROOF. We discuss in detail the case of m even; the case of m odd is covered by a similar argument. We use Theorem 3.1 to evaluate the difference

$$b = \log Z^{(l)}(V) - \log Z^{(m/2)}(V).$$

Condition (3.2) in Theorem 3.1 is valid, with $a(\varepsilon) = \|\{\widetilde{\omega}_j(\varepsilon)\}\| + |\sigma(\varepsilon)|$, by virtue of Lemma 4.1.

In view of the restriction diam $\widetilde{\omega}_j(\varepsilon) \leq 3$ (see (4.19)), it is easy to list all possible shapes of $\widetilde{\omega}_j(\varepsilon)$. It can be either

(i) the collection of $2d$ plaquettes forming the surface of a unit cube on $\widetilde{\mathbb{Z}}^d$ centered at $x \in \mathbb{Z}^d$ (we denote such an interface by $\vartheta_0(x)$), or

(ii) the collection of $2d(2d-1)$ plaquettes forming a boundary of a subset of $\mathbb{Z}^d$ which consists of some site $x \in \mathbb{Z}^d$ united with all its nearest-neighbor sites (we denote such an interface by $\vartheta_1(x)$).

In the case where the reference to a particular site x is not important, we simply write ϑ_0 instead of $\vartheta_0(x)$ and ϑ_1 instead of $\vartheta_1(x)$. Observe that the partition function inside the volume enclosed by ϑ_1 is not trivial (it contains terms corresponding to collections of the type one interfaces). Therefore, to fit the scheme of the previous section (see comments below (3.42)), we have to modify the statistical weight of the short quantum contour ε with $\widetilde{\omega}(\varepsilon) = \vartheta_1$, by multiplying it by a ratio of the conditioned partition functions

$$\frac{Z^{(I(\omega(\varepsilon)))}\left(\operatorname{Int}\widetilde{\omega}(\varepsilon)|\varepsilon\right)}{Z^{(E(\omega(\varepsilon)))}\left(\operatorname{Int}\widetilde{\omega}(\varepsilon)|\varepsilon^*\right)}, \tag{4.24}$$

which are defined as in (3.67), replacing $\omega_j(\Omega_n)$ by ε. It is easy to see that ratio (4.24) is less than 2.

To estimate b, we begin with analysing the contribution to $\ln Z^{(l)}(V)$ which comes from polymers π consisting of a single short quantum contour ε, with the set of bold plaquettes $\widetilde{\omega}(\varepsilon)$ coinciding with $\vartheta_0(x)$, and with $|E(\omega(\varepsilon)) - I(\omega(\varepsilon))| = 1$. The statistical weight of such a polymer coincides with the statistical weight of ε and is

$$w(\pi) = w(\varepsilon) = \exp(-\beta(\mu-1))\left(\frac{\beta}{2d}\right)^{n(\varepsilon)} \frac{1}{(n(\varepsilon))!}\left(l\cdot(m-l)\right)^{n(\varepsilon)/2}. \tag{4.25}$$

Obviously, the right-hand side of (4.25) is maximal for $l = \frac{m}{2}$. Observe now that there exists exactly $(2d)^{n(\varepsilon)}$ different short quantum contours ε, with $\widetilde{\omega}(\varepsilon) = \vartheta_0(x)$ and $|E(\omega(\varepsilon)) - I(\omega(\varepsilon))| = 1$ and fixed $n(\varepsilon)$. [These quantum contours correspond to $n(\varepsilon)$ independent jumps in any of $2d$ directions.] Hence, the contribution to b coming from the polymers just described is

$$b_1 = \sum_{x\in V}\left(\sum_{\substack{\varepsilon:\,\widetilde{\omega}(\varepsilon)=\vartheta_0(x),\\ E(\omega(\varepsilon))=l}} w(\varepsilon) - \sum_{\substack{\varepsilon:\,\widetilde{\omega}(\varepsilon)=\vartheta_0(x),\\ E(\omega(\varepsilon))=\frac{m}{2}}} w(\varepsilon)\right). \tag{4.26}$$

It is not hard to see that

$$
\begin{aligned}
b_1 &\le |\,[V]_v\,| \sum_{n=1}^{v} \exp(-\beta(\mu-1)) \left(\frac{\beta}{2d}\right)^n \frac{(2d)^n}{n!} \\
&\quad \times \left(\left(l \cdot (m-l)\right)^{n/2} - \left(\frac{m}{2}\cdot\frac{m}{2}\right)^{n/2} \right) \\
&\le |\,[V]_v\,| \exp(-\beta(\mu-1)) \left(\exp\left(\beta\sqrt{l\cdot(m-l)}\right) - \frac{1}{3}\exp\left(\beta\frac{m}{2}\right)\right) \\
&\le -|\,[V]_v\,| \exp\left(-\beta(\mu-1-m/2)\right)\left(1/3 - \exp(-\beta/m)\right).
\end{aligned}
\tag{4.27}
$$

Second, we take into account the contribution to b which comes from polymers π consisting of a single short quantum contour ε, with the set of bold plaquettes $\widetilde{\omega}(\varepsilon)$ coinciding with $\vartheta_0(x)$, and with $|E(\omega(\varepsilon)) - I(\omega(\varepsilon))| \ge 2$. As before, the statistical weight of such a polymer coincides with the statistical weight of ε and obeys

$$
\begin{aligned}
w(\pi) = w(\varepsilon) &\le \exp(-\beta(\mu-1)\cdot|\,E(\omega(\varepsilon)) - I(\omega(\varepsilon))\,|) \\
&\quad \times \left(\frac{\beta}{2d}\right)^{n(\varepsilon)} \frac{1}{(n(\varepsilon))!}\left(\frac{m+1}{2}\right)^{n(\varepsilon)},
\end{aligned}
\tag{4.28}
$$

as follows from (2.15) and (4.9). Observe that the number of different short quantum contours ε, with $\widetilde{\omega}(\varepsilon) = \vartheta_0(x)$ and with fixed $n(\varepsilon)$, does not exceed

$$
\left(2d|\,E(\omega(\varepsilon)) - I(\omega(\varepsilon))\,|\right)^{n(\varepsilon)}.
$$

Hence, the contribution to b coming from the polymers of this type is

$$
\begin{aligned}
b_2 &\le 2|\,[V]\,| \sum_{n=1}^{\infty} 2 \sum_{k=2}^{m} \exp(-\beta(\mu-1)k) \left(\frac{\beta}{2d}\right)^n \frac{(2dk)^n}{n!}\left(\frac{m+1}{2}\right)^n \\
&\le 4|\,[V]\,| \sum_{k=2}^{m} \exp\left(-\beta\left(\mu-1-\frac{m+1}{2}\right)k\right) \\
&\le 8|\,[V]\,| \exp\left(-\beta\left(\mu-1-\frac{m+1}{2}\right)2\right);
\end{aligned}
\tag{4.29}
$$

the last inequality holds for $\exp\left[-\beta\left(\mu-1-\frac{m+1}{2}\right)2\right] < \frac{1}{2}$.

The next contribution to b to be assessed comes from polymers π consisting of a single short quantum contour ε described in (ii). The statistical weight of such a polymer coincides with the modified statistical weight of ε and obeys

$$
\begin{aligned}
w(\pi) = 2w(\varepsilon) &\le 2\exp\Big[-\beta(\mu-1)(2d-1)\cdot|E(\omega(\varepsilon)) - I(\omega(\varepsilon))|\Big] \\
&\quad \times \left(\frac{\beta}{2d}\right)^{n(\varepsilon)} \frac{1}{(n(\varepsilon))!}\left(\frac{m+1}{2}\right)^{n(\varepsilon)}.
\end{aligned}
\tag{4.30}
$$

Bound (4.30) follows from (2.15), (4.9), and the equality $|\vartheta_1(x)| = 2d(2d-1)$. Observe that the number of different short quantum contours ε, with $\widetilde{\omega}(\varepsilon) = \vartheta_1(x)$ and fixed $n(\varepsilon)$, does not exceed $(2d\ 2d\ |E(\omega(\varepsilon)) - I(\omega(\varepsilon))|)^{n(\varepsilon)}$. Hence, the contribution to b coming from polymers of this type is

$$
\begin{aligned}
b_3 &\leq 2 \cdot 2|\,[V]\,|\sum_{n=1}^{\infty} 2 \sum_{k=1}^{m} \exp(-\beta(\mu-1)(2d-1)k) \\
&\qquad \times \left(\frac{\beta}{2d}\right)^n \frac{(2d2dk)^n}{n!} \left(\frac{m+1}{2}\right)^n \\
&\leq 8|\,[V]\,|\sum_{k=1}^{m} \exp\left[-\beta\left(\mu-1-\frac{m+1}{2}\right)(2d-1)k + \beta\frac{m+1}{2}k\right] \\
&\leq 16|\,[V]\,|\exp\left[-\beta\left(\mu-1-\frac{m+1}{2}\right)(2d-1) + \beta\frac{m+1}{2}\right];
\end{aligned}
\tag{4.31}
$$

the last inequality holds for $\exp\left[-\beta\left(\mu-1-\frac{m+1}{2}\right)(2d-1)+\beta\frac{m+1}{2}\right] < \frac{1}{2}$.

Finally, consider the contribution to b which comes from polymers $\pi = [\varepsilon_j^{\alpha_j}]$ containing at least two short quantum contours. Denoting by $[\varepsilon', \varepsilon'']$ the polymer contaning a pair of distinct (incompatible) short quantum contours $\varepsilon', \varepsilon''$, with multiplicity one, this contribution is

$$
\begin{aligned}
b_4 &\leq \sum_{\varepsilon',\varepsilon'':\ \varepsilon' \not\sim \varepsilon''} \ \sum_{\pi:\ \pi \supseteq [\varepsilon',\varepsilon'']} w(\pi) \\
&\leq \sum_{\varepsilon',\varepsilon'':\ \varepsilon' \not\sim \varepsilon''} w(\varepsilon')w(\varepsilon'')\exp(a(\varepsilon') + a(\varepsilon'')).
\end{aligned}
\tag{4.32}
$$

The last sum does not exceed

$$
\begin{aligned}
&\sum_{n'=0}^{\infty}\sum_{n''=0}^{\infty} \sum_{\substack{x',x'' \in V: \\ \text{dist}\,(x',x'') \leq n'+n''}} \left(\sum_{\substack{\varepsilon':\ \tilde{\omega}(\varepsilon')=\vartheta_0(x'), \\ n(\varepsilon')=n'}} w(\varepsilon')\exp(a(\varepsilon')) \right. \\
&\qquad\qquad \left. + \sum_{\substack{\varepsilon':\ \tilde{\omega}(\varepsilon')=\vartheta_1(x'), \\ n(\varepsilon')=n'}} w(\varepsilon')\exp(a(\varepsilon')) \right) \\
&\qquad \times \left(\sum_{\substack{\varepsilon'':\ \tilde{\omega}(\varepsilon'')=\vartheta_0(x''), \\ n(\varepsilon'')=n''}} w(\varepsilon'')\exp(a(\varepsilon'')) \right. \\
&\qquad\qquad \left. + \sum_{\substack{\varepsilon'':\ \tilde{\omega}(\varepsilon'')=\vartheta_1(x''), \\ n(\varepsilon'')=n''}} w(\varepsilon'')\exp(a(\varepsilon'')) \right).
\end{aligned}
\tag{4.33}
$$

The number of pairs $x', x'' \in V$ with $\operatorname{dist}(x', x'') \leq n' + n''$ is $\leq |V|(n'+n'')^d \leq |V| \exp(d(n'+n''))$. Consequently, the right-hand side of (4.33) is

$$\leq |V| \left(\sum_{n=0}^{\infty} \exp(dn) \sum_{\substack{\varepsilon:\, \tilde{\omega}(\varepsilon)=\vartheta_0(0),\\ n(\varepsilon)=n}} w(\varepsilon) \exp(a(\varepsilon)) + \sum_{n=0}^{\infty} \exp(dn) \sum_{\substack{\varepsilon:\, \tilde{\omega}(\varepsilon)=\vartheta_1(0),\\ n(\varepsilon)=n}} w(\varepsilon) \exp(a(\varepsilon)) \right)^2 . \tag{4.34}$$

Arguing as before (see (4.12), (4.29), and (4.31)), we bound (4.34) by

$$\begin{aligned}
&|V| \Bigg(\sum_{n=0}^{\infty} \sum_{k=1}^{m} \Bigg(2 \exp(-\beta(\mu-1)k + 2dk) \\
&\qquad \times \left(\frac{\beta}{2d}\right)^n \frac{(2dk)^n}{n!} \left(\frac{m+1}{2}\right)^n \exp(dn) \\
&\qquad + 2 \exp(-\beta(\mu-1)(2d-1)k + 2d(2d-1)k) \\
&\qquad \times \left(\frac{\beta}{2d}\right)^n \frac{(2d2dk)^n}{n!} \left(\frac{m+1}{2}\right)^n \exp(dn) \Bigg)\Bigg)^2 \\
&\leq 4|V| \Bigg(\sum_{k=1}^{m} \Bigg(\exp\left(-\beta\left(\mu - 1 - e^d \frac{m+1}{2}\right)k + 2dk\right) \\
&\qquad + \exp\Bigg(-\beta\left(\mu - 1 - e^d \frac{m+1}{2}\right)(2d-1)k \\
&\qquad + \beta e^d \frac{m+1}{2} k + 2d(2d-1)k + \exp(3d)\Bigg)\Bigg)\Bigg)^2 \\
&\leq 16|V| \left(\sum_{k=1}^{m} \exp\left(-\beta\left(\mu - 1 - e^d \frac{m+1}{2}\right)k + 2dk\right)\right)^2 \\
&\leq 64|V| \exp\left(-2\beta\left(\mu - 1 - e^d \frac{m+1}{2}\right) + 4d\right);
\end{aligned} \tag{4.35}$$

the last three inequalities hold for $\exp\left(-2\beta(\mu - 1 - 2e^d \frac{m+1}{2}) + 4d\right) < \frac{1}{2}$.

For $\mu \geq 2e^d(m+1)+1$ and $\beta \geq m \log 12$, we can combine (4.27), (4.29), (4.31), and (4.35) and obtain

$$b \leq -\frac{1}{8} | [V]_v | \exp\left[-\beta\left(\mu - 1 - \frac{m}{2}\right)\right]. \tag{4.36}$$

Now it is easy to see that, for

$$\begin{gathered}
\bar{\mu} = 6(2\log(2d)+1)(2ed)^{4d}, \quad \bar{\beta} = 32\exp(2d), \\
\mu \geq \bar{\mu}\, m, \quad \beta \geq \bar{\beta}\, m, \quad u = 4, \quad v = e\beta m, \\
p = \frac{\beta}{2d}\left[(\mu-1) - \frac{(m+1)}{4d} - (2\log(2d)+1)(2ed)^{4d}(m+1)\right],
\end{gathered} \tag{4.37}$$

we have

$$(p - 2\log(2d) - 2)(du - u + 1) \geq \beta\left(\mu - 1 - \frac{m}{2}\right) + \log(128d), \tag{4.38}$$

which gives the lemma. □

Note that the value $\bar{\mu}(d)$ is of order d^d and not an optimal one. In fact, our method allows us to reduce $\bar{\mu}(d)$, after a tedious calculation, to a quantity of order d.

As was noted before, Lemma 4.2 allows us to prove, by means of a standard argument, the existence of the limiting states φ^0 and $\varphi^{\pm}$. More precisely, it implies the existence of the limits (1.16), for $Y = Y^{(m/2)}$ in the case of m even and for $Y = Y^{(m/2\pm 1/2)}$ in the case of m odd; in the last case it also implies that the limiting reduced density matrices $\pi^{\pm}_{V^0}$ are distinct. The translation-invariance, periodicity and symmetry properties of the families of the limiting reduced density matrices are valid by construction. The correlation-decay property needed for ergodicity is established in a standard way, on the base of the polymer expansions obtained.

To complete the proof of Theorem 1, we have to establish the uniqueness of states φ^0 and $\varphi^{\pm}$ in the sense stated in Theorem 1. For this purpose, consider a cube V on $\mathbb{Z}^d$, with boundary condition Y_{V^c}, where Y is an arbitrary configuration. As in the previously discussed case of $Y = Y^{(l)}$, we can define classical contours and quantum contours under boundary condition Y. The difference with the previous case is that now the classical (and hence quantum) contours that "touch" the boundary ∂V may be "open", in the sense that the interface of such contours, still being connected sets of plaquettes of $\widetilde{\mathbb{Z}}^d$, do not neccessarily enclose a subset of V.

Let $\widetilde{Y}_V$ be a configuration in V such that

(i) $\widetilde{Y}_V + Y_{V^c}$ is admissible,

(ii) $\widetilde{Y}_{[V]_m} = Y^{(m/2)}_{[V]_m}$ for m even and $\widetilde{Y}_{[V]_m} = Y^{((m\pm 1)/2)}_{[V]_m}$ for m odd,

(iii) the compatible collection of classical contours $\{\gamma_i\}$ corresponding to $\widetilde{Y}_V$ has a minimal value of $\|\{\gamma_i\}\|$ (in the class of the configuration in V which satisfy conditions (i), (ii)).

It is not hard to check that such $\widetilde{Y}_V$ exists, and $\|\{\gamma_i(\widetilde{Y}_V)\}\| \leq m|\partial V|$. Using, in (2.5) and (2.6), $\widetilde{Y}_V$ instead of Y_V, it is possible to repeat the previous construction and define bulky contours Ω similar to those in Propositions 3.8 and 3.10. These bulky contours have the same statistical weight as before (which satisfies bound (3.97)), provided that they do not touch ∂V. Otherwise, their statistical weight satisfies a weaker bound

$$|w(\Omega)| \leq \exp\Bigg(-\sum_n ((p-1)\|\{\omega_j(\zeta_n(\Omega))\}\| + t|\sigma(\zeta_n(\Omega))|) \\ - q'\sum_k |\widetilde{\Omega}^{(k)}| + \beta(\mu - 1)\frac{m}{2d}|\partial V \cap \partial\widetilde{\Omega}| + |\partial V \cap \partial\widetilde{\Omega}|\Bigg). \tag{4.39}$$

Compared to (3.97), there are two new summands here, $\beta(\mu-1)\frac{m}{2d}|\partial V \cap \partial\widetilde{\Omega}|$ and $|\partial V \cap \partial\widetilde{\Omega}|$. The first comes from term $\beta(\mu - 1)|\widetilde{Y}_V|$ in (2.5) and (2.6). The second arises because, instead of (3.86), we now have, for the $\widetilde{\Omega}$'s under consideration, a bound (4.40)

$$\frac{Z^{(k)}(\widetilde{\Omega})}{Z^{(0)}(\widetilde{\Omega})} \leq \exp\Big(-(|\,[\widetilde{\Omega}]_v\,| - v|\partial\widetilde{\Omega}|) \times 16d \exp\big(-(p - 2\log(2d) - 2)(du - u + 1)\big) + |\partial V \cap \partial\widetilde{\Omega}|\Big).$$

In turn, the term $|\partial V \cap \partial\widetilde{\Omega}|$ appears in (4.40) and (4.39) in order to compensate the contribution coming from the short quantum contours in $\widetilde{\Omega}$ touching $\partial V \cap \partial\widetilde{\omega}$. [These short quantum contours have a modified statistical weight which we do not write in an exact form.]

Bound (4.39) implies that for any $\widetilde{\Omega}$ that touches ∂V and satisfies $|\widetilde{\Omega}| \geq |V|^{1-1/d+\delta}$ with $0 < \delta < \frac{1}{d}$, the probability of having such $\widetilde{\Omega}$, in the ensemble with boundary condition Y_{V^c}, tends to zero, with $V \nearrow \mathbb{Z}^d$. This observation implies, by means of a standard argument (see, e.g. **[Z]**), the uniqueness of the limiting states φ^0 and $\varphi^{\pm}$ (in the class of possible ergodic limiting states, with the boundary conditions given by the translation-periodic classical configurations). Hence, any limiting state (1.11) (actually, any limit point) coincides with φ^0 for m even and is a mixture of $\varphi^{\pm}$ for m odd.

Finally, using the same standard argument, it is not hard to check that for m odd and $Y = Y^{(l)}$, with $l = 0, \ldots, \frac{m-1}{2} (l = \frac{m+1}{2} \ldots, m)$, the configuration $\widetilde{Y}_V = Y_V^{(m/2-1/2)}$ (respectively, $\widetilde{Y}_V = Y_V^{(m/2+1/2)}$). This gives relations (1.12 a,b) and completes the proof of Theorem 1. □

Relations (1.18), (1.19), which in turn imply Theorem 2, follow immediately from the fact that the probability of having a quantum contour ζ with $\operatorname{Int}\widetilde{\omega}_j(\zeta) \ni 0$ tends to 0 as $\beta \to \infty$. □

References

[BR] O. Bratteli and D. W. Robinson,, *Operator algebras and quantum statistical mechanics*, vol. 1, 2, Springer-Verlag, Berlin, 1979, 1981.

[DMaSin] E.A. Dinaburg, A. E. Mazel, and Ya. G. Sinai, *The ANNNI model and contour models with interaction*, Mathematical Physics Reviews C (S. P. Nolvikov, ed.), vol. 6, Gordon and Breach, New York, 1987, pp. 113–168.

[DyLSi] F. Dyson, E. Lieb and B. Simon, *Phase transitions in quantum spin systems with isotropic and anisotropic interactions.*, Journ. Stat. Phys. **18** (1978), 335–383.

[FSiSp] J. Fröhlich, B. Simon and T. Spencer, *Infrared bounds, phase transitions and continuous symmetry breaking.*, Comm. Math. Phys. **50** (1976), 79–85.

[G1] J. Ginibre, *Existence of phase transitions for quantum lattice systems*, Comm. Math. Phys. **14** (1969), 205-234.

[G2] ______, *Some applications of functional integration in statistical mechanics*, Statistical Mechanics and Field Theory (C. de Witt and R. Stora, eds.), Gordon and Breach, New York, 1970, pp. 327–428.

[KL] T. Kennedy and E. H. Lieb, *An itinerant electron model with crystalline or magnetic long range order*, Physica A **138** (1986), 320–358.

[KoPr] R. Kotecky and D. Preiss, *Cluster expansion for abstract polymer models*, Comm. Math. Phys. **103** (1986), 491–498.

[M1] V. A. Malyshev, *Cluster expansions in lattice models of statistical physics and the quantum theory of fields.*, Uspekhi Mat. Nauk **35** (1980), no. 2, 3–53; English transl. Russian Math. Surveys **35** (1980).

[M2] ______, *Soliton sectors in lattice models with continuous time*, Funktsinal. Anal. Prilozhen. **13** (1979), no. 1, 31–41; English transl. Functional Anal. Appl. **13** (1979).

[MMi] V. A. Malyshev and R. A. Minlos, *Gibbs random fields. The method of cluster expansions*, "Nauka", Moscow, 1985; English transl., Kluwer, Dordrecht, 1991.

[MaSu] A. E. Mazel and Yu. M. Suhov, *Random surfaces with two-sided constarints: an application of the theory of dominant ground states.*, J. Stat. Phys. **64** (1991), no. 1/2, 111–134.

[MeMir-So] A. Messager and S. Miracle-Sole, in preparation.

[OPiSu] E. Olivieri, P. Picco, and Yu. M. Suhov, *On the Gibbs states for one-dimensional lattice boson systems with a long-range ineraction*, J. Stat. Phys. **70** (1993), no. 3/4, 985–1028.

[P] Y. M. Park, *Quantum statistical mechanics for superstable interactions: Bose – Einstein statistics.*, J. Stat. Phys. **40** (1985), 259–302.

[PS] S. A. Pirogov and Ya. G. Sinai, *Phase diagrams of classical lattice systems*, Teor. i Mat. Fizika **25** (1975), 358–369, 1185–1192.

[R] D. W. Robinson A proof of the existence of the phase transition in the anisotropic Heisenberg model, Comm. Math. Phys. **14** (1969), 195–204.

[S] E. Seiler, *Gauge theories as a problem of constructive quantum field theory and statistical mechanics*, Lecture Notes in Physics, vol. 159, Springer-Verlag, Berlin, 1982.

[Si] Ya. G. Sinai, *Theory of phase transitions*, Academia Kiado and Pergamon Press, London, Budapest, 1982.

[Su] Yu. M. Suhov, *Limit Gibbs state for a class of one-dimensional systems of quantum statistical mechanics.*, Comm. Math. Phys. **62** (1978), 119–136.

[Z] M. Zahradnik, *An alternate version of Pirogov–Sinai theory*, Comm. Math. Phys. **93** (1984), 559–581.

International Institute for Earthquake Prediction Theory and Mathematical Geophysics, The Russian Academy of Sciences, Warshavskoe sh., 79-2, Moscow 113556, Russia
St John's College, Cambridge CB2 1TP and Isaac Newton Institute for Mathematical Sciences, Cambridge, CB3 0EH, UK

St John's College, Cambridge CB2 1TP and Isaac Newton Institute for Mathematical Sciences, Cambridge, CB3 0EH, UK
Statistical Laboratory, Department of Pure Mathematics and Mathematical Statistics, University of Cambridge, Cambridge CB2 1SB, England, UK
Institute for Problems of Information Transmission, The Russian Academy of Sciences, GSP-4 Moscow 101447, Russia

Amer. Math. Soc. Transl.
(2) Vol. **171**, 1996

Fast "Turbulent" Dynamo for Smooth Maps on the Two-Torus

V. Oseledets

§1. Introduction

The concept of the fast dynamo was introduced by Zel'dovich. The main problem in the fast dynamo theory is to show that the rapid magnetic field growth can persist at arbitrarily low diffusivities. One can define the notion of the fast dynamo action for any dynamical system (flows and mappings) and try to calculate the limit growth rate of the dynamo action (i.e., the fast dynamo exponent) in terms of the geometry of the flow. This can be also used to verify some conjectures about the fast dynamo action, for example, the following one: for a "typical" dynamical system the fast dynamo exponent is the growth rate of the difference $N(t)$ between the number of untwisted and twisted periodic orbits on the support of the invariant measure that have the largest entropy among all such orbits with the length bounded by t. The eigenfunctions of the dynamo operator have a fine-scale structure and become very complicated in the limit of weak diffusion. From this viewpoint, models amenable to mathematical analysis are very interesting.

We recall that the first dynamo model to be analyzed mathematically as a smooth flow on a Riemannian manifold was the suspension of the cat map (Arnold, Zel'dovich, Ruzmaikin and Sokolov, 1981). One can show that the fast dynamo problem in this example is the same as the fast dynamo problem for the cat map. In (Oseledets, 1993) we analyzed similar models that are based on a smooth map on the two-torus. In particular, we proved that the fast dynamo exponent is equal to the asymptotic Lefschetz number. We also found the form of the eigenfunctions that have self-similar structure over different scales.

The kinematic turbulent dynamo problem was studied by many authors (Baxendale and Rozovskii, 1993). Despite the success in understanding the turbulent dynamo problem, no fast turbulent dynamo solution was found. In this paper we investigate the fast turbulent dynamo problem for a random smooth map on the two-torus. We calculate the fast dynamo exponent and obtain the "leading eigenvalue" and the "leading eigenfunction" for the family of random dynamo operators.

I would like to thank Susan Friedlander for helpful discussions.

This work was partially supported by Russian Foundation for Fundamental Sciences Grant No. 93-01-00239 and NSF Grant DMS-9300752.

§2. Random map of the two-torus

Let (Ω, ν) be a probability space with a probability measure ν. Let S be an ergodic automorphism (i.e., S and S^{-1} are measure-preserving transformations of Ω). A random smooth map $T(\omega)$, $\omega \in \Omega$, of the two-torus is defined by

$$T(\omega)x = A(\omega)x + \varphi(x, \omega), \tag{1}$$

where $A(\omega)$ is a measurable function with values in integer matrices and $\varphi(x, \omega)$ is a smooth function on the two-torus X (the set of points $x = (x_1, x_2) \bmod 1$) that depends measurably on ω. We assume that

$$\begin{gathered} \det \frac{\partial T(\omega)}{\partial x} = 1, \qquad \det A(\omega) = 1, \\ \int \left| \ln^+ \max_x \left(\det \frac{\partial T(\omega)}{\partial x} \right) \right| d\nu(\omega) < \infty, \end{gathered} \tag{2}$$

where $\frac{\partial T(\omega)}{\partial x}$ is the Jacobi matrix of the random map $T(\omega)$. We introduce fast "turbulent" dynamo models that comprise two operations: a random mapping $T(\omega)$ and diffusion. The evolution of a vector field $b(x)$, $x \in X$, under the random map operation is defined as follows:

$$(G_0(\omega)b)(x) = \frac{\partial T(\omega)}{\partial x}(T^{-1}(\omega)x) b(T^{-1}(\omega)x).$$

The diffusion is defined by the following operation:

$$(\mathcal{P}_\varepsilon b)(x) = \int p_\varepsilon(x - y) b(y)\, dy,$$

where $\varepsilon > 0$ and $p_\varepsilon(x - y)$ is the heat kernel for the two-torus, i.e., the kernel of the operator $\exp(\varepsilon\Delta)$, with Δ to be the Laplacian,

$$\Delta = \frac{\partial^2}{\partial x_1^2} + \frac{\partial^2}{\partial x_2^2}.$$

We are interested in the limit as $\varepsilon \to +0$. The two operations are taken in the following order: first the mapping and then the diffusion. In other words, starting with a vector field $b(x)$, we write the resulting vector field $G_\varepsilon(\omega)b$ as

$$(G_\varepsilon(\omega)b)(x) = \mathcal{P}_\varepsilon(G_0 b)(x).$$

The operator $G_\varepsilon(\omega)$ is called the random dynamo operator. The dynamo exponent is defined by

$$\chi(\varepsilon) = \lim_{t \to +\infty} \frac{1}{t} \ln \|G_\varepsilon(S^{t-1}\omega) \ldots G_\varepsilon(\omega)\|,$$

(this exponent is constant because S is ergodic), where $\|\cdot\|$ denotes the norm of an operator in the space L_2 of vector fields. It follows from Kingman's subadditive ergodic theorem that the above limit exists on an S-invariant subset of Ω of full measure.

The *fast dynamo exponent* is defined by

$$\chi_0 = \liminf_{\varepsilon \to +0} \chi(\varepsilon).$$

The random map $T(\omega)$ is the *fast dynamo* if the fast dynamo exponent is positive:

$$\chi_0 > 0.$$

We define the *Lyapunov exponent* χ_L by

$$\chi_L = \lim_{t\to+\infty} \frac{1}{t} \ln \|B(S^{t-1}\omega)\dots B(\omega)\|,$$

where $B(\omega) = (A^{-1}(\omega))^*$.

THEOREM. *The fast dynamo exponent χ_0 is equal to the Lyapunov exponent χ_L.*

PROOF. Define

$$G(t,\varepsilon,\omega) = G_\varepsilon(S^{t-1}\omega)\dots G_\varepsilon(\omega).$$

Let H be the space of smooth vector fields $B(x)$ with $\div B = 0$ and $H_1 \subset H$ be the subspace consisting of those B for which $\int B(x)dx = 0$. For $B \in H_1$, we have

$$B(x) = \Big(\frac{\partial a}{\partial x_2}, -\frac{\partial a}{\partial x_1}\Big)$$

and

$$G(t,\varepsilon,\omega)B(x) = \Big(\frac{\partial a_t}{\partial x_2}, \frac{\partial a_t}{\partial x_1}\Big),$$

where $a_t = d(t,\varepsilon,\omega)a$ and the operator $d(t,\varepsilon,\omega)$ is defined as

$$d(t,\varepsilon,\omega) = d_\varepsilon(S^{t-1}\omega)\cdots d_\varepsilon(\omega),$$

with $d_\varepsilon(\omega) = \exp(\varepsilon\Delta)d_0(\omega)$ and $d_0(\omega)a(x) = a(T^{-1}(\omega)x)$. Hence we obtain

$$\begin{aligned}(G(t,\varepsilon,\omega)B, C) &= \int (G(t,\varepsilon,\omega)B(x))C(x)dx \\ &= \int (d(t,\varepsilon,\omega)a(x))\Big(\frac{\partial C_2}{\partial x_1} - \frac{\partial C_2}{\partial x_2}\Big)(x)dx,\end{aligned}$$

where $C = (C_1, C_2)$. Since $\|d(t,\varepsilon,\omega)\|_2 = 1$, we have

$$|(G(t,\varepsilon,\omega)B, C)| \le \text{const}$$

for smooth fields B and C. The multiplicative ergodic theorem (Ruelle, 1982) implies that

$$\lim_{t\to+\infty} \frac{1}{t} \log \|(G(t,\varepsilon,\omega) \mid H_1)\| = 0.$$

We can write

$$B(x) = \Big(\int B_1 dx, \int B_2 dx\Big) + B^1(x),$$

where $\int B_1(x)dx = 0$. Hence, $H = H_0 \bigoplus H_1$, where H_0 is a two-dimensional vector space.

The operator $G(t,\varepsilon,\omega)$ in the space H is represented by the triangular matrix

$$\begin{pmatrix} G_{00}(t), & 0 \\ G_{10}(t), & G_{11}(t) \end{pmatrix},$$

for which

$$\begin{aligned}&\lim_{t\to\infty} \frac{1}{t} \log \|G_{00}(t)\| = \chi_L, \\ &\lim_{t\to\infty} \frac{1}{t} \log \|G_{11}(t)\| = 0, \\ &\lim_{t\to\infty} \sup \frac{1}{t} \log \|G_{10}(t)\| \le \chi_L.\end{aligned}$$

We obtain

$$\lim_{t\to\infty} \frac{1}{t} \log \|(G(t,\varepsilon,\omega) \mid H)\| = \chi_L,$$

for ν-a.s. $\omega \in \Omega$.

Let us now consider a smooth field B. Set

$$b_t(x) = \div(G(t,\varepsilon,\omega)B(x)).$$

One can easily check that

$$b_t(x) = d(t,\varepsilon,\omega)b_0(x), \qquad \|b_t(x)\|_2 \le \text{const}.$$

Let

$$G(t,\varepsilon,\omega)B(x) = \operatorname{grad} f_t(x) + B_t,$$

where $B_t \in H$, $f_t(x)$ is a smooth function on the two-torus. Then

$$\|\operatorname{grad} f_t\|_2 \le \text{const}.$$

We also have

$$\chi(\varepsilon) = \lim_{t\to\infty} \frac{1}{t} \log \|(G(t,\varepsilon,\omega)\| = \lim_{t\to\infty} \frac{1}{t} \log \|(G(t,\varepsilon,\omega) \mid H)\| = \chi_L.$$

This implies that $\chi_0 = \chi_L$, which completes the proof of the theorem.

§3. The "leading eigenfunction"

Let $\Lambda_\varepsilon(\omega)$ be a "leading eigenvalue" of $G_\varepsilon(\omega)$, i.e., for some function $b_\varepsilon(x,\omega) \ne 0$ we have

$$G_\varepsilon(\omega)b_\varepsilon(x,\omega) = \Lambda_\varepsilon(\omega)b_\varepsilon(x,S\omega),$$
$$\int \ln|\Lambda_\varepsilon(\omega)|\,d\nu(\omega) = \chi(\varepsilon).$$

We call such a function $b_\varepsilon(x,\omega)$ a "leading eigenfunction".

It is interesting to analyze the limit of the "leading eigenfunction" as $\varepsilon \to 0$. Let $\widetilde{D}_\varepsilon = -IG_\varepsilon I$, where

$$I = \begin{pmatrix} 0 & -1 \\ 1 & 0 \end{pmatrix}.$$

We can write

$$\widetilde{D}_\varepsilon b = P_\varepsilon \left[\left(\frac{\partial T^{-1}(\omega)}{\partial x} \right)^* (x) b(T^{-1}(\omega)x) \right].$$

If b_ε is the "eigenfunction" for G_ε with an "eigenvalue" Λ_ε, then, putting $\widetilde{b}_\varepsilon(x,\omega) = (-Ib_\varepsilon)(x,\omega)$, we have

(3)
$$\widetilde{D}_\varepsilon(\omega)\widetilde{b}_\varepsilon(x,\omega) = P_\varepsilon \left[\left(\frac{\partial T^{-1}(\omega)}{\partial x} \right)^* (x)\widetilde{b}_\varepsilon(x,\omega)(T^{-1}(\omega)x) \right] = \Lambda_\varepsilon(\omega)\widetilde{b}_\varepsilon(x,S\omega).$$

Let $b^0(\omega)$ be the "eigenvector" of $(A^{-1}(\omega))^*$ corresponding to the "eigenvalue" $\lambda(\omega)$. It satisfies the conditions

$$(A^{-1}(\omega))^* b^0(\omega) = \lambda(\omega) b^0(S\omega), \qquad \int \ln|\lambda(\omega)|\,d\nu(\omega) > 0.$$

Note that

$$\int \ln|\lambda(\omega)|\, d\nu(\omega) = \chi_0.$$

Let

$$\widetilde{b}_\varepsilon(x,\omega) = b^0(\omega) + \operatorname{grad} g_\varepsilon(x,\omega)$$

be a representation of the "eigenfuncton". We obtain

$$\begin{aligned}\widetilde{D}_\varepsilon(\omega)b^0(\omega) &= P_\varepsilon[(A^{-1}(\omega))^* b^0(\omega)] + \left(\frac{\partial\psi}{\partial x}\right)^*(x,\omega)b^0(\omega)\\ &= \lambda(\omega)b^0(S\omega) + \operatorname{grad} P_\varepsilon[b^0\psi(x,\omega)],\\ \widetilde{D}_\varepsilon \operatorname{grad} g_\varepsilon(x) &= \operatorname{grad}(P_\varepsilon[g_\varepsilon(T^{-1}(\omega)x)]),\end{aligned}$$

where

$$T^{-1}(\omega)x = A^{-1}(\omega)x + \psi(x,\omega).$$

Comparing with (3), we have

$$P_\varepsilon[g_\varepsilon(T^{-1}(\omega)x,\omega)] + P_\varepsilon[b^0(\omega)\psi(x,\omega)] = \lambda(\omega)g_\varepsilon(x,S\omega). \tag{4}$$

Hence

$$g_\varepsilon(x) = \widetilde{S}^{-1}(1-\widetilde{S}^{-1}\widetilde{D}_{\varepsilon,0})^{-1}P_\varepsilon[b^0\psi] = \widetilde{S}^{-1}\sum_{k=0}^{\infty}(\widetilde{S}^{-1}\widetilde{D}_{\varepsilon,0})^k P_\varepsilon[b^0\psi(x)], \tag{5}$$

where

$$\widetilde{D}_{\varepsilon,0}g = P_\varepsilon[g(T^{-1}(\omega)x,\omega)]$$

and

$$\widetilde{S}g = \lambda(\omega)g(x,S\omega).$$

Finally, we obtain the "leading eigenvalue" $\Lambda_\varepsilon(\omega) = \lambda(\omega)$ and the "leading eigenfunction" of the dynamo operator G_ε:

$$b_\varepsilon(x,\omega) = Ib^0(\omega) + I \operatorname{grad}\left[\lambda\widetilde{S}^{-1}\sum_{k=0}^{\infty}(\lambda\widetilde{S}^{-1}\widetilde{D}_{\varepsilon,0})^k P_\varepsilon[b^0(\omega)\psi(x,\omega)]\right], \tag{6}$$

where

$$\widetilde{D}_{\varepsilon,0}g = P_\varepsilon(g(T^{-1}(\omega)x))$$

and

$$T^{-1}(\omega)x = A^{-1}(\omega)x + \psi(x,\omega).$$

As $\varepsilon \to 0$, the function $g_\varepsilon(x,\omega)$ tends to the function $g_0(x,\omega)$ defined by

$$g_0(x,\omega) = \lambda\widetilde{S}^{-1}\sum_{k=0}^{\infty}(\lambda\widetilde{S}^{-1}\widetilde{D}_{0,0})^k[b^0(\omega)\psi(x,\omega)]. \tag{7}$$

In the case of zero diffusion the "leading eigenfunction" is given by the following equality:

$$b_0(x,\omega) = Ib^0(\omega) + I \operatorname{grad} g_0(x,\omega). \tag{8}$$

In a typical case $b_0(x,\omega)$ is a distribution.

References

1. V. I. Arnold, Ya. B. Zeldovich, A. A. Ruzmaikin, and D. D. Sokoloff, *A magnetic field in a stationary flow with stretching in Riemannian space*, Zh. Eksp. Teor. Fiz. **81** (1981), 2052–2058; English transl., Sov. Phys. JETP **54** (1981), 1083–1086.
2. P. H. Baxendale and B. L. Rozovskii, *Kinematic dynamo and intermittence in a turbulent flow*, Geophys. Astrophys. Fluid Dynamics **73** (1993), 33–60.
3. V. I. Oseledets, *L-entropy and antidynamo theorem*, Proc. 6th Internat. Symp. on Information theory, vol. III, Moscow-Tashkent, 1984, pp. 162–163.
4. ______, *Fast dynamo problem for a smooth map on two-torus*, Geophys. Astrophys. Fluid Dynamics **73** (1993), 133–145.
5. D. Ruelle, *Characteristic exponents and invariant manifolds for compact in Hilbert space*, Ann. Math. (2) **115** (1982), 243–290.

Rogozhsky val 15, apt.41, 109147, Moscow, Russia

Amer. Math. Soc. Transl.
(2) Vol. **171**, 1996

Mechanical Background of Brownian Motion

M. Soloveitchik

§0. Introduction

We begin with the brief description of the theory of Brownian Motion. We follow the approach developed by Einstein and Smoluchovski. Consider a massive charged ball moving in the Euclidean space $\mathbb{R}^d$ that is filled in by a neutral gas of identical particles. We assume that the ball undergoes on action of an external electrical field f. The ball interacts with particles of the gas by a suitable repulsive potential. (The reader may think of a gas of point-like particles colliding elastically with the ball). It is assumed that the system is in the thermodynamic equilibrium. Namely, we set $f = 0$ and fix a Gibbs measure on the space of all possible configurations of the system. This Gibbs measure corresponds as usual to certain values of the inverse temperature β and the average particle density. It is known that if $f = 0$, this measure is invariant with respect to the dynamics. Denote by $Q(t)$ the position of the ball at time t. The following fundamental relations are accepted as main assumptions of the theory of Brownian Motion. They were discovered almost 90 years ago and agree well with numerous physical experiments.

1. Drift and diffusion.

$$Q(t) = D(f)t + \Sigma(f)W_t + o(\sqrt{t}), \quad \text{as } t \to \infty.$$

Here $D(f)$ is a vector called the mean drift and $\Sigma(f)$ is a positive definite operator; Σ^2 is called the diffusivity of the Brownian particle, W_t denotes the standard Wiener process.

2. Einstein relation.

$$D(f) = \frac{\beta}{2}\Sigma^2(0)f + o(f), \quad \text{as } f \to 0.$$

The Brownian Motion became a stimulating subject for various mathematical theories. One of the most important problems of modern statistical mechanics is to

1991 *Mathematics Subject Classification*. Primary 82C70; Secondary 60F17.

I am grateful to all mathematicians who made this paper possible. I would especially like to emphasize that without the stimulating influence of Ya. G. Sinai — not only as a scientist but also as a teacher — many of the results presented here would never be achieved. He helped me get started in mathematics and I am deeply indebted to him.

give a rigorous mathematical justification of axioms (1) and (2). More precisely, this means the following.

1. We must consider a mechanical system of the type "classical test object + constant external force acting on the test object + gas of classical particles". The underlining dynamics of the system have to be deterministic and Hamiltonian. No resorting to stochastic evolution is allowed.

2. Randomness may enter only through initial data. A measure on the space of initial data has to be of the Gibbs type. This measure should be invariant under the dynamics corresponding to the trivial external field $f = 0$.

3. The only "large parameter" is time. All physical parameters, for instance, density and masses of particles, temperature, etc. should be constant.

The problem of justification seems difficult even if one resorts to stochastic evolution or to additional parameter scaling. (See for example **[SzTo2]**, **[DuGoLe]**, **[CaDu]**, **[LeRo]**). In this paper we give a detailed discussion of the "pure mechanical" case. Our aim is to outline recent progress in this area. The paper is organized as follows. In §1 we give a rigorous mathematical description of our main models: the one-dimensional Rayleigh gas, the modified Rayleigh gas, and the periodic Lorentz gas. In §2 we present results concerning the first model. Our exposition is based mostly on the paper **[SiSo]**. In §3 we deal with the modified version of Rayleigh gas suggested by Lebowitz. These results have been recently announced in the author's joint works with Boldrighini **[BoSo1, BoSo2]**. The detailed proofs will be given in forthcoming publications. In §4 we briefly discuss results of **[ChEyLeSi]**. As a rule, we do not provide detailed proofs but explain basic ideas.

§1. Description of the main models

So far, there have been no rigorous mathematical models based only on the above physical postulates. A significant progress has been made only under some "artificial" additional assumptions. Below we describe our main models.

(i) *One-dimensional Rayleigh gas.* This model consists of a massive (charged) particle (M.P.) of mass M on a line immersed in ideal gas of identical point-like particles each of mass m. Gas particles do not interact and move according to the free dynamics until a collision with the massive particle occurs. Collisions of M.P. with gas particles are elastic. More precisely, let us define the *extended phase space* of the system by $\hat{\Omega} = \mathbb{R}^2 \times \mathbf{Y}$, where $\mathbf{Y} = \{Y \subset \mathbb{R}^2 : \operatorname{card}(Y \cap K) < \infty$ for any compact $K \subset \mathbb{R}^2\}$. Any point in Y is called a particle of the ideal gas and is denoted by $(q, v) \in Y \in \mathbf{Y}$. Here q and v are the position and the velocity of the particle.

We will denote a point $\hat{\omega} \in \hat{\Omega}$ by $\hat{\omega} = ((Q, V), Y)$, where Q and V are the position and the velocity of the massive particle. Let us define the action of the one-dimensional translation group on $\hat{\Omega}$ by the formula:

$$R_a(\hat{\omega}) = \hat{\omega}_1 : (q, v) \in Y(\hat{\omega}_1) \Leftrightarrow (q - a, v) \in Y(\hat{\omega}),$$
$$Q(\hat{\omega}_1) = Q(\hat{\omega}) + a, \qquad V(\hat{\omega}_1) = V(\hat{\omega}).$$

The *phase space* Ω is the quotient of $\hat{\Omega}$ with respect to this action. We identify Ω with the subset of $\hat{\Omega}$ corresponding to $Q = 0$. A point $\omega \in \Omega$ is a pair $\omega = (V, Y)$. We endow Ω (and $\hat{\Omega}$) with the standard topology and the corresponding σ-algebra of Borel sets and define a fundamental set of neighbourhoods of a point $Y_0 \in \mathbf{Y}$ by $U_S(Y_0) = \{Y \subset \mathbb{R}^2 : \operatorname{card}(Y \cap S) = \operatorname{card}(Y_0 \cap S)\}$, where $S \subset \mathbb{R}^2$ is bounded and open. The space $\mathbf{Y}$ with the given topology is a polish space. The topology on both

Ω and $\hat{\Omega}$ is the product topology. We denote by $\mathbf{B}$ and $\hat{\mathbf{B}}$ the corresponding Borel σ-algebras. Let us introduce the basic reference measure on $(\Omega, \mathbf{B})$ by the formula:

$$\mu(d\omega) = \sqrt{\frac{\beta M}{2\pi}} \exp(-\beta M V^2/2) dV \, P(dY),$$

where P denotes the Poisson field on the one particle phase space $\mathbb{R}^2$ with the intensity measure

$$n(dq\, dv) = n_0 \sqrt{\frac{\beta m}{2\pi}} \exp(-\beta m v^2/2)\, dv\, dq.$$

Here n_0 and β are positive constants corresponding to the density of the particle and to the inverse temperature of the system. Note that any P-typical configuration of particles is locally finite with respect to the q coordinate. The main reference measure for the extended phase space is $\mu \times dQ$. The dynamics $\{\hat{T}^t\}$ on $\hat{\Omega}$ corresponds to elastic collisions. All particles keep their velocities until they collide with M.P., which moves between collisions with constant acceleration f/M, where $f \geq 0$ is the force acting on M.P. Since $\hat{T}^t R_a = R_a T^t$, we may define the corresponding dynamics $\{T^t\}$ on the quotient space Ω. Note that μ is invariant under $\{T^t\}$ only by $f = 0$.

Our main object is the displacement of M.P. defined on Ω by

$$Q(t, \omega) = \int_0^t V(T^s(\omega))\, ds.$$

(ii) *Modified Rayleigh gas.* This system is a modification of one-dimensional Rayleigh gas proposed by Lebowitz.

We denote points in the plane $\mathbb{R}^2$ by $q = (q_1, q_2)$ and use the adjectives "horizontal" and "vertical" for the first and the second coordinate axis, respectively. The system is two-dimensional and consists of a rod of mass M and a gas of infinitely many point-like particles with equal masses $m < M$. The rod has length ℓ, and is infinitely thin. Its center point is constrained to move on the horizontal q_1-axis, and the rod keeps a vertical position, i.e., its orientation is fixed and orthogonal to the q_1-axis. The horizontal velocity of the rod is denoted by V and its coordinate on the q_1-axis by Q. The ideal gas is described by a locally (in q) finite subset Y of one particle phase space $\mathcal{M} = \mathbb{R}^2 \times \mathbb{R}^2$. A configuration of the whole system is described by a point in the *extended phase space* $\hat{\Omega}$. A point $\hat{\omega} \in \hat{\Omega}$ will be written as $\hat{\omega} = ((Q, V), Y)$, where Q, V denote the position and the velocity of the rod, respectively, and Y is the particle configuration. We define Ω as the quotient of $\hat{\Omega}$ with respect to the group of horizontal space translations. A point $\omega \in \Omega$ can be written in obvious notations as a pair $\omega = (V, Y)$. One can introduce Borel σ-algebras $\hat{\mathbf{B}}$ and $\mathbf{B}$ on $\hat{\Omega}$ and Ω, respectively, as above. The main reference measure on $(\Omega, \mathbf{B})$ in this case is

$$\mu(d\omega) = \sqrt{\frac{\beta M}{2\pi}} \exp(-\beta M V^2/2)\, dV \, P(dY),$$

where P is the Poisson field on the one particle phase space $\mathcal{M}$ with the intensity measure

$$n(dq\, dv) = n_0 \sqrt{\frac{\beta m}{2\pi}} \exp(-\beta m {v_1}^2/2) h(dv_2)\, dv_1\, dq,$$

where $h(\cdot)$ is a distribution of the vertical velocity. We assume that

$$\int_{-\infty}^{+\infty} h(dw)|w| < \infty,$$

and that there exists $a > 0$ such that $h([-a\,,\,a]) = 0$.

The last condition seems artificial. It ensures that there are no particles whose vertical velocities are too small and thus any particle inside the region available to the rod will leave it after a renewal time. This property is crucial since it removes, at least partially, the conservation of memory provided by recollisions of particles with the rod. The dynamics on $\hat{\Omega}$ (and on Ω) is the usual dynamics of elastic collisions. All particles keep their velocities until they collide with the rod, and upon collisions, the vertical velocities do not change. The rod moves between collisions according to the motion with constant acceleration f/M, where $f \geq 0$ is the force acting on the rod. The main problem is to investigate the displacement of the rod in the horizontal direction.

(iii) *Periodic Lorentz gas in* $\mathbb{R}^2$. This system corresponds to the motion of a single point-like particle under the action of elastic collisions with a finite number of fixed convex scatterers in a periodic domain of the plane $\mathbb{R}^2$. The following equations ("Gaussian dynamics") govern the motion between collisions:

$$\dot{q} = v, \quad \dot{v} = \frac{1}{m}\left(f - \frac{\langle f, v\rangle}{\langle v, v\rangle}v\right).$$

Here q denotes the position of the particle, v the corresponding velocity, f a constant vector of external force field, and m the mass of the particle. The brackets $\langle f, v\rangle$ denote the standard scalar product. Note that the kinetic energy of the particle is conserved. Set $\beta^{-1} = m|v|^2$.

Due to periodicity, the suitable configuration space of the system is the region G on the two-dimensional torus $\mathbb{T}^2$ that is the complement to the union of the scatterers. The phase space of the system is $\Omega = G \times \mathbf{S}^1_\beta$, where $\mathbf{S}^1_\beta = \{v : \beta^{-1} = m|v|^2\}$. The main reference (probability) measure on Ω is the Liouville measure $\mu = dq\,ds$. The important technical assumption is the *finite horizon* condition. To explain what it means let us define the time of the next collision by

$$\tau(\omega) = \inf\{t > 0 : q(T^t(\omega)) \in \partial G\}.$$

We require that $\tau(\omega) < \text{const}$. This model is different from the previous ones. Rigorously speaking, it does not satisfy our conditions formulated in the Introduction, since the dynamics is not hamiltonian (if f is nonzero). Absorbtion of energy by the medium is artificially simulated by the friction term in the equations of motion. However, one can accept that this simulation is sufficiently good to describe the contribution of moving scatterers. The results that have been obtained for this model are surprisingly complete and essentially multi-dimensional.

In order to make our consideration mathematically rigorous we will have to discuss the dynamics associated with the above models. In cases i) or ii) there are examples of initial configurations for which the dynamics can be defined only up to some finite time t. A more detailed analysis reveals the following reasons for that:

(a) there are infinitely many particles coming into some bounded region in a finite time;

(b) a non-transversal collision may occur. This means that the collision is multiple (two or more gas particles and M.P. (or the rod) collide simultaneusly) or M.P. (or the rod) and the colliding particle have equal (horizontal) velocities at the time of collision;

(c) infinitely many collisions may occur during a finite interval of time.

We will show that all these situations may occur only for a "small" set of initial conditions (i.e., for a set of zero measure). We consider only the one-dimensional Rayleigh gas. The modified Rayleigh gas can be studeid in a similar way. The case (a) easily follows from the observation that

$$\lim_{\substack{|q|\to\infty \\ (q,v)\in Y(\omega)}} \frac{|v|}{\log|q|} = 0.$$

μ-a.e.

Now we must construct the dynamics for a closed subsystem containing finite number of particles. This problem is finite-dimensional but nontrivial. Let us briefly discuss it in a more general setting. Consider a finite number of particles (on a line) given by their coordinates and velocities (q_i, v_i), $i = 1, \ldots N$, and endowed with masses m_i. Assume that particles collide elastically and undergo the action of external forces. Set $r_i = q_i\sqrt{m_i}$, $w_i = v_i\sqrt{m_i}$. In these new coordinates the problem can be stated as follows. In the Euclidean space $\mathbb{R}^N$ we consider a polyhedral cone defined by a finite number of linear inequalities $\theta_j(r) > 0$. The motion of a point-like particle inside the cone is governed by the equations $\dot{r} = w$, $\dot{w} = F(r)$ and on the regular part of the boundary by the elastic reflection rule. We show that the corresponding dynamics is well defined at least for a set of full $\nu = dr\, dw$ measure. It is easy to see that the initial conditions leading to the case (b) (i.e., to a non-transversal reflection) correspond to a subset of codimension at least 1. Thus we should estimate the set Z_∞ of initial conditions leading to the case (c). Let us present some related results.

PROPOSITION 1.1. *If the external force F is equal to zero, then Z_∞ is empty. Moreover, the number of possible reflections is uniformly bounded for all initial conditions.*

The proof is given in **[Si]** or **[Ga]**.

PROPOSITION 1.2. *If $F \in C^1$ and $|F(r)| < \text{const}(1 + |r|)$, then Z_∞ is a set of ν-measure zero and of the first topological category.*

An elementary proof of this fact is contained in **[So1]**. The above statements allow one to construct the required dynamics on some invariant subset $\Omega_* \subset \Omega$ for all the systems that we consider. We identify the set Ω_* with Ω.

§2. Rayleigh gas. Balance equations.

We begin with the case of one-dimensional Rayleigh gas. The first problem we must deal with is the lack of a suitable stationary distribution when the external field f is not equal to zero. The natural way to overcome this difficulty is to construct the distribution as the limit of the main reference measure under the time evolution. This has been rigorously proved only for the cases (ii) (Modified Rayleigh gas) and (iii) (Lorentz gas). For one-dimensional Rayleigh gas the problem is still open and we will consider only the equilibrium case ($f = 0$). We are interested in proving the Central Limit Theorem for the displacement $Q(t, \omega)$ of M.P. The first result deals with the degenerate case $M = m$. The dynamics in this case is relatively simple since the motion of the M.P. simply copies the trajectory of the last collided gas particle. In **[Sp]**, the displacement of the M.P. was explicitly represented as a functional of gas configuration.

THEOREM 2.1 ([**Sp**]). *Assume that $M = m$. Then the random process $\frac{1}{\sqrt{tN}} Q(tN, \cdot)$ weekly converges (as $N \to \infty$) to the Wiener process $W_{\sigma^2 t}$. The limit variance*

$$\sigma^2 = \sqrt{\frac{\pi}{2}} \frac{1}{n_0 m \sqrt{\beta}}.$$

The case $M > m$ is much harder. The main technique is based on "the balance equations". The following heuristic argument explains the meaning of these equations. Assume first that $Q(t, \omega) > 0$ is of order $\sqrt{t}$. Consider $A_t^+ = \{(q, v) \in Y(\omega) : q < 0,\ q + tv > 0\}$. This is the set of gas particles that lie to the left of M.P., have positive velocities, and intersect point zero under the free dynamics up to the time $t > 0$. What happens with these particles at time t? One must consider two different cases:

1. A particle collides at least once with M.P. That means that this particle is contained in

$$B_t = \{(q, v) \in Y(\omega) : q < 0,\ q + \tau v = Q(\tau, \omega) \text{ for some } \tau \in (0, t]\}.$$

Let us accept as a reasonable heuristic conjecture that the number of particles in $B_t \setminus A_t^+$ is small in probability as compared to $\sqrt{t}$.

Particles in $A_t^+ \setminus B_t$ realise the following second possibility.

2. A particle does not collide with M.P. up to time t. At time t such particles are contained in the segment $[0, Q(t, \omega)]$ and have positive velocities. Again, let us accept as a reasonable conjecture that up to terms of smaller order *only* such particles satisfy these conditions at time t. The number of such particles may be thus approximately estimated by $\frac{n_0}{2} Q(t, \omega)$, where, as above, n_0 is the density of gas particles. Denoting $b_t = \text{card}(B_t)$, $a_t^+ = \text{card}(a_t^+)$, we come to the following conjecture:

$$a_t^+ \approx b_t + \frac{n_0}{2} Q(t, \cdot).$$

The precise statement is given by the following proposition.

PROPOSITION 2.1 ([**SiSo**]). *We have*

$$a_t^+(\omega) = b_t(\omega) + \frac{n_0}{2} Q(t, \omega) + \varepsilon(t, \omega),$$

where

$$\frac{\varepsilon(t, \cdot)}{\sqrt{t}} \to 0$$

in probability.

Let us remark that we have no assumptions concerning the sign of $Q(t, \omega)$ and that in fact the normalized correction $\frac{\varepsilon(t, \cdot)}{\sqrt{t}} \to 0$ in L_2 ([**SzTo1,2**]). Consider the involution $\varphi \colon \Omega \to \Omega$ which changes the signs of all velocities. Note that

$$b_t(\varphi(T^t(\omega))) = b_t(\omega),$$
$$Q(t, \varphi(T^t(\omega))) = -Q(t, \omega).$$

Setting $a_t^-(\omega) = a_t^+(\varphi(T^t(\omega)))$ we obtain (see [**SiSo**])

$$(*) \qquad Q(t, \omega) = \frac{1}{n_0}(a_t^+(\omega) - a_t^-(\omega)) + \delta(t, \omega),$$

where $\frac{\delta(t, \cdot)}{\sqrt{t}} \to 0$ in probability (in L_2).

This is the balance equation that represents the conservation law of particles. The reader may find similar equations based on momentum and energy conservation in **[Sz-To1,2]**. Note, that the values a_t^+, a_t^- have the same Poisson distribution. Thus if we set

$$\alpha_t = E_\mu(a_t^+) = E_\mu(a_t^-),$$

and $\hat{a}_t^+ = a_t^+ - \alpha_t$, $\hat{a}_t^- = a_t^- - \alpha_t$, then

$$\frac{1}{\sqrt{t}} Q(t,\omega) = \frac{1}{n_0\sqrt{t}}(\hat{a}_t^+(\omega) - \hat{a}_t^-(\omega)) + \frac{1}{\sqrt{t}}\delta(t,\omega).$$

The variables $\frac{1}{\sqrt{t}}\hat{a}_t^\pm$ all have the same distribution, which converges to a Gaussian distribution. Now we state a stronger result than the convergence of the normalized displacement of M.P.

THEOREM 2.2 (**[SiSo]**). *The distributions of $\frac{1}{\sqrt{t}}Q(t,\omega)$ are tight and any limit point can be represented as a distribution of a sum of two Gaussian variables.*

"Balance" arguments allow us also to prove the "almost everywhere" type convergence.

THEOREM 2.3 (**[SiSo]**). *For each $\varepsilon > 0$ we have $\lim_{t\to\infty} Q(t,\omega)/t^{\frac{1}{2}+\varepsilon} = 0$ for μ-a.e. ω.*

More detailed analysis shows that $\sqrt{t}$ is actually the appropriate scale for this model. The "lower bounds" for the normalized displacement are given by

THEOREM 2.4 (**[SiSo]**). *For each $t > 0$ the normalized displacement $\frac{Q(t,\omega)}{\sqrt{t}}$ has the following representation:*

$$\frac{Q(t)}{\sqrt{t}} = \gamma(t) + \xi(t) + \delta(t),$$

where the random variable $\gamma(t)$ has a Gaussian distribution with the expectation 0 and the variance $\sigma_-^2 = \sqrt{\frac{\pi}{8}}\frac{1}{n_0 m\sqrt{t}}$ not depending on M, $\xi(t)$ is a random variable independent of $\gamma(t)$ and $\delta(t) \to 0$ in probability.

COROLLARY. *Any limit distribution for the tight family $\frac{1}{\sqrt{t}}Q(t,\omega)$ has the following properties:*

1. *It may be represented as a distribution of the random variable $\gamma + \xi$, where γ is the Gaussian variable with the expectation 0 and the variance σ_-^2, ξ is a random variable independent of γ.*
2. *It is absolutely continuous and its density is analytic.*
3. *The variance σ_∞^2 of the limit distribution satisfies the inequalities:*

$$(**)\qquad \sigma_-^2 = \sqrt{\frac{\pi}{8}}\frac{1}{n_0 m\sqrt{t}} \le \sigma_\infty^2 \le \sigma_+^2 = \sqrt{\frac{\pi}{2}}\frac{1}{n_0 m\sqrt{t}}.$$

Note that (1) is the straightforward consequence of Theorem 2.4, and (2) follows from (1) since the characteristic function of this distribution has to decay superexponentially. The lower bound in (3) follows from (1), the upper bound follows from the representation $(*)$ and coinsides with the limit variance in the case of $M = m$ (Theorem 2.1). Thus the limit distribution has an analytic density and may be represented as a sum of two Gaussian variables, the Gaussian variable γ, and independent one. These conditions are not sufficient to conclude that each limit distribution is Gaussian. The inequality $(**)$ is remarkable since the boundary values of the asymptotic variance do not depend on M. Let us note hat the same lower bound was obtained independently by Szasz and Toth **[SzTo1, SzTo2]** who used completely different methods. Moreover, they discovered the following surprising phenomenon. If the limit variance does not depend on M, then the normalized displacement converges in distribution to a Gaussian variable. Numerical simulations **[BoCoFr]** show that the required condition is probably false: the limit variance strongly depends on M. Moreover, the lower bound given above is also sharp and occurs when $M \to \infty$. Existence of the boundary values for the limit variance that is independent of M is, perhaps, a pure one-dimensional phenomenon. No similar bounds are known for the modified Rayleigh gas. Note that we were able to obtain the results concerning one-dimensional Reyleigh gas without the use of ergodic properties of the system. In fact, there are no results of this kind. Let us formulate the following *open problem.*

Prove that in the case of Reyleigh gas and $M \neq m$ the dynamical system is ergodic, mixing, has the K-property, etc.

The main reason this system is ergodic and the process $V(t)$ is "close to Markovian" is based on the following observation: fresh particles which, after a long time, achieve M.P. become statistically almost independent of their past trajectories since they come from well separated regions of the space. The best situation is the following: when a "new" particle comes, "old" particles escape forever and give no contribution to the velocity of M.P. Of course, it is not the case because recollisions are possible. Moreover, we have no a priori restrictions on the number of possible recollisions between a given gas particle and M.P. This number may be infinite as well. The following result shows that the situation is not so hopeless.

THEOREM 2.5 (**[SiSo]**). *For μ-almost every ω each particle has finitely many collisions with M.P.*

PROOF. We derive the statement from Theorem 2.3. Let us denote by $\Gamma^{\infty}(\omega)$ the set of all particles in $Y(\omega)$ that are at the left of M.P. and undergo infinitely many collisions as $t > 0$. It follows from Theorem 2.3 that the velocity of any particle contained in $\Gamma^{\infty}(\omega)$ has to be positive for each $t > 0$, because if it becomes negative, then the particle escapes to infinity. Define $(q^{+}(\omega), v^{+}(\omega)) \in \Gamma^{\infty}(\omega)$ by $q^{+}(\omega) = \max\{q : (q, v) \in \Gamma^{\infty}(\omega)\}$.

Set $W(\omega) = v^{+}(\omega)$ if $\Gamma^{\infty}(\omega) \neq \varnothing$ and $W(\omega) = 0$ otherwise. Evidently, $W(\omega) \geq 0$. It is easy to see that

$$\int_0^t W(T^s(\omega))\, ds \leq Q(t, \omega) - q^{+}(\omega).$$

It follows from Theorem 2.3 that

$$\lim_{t \to \infty} \frac{1}{t} Q(t, \omega) = 0 \text{ a. e.}$$

Thus

$$\lim_{t \to \infty} \frac{1}{t} \int_0^t W(T^s(\omega))\, ds = 0 \text{ a.e.},$$

and due to the ergodic theorem $W = 0$ a.e. This is equivalent to $\Gamma^{\infty} = \varnothing$ a.e. The same arguments work for particles at the right of M.P. The proof is complete.

Consider a particle $x \in Y(\omega)$. Denote by $\Delta(x)$ the time interval between the first and the last collision of x with M.P. It follows from the above that this interval is finite and nonempty. Set $D(\omega) = \bigcup_{x \in Y(\omega)} \Delta(x)$. We say that ω provides a cluster decomposition if $D(\omega) = \bigcup J_k$, where each J_k is a finite interval and $J_k \cap J_i = \varnothing$ for $i \neq k$. An abundance of ergodic properties of the system is due to the fact that cluster decomposition is a typical property. This fact has been established for slightly different systems, where M.P. is localized by an external potential increasing to infinity, or by elastic barriers. It can also be used to prove K- and Bernoulli regularity. We refer the reader to **[GoLeRa, BoPePrSiSo, So2, So3]**.

In the case of Rayleigh gas, the problem of recollisions seems too difficult. In the next section we demonstrate how to overcome this obstacle for the modified Rayleigh gas.

§3. Drift and diffusion for the modified Rayleigh gas

Let us return to the general non-equilibrium case and consider the modified Rayleigh gas described above as our main model.

Our approach to the problem is based on two simple but very usefull observations. The first one allows us to deal with Markov processes. Consider the part of one-particle configuration and phase space available to the rod:

$$\mathcal{S} = \{q \in \mathbb{R}^2 : |q_2| < \ell/2\}, \qquad \mathcal{M}_S = \{(q, v) : q \in \mathcal{S}\}.$$

The subsystem in $\mathcal{S}$ (in the fixed and in the moving reference frame) is described by

$$\hat{X} = \hat{X}(\hat{\omega}) = (Q(\hat{\omega}), V(\hat{\omega}), Y(\hat{\omega}) \cap \mathcal{M}_S),$$
$$X = X(\omega) = (Q(\omega), V(\omega), Y(\omega) \cap \mathcal{M}_S).$$

We denote by π the measure induced by μ on $\mathcal{X} = \{X(\omega), \omega \in \Omega\}$. More precisely, we set $\pi(A) = \mu\{\omega : X(\omega) \in A\}$.

Consider then $X_t(\omega) = X(T^t(\omega))$, $t \geq 0$ — the evolution of the configuration inside the strip $\mathcal{S}$ in terms of the entire configuration ω. Clearly, X_t is a Markov process with transition probabilities given in terms of the Poisson measure:

$$\mathcal{P}^t(X, A) = P(\{Y : T^t(X \cup (Y \setminus \mathcal{M}_S)\}).$$

This reduction to a Markov process has appeared for a similar system in **[GoLeRa]** and later in **[ErTu]**. Formally, it is similar to Markovian partitions in the theory of dynamical systems.

The second observation is the following.

PROPOSITION 3.1 (**[BoSo1]**). *Let $\mu_t = \mu(T^{-t}(\cdot))$ be the family of the measures generated by the dynamics. Then for each t the measure μ_t is equivalent to measure μ and the Radon-Nikodym derivative is equal to*

$$\frac{d\mu_t}{d\mu} = \exp\Big(\beta f \int_0^t V(T^{-t+s}(\omega))\, ds\Big).$$

PROOF. Let us consider the dynamics in the extended phase space. In the coordinates $\hat{\omega} = (Q, \omega)$ we get

$$\hat{T}^t(\hat{\omega}) = \Big(Q + \int_0^t V(T^s(\omega))\, ds, T^t(\omega)\Big).$$

On the other hand, the Gibbs measure $\hat{\mu}_f = \exp(\beta f Q)\, dQ \mu(d\omega)$ is invariant under $\{\hat{T}^t\}$. Define $\psi(\hat{\omega}) = \mathbf{I}_J(Q)\varphi(\omega)$, where φ is a bounded measurable function and $\mathbf{I}_J$ is the indicator of the finite interval (a, b). Invariance of the Gibbs measure implies that $\hat{\mu}_f(\psi) = \hat{\mu}_f(\psi(\hat{T}^t(\cdot))$. This yields

$$\int_\Omega \mu(d\omega)\varphi(\omega) \cdot \int_a^b \exp(\beta f Q)\, dQ$$
$$= \int_\Omega \mu(d\omega)\varphi(T^t(\omega)) \cdot \int_a^b \exp\Big(\beta f \Big(Q - \int_0^t V(T^s(\omega))\, ds\Big)\Big)\, dQ.$$

The right-hand side is equal to

$$\int_\Omega \mu(d\omega)\varphi(T^t(\omega)) \exp\Big(- \beta f \int_0^t V(T^s(\omega))\, ds\Big) \cdot \int_a^b \exp(\beta f(Q))dQ.$$

Thus

$$\int_\Omega \mu(d\omega)\varphi(T^t(\omega)) \exp\Big(- \beta f \int_0^t V(T^s(\omega))\, ds\Big) = \int_\Omega \mu(d\omega)\varphi(\omega),$$

and the result follows if we replace ω by $T^{-t}(\omega)$.

Note that the same arguments work for the Rayleigh gas as well. The previous proposition implies the following *summation rule* :

$$\int_\Omega \mu(d\omega) \exp\Big(- \beta f \int_0^t V(T^s(\omega))\, ds\Big) = 1$$

independently of t. This equality is an important technical tool and provides a basis to derive the Einstein relation heuristicly. Suppose that the "drift + diffusion" representation (see Introduction) is valid:

$$\int_0^t V(T^s(\omega))\, ds) = td(f) + \sigma(f) W_t.$$

Substituting this equation into the summation rule we obtain that

$$E(\exp(td(f) + \sigma(f)W_t)) = 1.$$

Thus $d(f) = \frac{\sigma^2(f)}{2}\beta f = \frac{\sigma^2(0)}{2}\beta f + o(f)$ under the assumption that the variance is continuous with respect to values of external field. Note that the above arguments are a contemporary interpretation of original Einstein's arguments, which he conducted on a level of physical evidence long before the corresponding mathematical notions were introduced. Another important fact that easily follows from Proposition 3.1 is the following.

PROPOSITION 3.2. *For any $t \geq 0$ the measures $\pi\mathcal{P}^t$ and π are equivalent on $\mathcal{X}$, and the Radon-Nikodym derivative of $\pi\mathcal{P}^t$ with respect to π is given by the formula*

$$\frac{d\pi\mathcal{P}^t}{d\pi}(X) = \int P(dY) \exp\Big(\beta f \int_0^t V(T^{-s}(X \cup (Y \setminus \mathcal{M}_S)))\, ds\Big).$$

We have not really used specific properties of modified Rayleigh gas such as the restrictions on vertical velocities of particles. We recall that there exists the renewal time $\tau > 0$, which is equal to the maximal time a given particle spends inside the strip available to the rod. Using this property we obtain the following statement.

PROPOSITION 3.3. *If $t > 2\tau$, then the measure π is absolutely continuous with respect to $\mathcal{P}^t(X_0, dX)$ for π-typical X_0.*

The proof of this proposition is based on the following observation. If for a time τ no particles come into some large region of the strip containing the rod, then at time τ this region contains only the rod, whose velocity depends on the initial configuration. Using new incoming particles we may produce any other configuration chosen in advance inside the strip at any other time τ. The reader may find details in [**BoSo1**]. Note that the measure $\mathcal{P}^t(X_0, dX)$ is not absolutely continuous with respect to π.

For technical reasons, it is more convenient to consider a Markov chain instead of a continuous time process. For this purpose, set $\mathbf{P} = \mathcal{P}^{4\tau}$ and consider the chain $(\mathcal{X}, \mathbf{P})$. This chain has the following properties :

1. the measures $\pi\mathbf{P}$ and π are equivalent;
2. π is absolutely continuous with respect to $\mathbf{P}(X, \cdot)$ for π-almost all X.

Using the standard results on Markov chains [**Nu**] we may conclude that there is an absorbing set $\mathcal{X}_1 \subset \mathcal{X}$ such that $\pi(\mathcal{X}_1) = 1$, and the restriction of the chain to this set is π-irreducible and aperiodic. Moreover, π is a maximal irreducibility measure. Without loss of generality, we may restrict ourselves to the chain $(\mathcal{X}_1, \mathbf{P})$ and write $\mathcal{X}$ instead of $\mathcal{X}_1$. It turns out that this Markov chain has strong ergodic properties. Let us briefly explain the basic reasons for that. Set

$$w(B) = \sup\{|v_1| : (q, v) \in B, \tau|v_1| \geq (1/4)|q_1|\},$$

where $B \subset \mathcal{M}$ and τ is the renewal time. Define

$$W(X) = \max\{|V(X)|, w(Y_S)\},$$

where Y_S denotes the particles inside the strip $\mathcal{S}$. Let us recall some notions of the theory of irreducible Markov chains (see [**Nu**]). A set $B \subset \mathcal{X}$ is said to be *small* (or minorant) if $\pi(B) > 0$, and there exists a positive measure λ and an integer positive number k_0 such that $\mathbf{P}^{k_0}(X, dY) \geq I_B(X) \cdot \lambda(dY)$. Evidently, the measure λ is absolutely continuous with respect to π. The role of the function W can be explained by the following result.

PROPOSITION 3.4 (Relaxation condition, see [**BoSo1**]).

1. *For any $U > 0$ the set $A_U = \{X : W(X) \leq U\}$ is a small.*
2. *For any choice of $\gamma \in (0, 1)$, sufficiently large U, and all $X_0 \in A_U$, the following inequalitiy holds*:

$$\mathbf{P}(X_0, \{X : W(X) > U^\gamma\}) < \exp(-cU^\delta),$$

where $c > 0$ is a constant and $\delta \in (0, \gamma)$.

The important observation [**BoSo1**] shows that Proposition 3.4 implies *geometric ergodicity* of the corresponding Markov chain. More precisely, this means the following:

1. There is an absorbing set $H \subset \mathcal{X}$ such that the restriction of the chain on to H is the Harris recurrent [**Nu**].
2. The unique $\mathbf{P}$-invariant measure ν concentrated on H is finite and equivalent to π. Moreover, if π is normalized to be a probability measure, the variational distance satisfies

$$||\mathbf{P}^n(X, \cdot) - \nu|| \leq g(X)\exp(-\chi n),$$

where $g \in L_1(\nu)$, $X \in H$, $\chi > 0$.

Returning to the initial time-continuous process we obtain

THEOREM 3.5 (Convergence to the invariant measure, see [**BoSo1**]). *There exists a probability measure ν_f invariant under $\{\mathcal{P}^t\}$ and equivalent to the measure π. Moreover*

$$||\mathcal{P}^t(X,\cdot) - \nu_f|| \le g(X)\exp(-\chi t)$$

for any X contained in the subset H of of full π-measure, with $g(X) \in L_1(\nu_f)$ and $\chi > 0$.

As shown in [**BoSo2**], the function g may be written as const $W(X)$. Proposition 3.4 implies that $W \in L_p(\nu_f)$, $p \ge 1$. Thus we can apply the ergodic theorem to the function $V = V(X)$ (since $|V(X)| \le |W(X)|$).

THEOREM 3.6 (Existence of a positive drift, see [**BoSo1**]). *The limit*

$$d_f = \lim_{t\to\infty} \frac{1}{t} Q(t,\omega) = \nu_f(V)$$

exists, is finite, and does not depend on ω for μ-a.a. ω. Moreover, it is strictly positive for $f > 0$.

The following results deal with the deviations from the drift.

THEOREM 3.7 (Diffusion, see [**BoSo1**]). *As $N \to \infty$, the process*

$$\xi_t^N(\omega) = \frac{Q(tN,\omega) - d_f \cdot tN}{\sqrt{tN}}$$

converges weakly in the space of continuous functions to the Wiener process $W_{\sigma_f^2 t}$ with nondegenerate diffusion constant $\sigma_f^2 > 0$.

THEOREM 3.8 (Deviations from the drift, see [**BoSo2**]). *For μ-a.a. $\omega \in \Omega$,*

$$\limsup_{t\to\infty} \frac{|Q(t,\omega) - d_f t|}{\sqrt{t}\log t} \le A < \infty.$$

Here A is finite and does not depend on ω.

We emphasize that the limit drift and diffusion in the Central Limit Theorem are shown to be strictly nondegenerate. Usually this fact is more difficult to obtain than the Limit Theorem itself. This assertion does not follow from arbitrarily strong mixing property of the process (excluding, perhaps, the case of independent increments) and its proof usually requires some very special arguments. Let me briefly explain them.

First we show that $\sigma_f^2 > 0$. To do this we introduce the σ-algebra Σ generated by the variables $X_0, X_1, \ldots$, where we set $X_k = X(T^{2\tau k}(\omega))$. Choose the initial distribution (i.e., the distribution of X_0) equal to ν_f. Consider the variance $\mathcal{D}(Q(2\tau N,\cdot))$ This value is estimated from below by the average conditional variance with respect to Σ as follows:

$$\mathcal{D}(Q) \ge E_{\nu_f}(E\,(Q - E(Q\,|\Sigma))^2|\Sigma) \stackrel{\text{def}}{=} E_{\nu_f}(\mathcal{D}_\Sigma(Q)).$$

Let us write

$$Q(2\tau N,\,\omega) = \sum_0^{N-1} \Delta Q_j,$$

where

$$\Delta Q_j = \int_{2\tau j}^{2\tau(j+1)} V(s)\,ds.$$

Since the process X_t is Markovian, the values ΔQ_j are Σ-conditionally independent. Hence

$$\mathcal{D}_\Sigma\Big(\sum_0^{N-1}\Delta Q_j\Big)=\sum_0^{N-1}\mathcal{D}_\Sigma(\Delta Q_j)=\sum_0^{N-1}\mathcal{D}_{\{X_j,X_{j+1}\}}(\Delta Q_j),$$

where $\mathcal{D}_{\{X_j,X_{j+1}\}}$ denotes the conditional variance with respect to the σ-algebra generated by X_j, X_{j+1}.

Due to stationarity, the variables $\mathcal{D}_{\{X_j,X_{j+1}\}}(\Delta Q_j)$ are identically distributed. Thus,

$$E(\mathcal{D}_{\{X_j,X_{j+1}\}}(\Delta Q_j))=E(\mathcal{D}_{\{X_0,X_1\}}(\Delta Q_0))=a.$$

Finally, we have

$$\mathcal{D}(Q(2\tau N,\cdot))\geq Na.$$

It remains to show that $a>0$. Suppose $a=0$. This implies that the displacement ΔQ_0 is a measurable function only of X_0, X_1. However, it is easy to construct a configuration of incoming particles which goes from X_0 to $X_{2\tau}$ and provides an arbitrarily chosen value of ΔQ_0. Thus $\sigma_f^2>a/2\tau>0$.

Let $f>0$. Suppose that $d_f\leq 0$. By the Central Limit Theorem, for any $\varepsilon>0$ we have

$$\begin{aligned}\mu\{\omega: Q(t,\omega)<-\varepsilon\sqrt{t}\}&\geq\mu\{\omega: Q(t,\omega)-d_f t<-\varepsilon\sqrt{t}\}\\&\to\frac{1}{\sqrt{2\pi}\sigma_f}\int_\varepsilon^\infty\exp(-\frac{s^2}{2\sigma_f^2})\,ds>0.\end{aligned}$$

On the other hand, by the summation rule

$$\int_\Omega\mu(d\omega)\exp\Big(-\beta f\int_0^t V(T^s(\omega))\,ds\Big)=1,$$

and hence by the Chebyshev inequality,

$$\mu\{\omega: Q(t,\omega)<-\varepsilon\sqrt{t}\}\leq\exp(-\beta f\varepsilon\sqrt{t})\to 0\ \text{ as } t\to\infty.$$

Thus the conditions $f>0$ and $d_f\leq 0$ are incompatible. This completes the proof.

Let us briefly discuss the equilibrium state in the phase space Ω. From Proposition 2.4 one can obtain that $\mu_t=\mu(T^{-t})\to\mu_f$ in the sense of weak convergence and in the variational norm on each σ-algebra $\mathbf{B}_L$ generated by configurations inside the horizontal strip $\{|q_2|\leq 0\}$. The measure μ_f is $\{T^t\}$-invariant. Its restriction to each $\mathbf{B}_L$ is equivalent to that of μ. In the whole space Ω the measure μ_f is *strictly singular* to μ since $f\neq 0$. Indeed, Theorem 3.6 implies that, as $t\to\infty$,

$$\frac{d\mu_t}{d\mu}=\exp(\beta f\int_0^t V(T^{-t+s}(\omega))\,ds)\to 0$$

for μ-a.e. ω. A similar fact has been established for the Lorentz gas (see [**ChEyLeSi**]). It is accepted that this fact is typical for systems with deterministic dynamics. For models with stochastic dynamics the stationary measure is typically equivalent to the initial one (see [**LeRo**]).

Let us note that the results presented here are concerned only with the first condition of the Einstein theory, namely the "drift + diffusion" representation. At present, this it is the only example of a pure mechanical system of this type for which the representation with nondegenerate parameters is rigorously established. The validity of the Einstein relation for the modified Rayleigh gas remains an open problem.

§4. Einstein relation for the periodic Lorentz gas

In this section we briefly describe the results of **[ChEyLeSi]** concerning the periodic Lorentz gas. We exclude the external magnetic field. It is the only known example of a deterministic system for which the Einstein relation is completely proved. We keep the notations of §1 and suppose that the absolute value of the external force $|f|$ is sufficiently small. (We recall that in this case f is a 2-dimensional vector.)

THEOREM 4.1. *The measures $\mu_t = \mu(T^{-t})$ converge weakly as $t \to \infty$ to the stationary ergodic measure μ_f. For f sufficiently small but nonzero, the measure μ_f is singular with respect to the initial measure μ. Its fractal dimension is strictly less than* 3.

Define the mean drift of the particle by $d_f = \mu_f(v)$.

Let us remark that there is a remarkable relation between the drift and dynamical characteristics of the system. Namely, $\beta\langle d_f, f\rangle = -(\lambda_f^s + \lambda_f^s)$, where $(\lambda_f^s, \lambda_f^s)$ are nonzero Lyapunov exponents of the measure μ_f (we assume that the Bolzmann constant $k_B = 1$).

In the case $f = 0$ we introduce the diffusion operator as follows.

THEOREM 4.2 (**[BuSi]**). *The operator σ^2 given by*

$$(\sigma^2)_{i,j} = 2\int_0^\infty dt \int_\Omega \mu(d\omega)(v_i v_j(T^t(\omega)))$$

is well defined (since the corresponding integral converges) and strictly positive. The process

$$\frac{1}{\sqrt{Nt}}\int_0^{Nt} v(T^s(\omega))\, ds$$

converges weakly, as $N \to \infty$, in the space of continuous functions to $\sigma \cdot W_t$.

THEOREM 4.3 (Einstein relation). *Under the previous assumptions,*

$$d_f = \frac{\beta}{2}\sigma^2 \cdot f + o(f).$$

The arguments leading to these results are based on the Markov partition techniques. We refer the reader to **[BuSi]**, **[ChEyLeSi]** for details.

References

[BoCoFr] C. Boldrighini, G. C. Cosimi, and S. Frigio, *Diffusion and Einstein relation for a massive particle in a one-dimensional free gas: numerical evidence*, J. Stat. Phys. **59** (1990), 1241–1250.

[BoPePrSiSo] C. Boldrighini, S. Pellegrinotti, E. Presutti, Ya. G. Sinai, and M. R. Soloveitchik, *Ergodic properties of a semi-infinite one-dimensional systems of classical statistical mechanics*, Comm. Math. Phys. **101** (1985), 363–382.

[BoSo1] C. Boldrighini and M. R. Soloveitchik, *Drift and diffusion for a mechanical system*, Preprint 94-37 (SFB 359) (1994), University of Heidelberg.

[BoSo2] ———, *On "large deviations" in a mechanical system*, Preprint no. 27 (1993), Landau Center for Research in Mathematical Analysis, Hebrew University of Jerusalem.

[BuSi] L. A. Bunimovich and Ya. G. Sinai, *Statistical properties of Lorentz gas with periodic configuration of scatterers*, Comm. Math. Phys. **78** (1981), 479–497.

[CaDu] P. Calderoni and D. Durr, *The Smoluchovsky limit for a simple mechanical Model*, Preprint (1986).

[ChEyLeSi] N. Chernov, G. L. Eyink, J. L. Lebowitz, and Ya. G. Sinai, *Steady state electrical conduction in the periodic Lorentz gas*, Comm. Math. Phys. **154** (1993), 569–601.

[DuGoLe] D. Durr, S. Goldstein, and J. L. Lebowitz, *Asymptotics of particle trajectories in infinite one -dimensional system with Collisions*, Comm. Pure Appl. Math. **38** (1985), 573–597.

[ErTu] L. Erdos and D. Tuyen, *Central Limit Theorems for the one-dimensional Rayleigh gas with semipermeable barriers*, Comm. Math. Phys. **143** (1992), 451–466.

[Ei1] A. Einstein, *Die von der molekularkinetischen Theorie der Wrme geforderte Bewegung von in ruhenden Flüssigkeit suspendierten Teilchen*, Annalen der Physik **17** (1905), 549–560.

[Ei2] ———, *Zur Theorie der Brownschen Bewegung*, Annalen der Physik **19** (1906), 371–381.

[Ga] G. A. Galperin, *Elastic collisions of particles on the line jour Uspekhi Mat. Nauk* **33** (1978), 211–212. (Russian)

[GoLeRa] S. Goldstein and J. L. Lebowitz, Ravishankar, *Ergodic properties of a system in Contact with a heat bath: a one-dimensional model*, Comm. Math. Phys. **85**, 419.

[LeRo] J. L. Lebowitz and H. Rost, *The Einstein relation for the displacement of a test particle in a random environment*, Preprint 94-15 (SFB 359) (1994), Heidelberg University.

[Nu] E. Nummelin, *General irreducible Markov chains and non-negative operators*, Cambridge Univ. Press, Cambridge, 1989.

[Si] Ya. G. Sinai, *Billiard trajectories in a polyhedral angle*, Uspekhi Mat. Nauk **33** (1978), 231–232.

[SiSo] Ya. G. Sinai and M. R. Soloveitchik, *One-dimensional classical massive particle in the ideal gas*, Comm. Math. Phys. **104** (1986), 423–443.

[So1] M. R. Soloveitchik, *Conservative dynamical system in a polyhedral angle. Existence of the dynamics*, Preprint 94-33 (SFB 359) (1994), University of Heidelberg.

[So 2] ———, *Ergodic properties of systems with an external potential in classical statistical mechanics*, Math. USSR–Izv. **34** (1990), no. 1, 181–201.

[So3] ———, *The sufficient condition of Bernoulli property for K-systems and its application in classical statistical mechanics*, Mat. Zametki **45** (1989), no. 2, 105–111.

[Sp] F. Spitzer, *Uniform motion with elastic collisions of an infinite particle system*, J. Math. Mech. **18** (1969), 973–989.

[SzTo1] D. Szasz and B. Toth, *Bounds on the limiting variance of the heavy particle*, Comm. Math. Phys. **104** (1986), 445–455.

[SzTo2] ———, *Towards a unified dynamical theory of the Brownian particle in an ideal gas*, Comm. Math. Phys. **111** (1987), 41–62.

DEPARTMENT OF APPLIED MATHEMATICS, UNIVERSITY OF HEIDELBERG, IM NEUENHEIMER FELD 294, 69120 HEIDELBERG, GERMANY